J. N. Mordeson, D. S. Malik, N. Kuroki

Fuzzy Semigroups

Springer

Berlin
Heidelberg
New York
Hong Kong
London
Milano
Paris
Tokyo

Studies in Fuzziness and Soft Computing, Volume 131
http://www.springer.de/cgi-bin/search_book.pl?series=2941

Editor-in-chief
Prof. Janusz Kacprzyk
Systems Research Institute
Polish Academy of Sciences
ul. Newelska 6
01-447 Warsaw
Poland
E-mail: kacprzyk@ibspan.waw.pl

Further volumes of this series can be found on our homepage

Vol. 113. A. Abraham, L.C. Jain and J. Kacprzyk (Eds.)
Recent Advances in Intelligent Paradigms and Applications", 2003
ISBN 3-7908-1538-1

Vol. 114. M. Fitting and E. Orowska (Eds.)
Beyond Two: Theory and Applications of Multiple Valued Logic, 2003
ISBN 3-7908-1541-1

Vol. 115. J.J. Buckley
Fuzzy Probabilities, 2003
ISBN 3-7908-1542-X

Vol. 116. C. Zhou, D. Maravall and D. Ruan (Eds.)
Autonomous Robotic Systems, 2003
ISBN 3-7908-1546-2

Vol 117. O. Castillo, P. Melin
Soft Computing and Fractal Theory for Intelligent Manufacturing, 2003
ISBN 3-7908-1547-0

Vol. 118. M. Wygralak
Cardinalities of Fuzzy Sets, 2003
ISBN 3-540-00337-1

Vol. 119. Karmeshu (Ed.)
Entropy Measures, Maximum Entropy Principle and Emerging Applications, 2003
ISBN 3-540-00242-1

Vol. 120. H.M. Cartwright, L.M. Sztandera (Eds.)
Soft Computing Approaches in Chemistry, 2003
ISBN 3-540-00245-6

Vol. 121. J. Lee (Ed.)
Software Engineering with Computational Intelligence, 2003
ISBN 3-540-00472-6

Vol. 122. M. Nachtegael, D. Van der Weken, D. Van de Ville and E.E. Kerre (Eds.)
Fuzzy Filters for Image Processing, 2003
ISBN 3-540-00465-3

Vol. 123. V. Torra (Ed.)
Information Fusion in Data Mining, 2003
ISBN 3-540-00676-1

Vol. 124. X. Yu, J. Kacprzyk (Eds.)
Applied Decision Support with Soft Computing, 2003
ISBN 3-540-02491-3

Vol. 125. M. Inuiguchi, S. Hirano and S. Tsumoto (Eds.)
Rough Set Theory and Granular Computing, 2003
ISBN 3-540-00574-9

Vol. 126. J.-L. Verdegay (Ed.)
Fuzzy Sets Based Heuristics for Optimization, 2003
ISBN 3-540-00551-X

Vol 127. L. Reznik, V. Kreinovich (Eds.)
Soft Computing in Measurement and Information Acquisition, 2003
ISBN 3-540-00246-4

Vol 128. J. Casillas, O. Cordón, F. Herrera, L. Magdalena (Eds.)
Interpretability Issues in Fuzzy Modeling, 2003
ISBN 3-540-02932-X

Vol 129. J. Casillas, O. Cordón, F. Herrera, L. Magdalena (Eds.)
Accuracy Improvements in Linguistic Fuzzy Modeling, 2003
ISBN 3-540-02933-8

Vol 130. P.S. Nair
Uncertainty in Multi-Source Databases, 2003
ISBN 3-540-03242-8

John N. Mordeson
Davender S. Malik
Nobuaki Kuroki

Fuzzy Semigroups

Springer

Prof. Dr. John N. Mordeson
E-mail: mordes@creighton.edu

Prof. Davender S. Malik
malik@creighton.edu

Creighton University
Dept. of Mathematics and Computer Science
Omaha, NE 68178
USA

Prof. Nobuaki Kuroki
E-mail: kuroki@juen.ac.jp

Joetsu University of Education
Dept. of Mathematics
Joetsu-shi,
943 Niigate-ken
Japan

ISSN 1434-9922
ISBN 3-540-03243-6 Springer-Verlag Berlin Heidelberg New York

Library of Congress Cataloging-in-Publication-Data

Fuzzy semigroups / John N. Mordeson, Davender S. Malik, Nobuaki Kuroki.
p. cm. -- (Studies in fuzziness and soft computing ; 131)
Includes bibliographical references and index.
ISBN 3-540-03243-6 (acid-free paper)
1. Semigroups. 2. Fuzzy sets. 3. Machine theory. I. Malik, D. S. II. Kuroki, Nobuaki,
1941 - III. Title. IV. Series.
QA182.M67 2003
512'.2--dc21

Springer-Verlag Berlin Heidelberg New York
a member of BertelsmannSpringer Science+Business Media GmbH
http://www.springer.de

Typesetting: data delivered by authors
Cover design: E. Kirchner, Springer-Verlag, Heidelberg
Printed on acid free paper 62/3020/M - 5 4 3 2 1 0

Preface

Lotfi Zadeh introduced the notion of a fuzzy subset of a set in 1965. His seminal paper has opened up new insights and applications in a wide range of scientific fields. Azriel Rosenfeld used the notion of a fuzzy subset to put forth cornerstone papers in several areas of mathematics, among other discplines. Rosenfeld is the father of fuzzy abstract algebra. Kuroki is responsible for much of fuzzy ideal theory of semigroups. Others who worked on fuzzy semigroup theory, such as Xie, are mentioned in the bibliography. The purpose of this book is to present an up to date account of fuzzy subsemigroups and fuzzy ideals of a semigroup. We concentrate mainly on theoretical aspects, but we do include applications. The applications are in the areas of fuzzy coding theory, fuzzy finite state machines, and fuzzy languages. An extensive account of fuzzy automata and fuzzy languages is given in [100]. Consequently, we only consider results in these areas that have not appeared in [100] and that pertain to semigroups.

In Chapter 1, we review some basic results on fuzzy subsets, semigroups, codes, finite state machines, and languages. The purpose of this chapter is to present basic results that are needed in the remainder of the book.

In Chapter 2, we introduce certain fuzzy ideals of a semigroup, namely, fuzzy two-sided ideals, fuzzy bi-ideals, fuzzy interior ideals, fuzzy quasi-ideals, and fuzzy generalized bi-ideals. We give some fundamental properties properties of these fuzzy ideals. We also characterize fuzzy ideals generated by fuzzy subsets.

In Chapter 3, we characterize a regular semigroup, a left (right) regular semigroup, a completely regular semigroup, an intra-regular semigroup, a quasi-regular semigroup, and a semisimple semigroup in terms of fuzzy ideals. For example, we show that a semigroup S is intra-regular if and only if $f \cap g \subseteq f \circ g$ for every fuzzy left ideal f and every fuzzy right ideal g of S.

In Chapter 4, we characterize a semigroup that is a semilattice of left (right) simple groups, of left (right) groups, and a semigroup that is a semilattice of groups in terms of fuzzy ideals. The remainder of Chapter 4 is concerned with the study of fuzzy normal semigroups and convexity and Green's relations.

In Chapter 5, we discuss the fuzzy congruences on a semigroup and a group. Let G be a group. Let $FN(G)$ and $FC(G)$ denote the set of all normal subgroups of G and the set of all fuzzy congruences on G, respectively. We show that there exists a one-to-one function from $FN(G)$ onto $FC(G)$. The remainder of Chapter 5 is concerned with the study of Homomorphism Theorems on fuzzy semigroups and fuzzy congruences on inverse semigroups, especially, idempotent-separating fuzzy congruences and group fuzzy congruences on inverse semigroups.

Chapter 6 is concerned with T^*-pure semigroups. We show that a semigroup S is regular and T^*-pure if and only if S is a semilattice of groups. Properties of semilattice fuzzy congruences and fuzzy congruences on T^*-pure semigroups are also given.

In Chapter 7, we consider different types of fuzzy prime ideals. We show that a nonconstant fuzzy prime ideal f of a semigroup S is prime if and only if f is two-valued, there exist x_0 of S such that $f(x_0) = 1$, and $\{x \in S \mid f(x) = 1\}$ is a prime ideal of S. Relations among types of fuzzy ideals are established. Notions such as fuzzy multiplication semigroups and fuzzy ideal extensions are also studied.

Applications are given in Chapters 8, 9, and 10. In Chapter 8, we consider fuzzy codes on a free monoid. We are particularly interested in fuzzy prefix codes and maximal fuzzy prefix codes. We study some of their algebraic properties. Much of what we present is due to Shen, Mo, and Pen. We close Chapter 8, with results due to Gerla. In Chapter 9, we present some results on fuzzy finite state machines due to Kim, Cho, Kim, and Lee. The approach here is a little different than the the one presented in [100]. We present a discussion of this approach by Kumbhojkar and Chaudari. Chapter 10 concludes the book with results by Santoz and Kandel. It concerns some results on regular fuzzy languages and codes which did not appear in [100].

The authors are grateful to the staffs of Springer-Verlag. We are indebted to Dr. Timothy Austin, Dean, Creighton College of Arts and Sciences, and to Dr. and Mrs. George Haddix for their support of our work. We also wish to thank Professors Paul P. Wang and Hu Cheng-ming for their support of fuzzy mathematics. The first author dedicates the book to his grandson Josh Wade. The second author dedicates the book to his wife Sadhana and daughter Shelly. The third author dedicated the book to his son Ryouta and daughter Maiko.

Contents

1
Introduction

1.1 Notation

In this section, we set down some notation. We let $\mathbb{N}$ denote the set of positive integers, $\mathbb{Z}$ the set of integers, $\mathbb{Q}$ the set of rational numbers, $\mathbb{R}$ the set of real numbers, $\mathbb{R}^+$ the set of positive real numbers, and $\mathbb{N}^0 = \mathbb{N} \cup \{0\}$.

We let $|S|$ denote the cardinality of S, i.e., the number of elements of S. If S is a set such that $|S| < \infty$, then S is called a **finite** set; otherwise S is called an **infinite** set.

A set A is said to be a **subset** of a set S if every element of A is an element of S. In this case, we write $A \subseteq S$ and say that A is contained in S. If $A \subseteq S$, but $A \neq S$, then we write $A \subset S$ and say that A is properly contained in S or that A is a **proper subset** of S.

The **empty set** or **null set** is denoted by $\emptyset$.

Given two sets A and B, the **relative complement** (or **set difference**) of B in A, written $A \backslash B$, is the set

$$A \backslash B = \{x \mid x \in A, \text{ but } x \notin B\}.$$

We sometimes write $\overline{A}$ for the relative complement of a set A in the universal set.

If A and B are sets, we let $A \times B$ denote the **Cartesian cross-product** of A and B, i.e., $A \times B$ is the set of all ordered pairs (a, b), where $a \in A$ and $b \in B$. We write A^n for the set of all ordered n-tuples of elements from A, where $n \in \mathbb{N}$. We let $\mathcal{P}(X)$ denote the **power set** of a set X, i.e., $\mathcal{P}(X)$ is the set of all subsets of X.

1.2 Relations

In this section, we review some basic properties of relations.

Definition 1.2.1 *A* ***binary relation*** *or simply a* **relation** *R from a set A into a set B is a subset of $A \times B$.*

Let R be a relation from a set A into a set B. We sometimes write xRy or $R(x) = y$ for $(x, y) \in R$. If $A = B$, then R is called a **binary relation** on A.

Definition 1.2.2 *Let R be a relation from a set A into a set B. Then the* ***domain*** *of R, written $Dom(R)$, is defined to be the set*

$$\{x \mid x \in A \text{ and there exists } y \in B \text{ such that } (x, y) \in R\}.$$

The ***image*** *of R, written $Im(R)$, is defined to be the set*

$$\{y \mid y \in B \text{ and there exists } x \in A \text{ such that } (x, y) \in R\}.$$

Definition 1.2.3 *Let R be a binary relation on a set A. Then R is called*
(1) ***reflexive*** *if for all $x \in A$, xRx,*
(2) ***symmetric*** *if for all $x, y \in A$, xRy implies yRx,*
(3) ***transitive*** *if for all $x, y, z \in A$, xRy and yRz imply xRz.*

Definition 1.2.4 *A binary relation E on a set A is called an* ***equivalence relation*** *on A if E is reflexive, symmetric, and transitive.*

Definition 1.2.5 *Let E be an equivalence relation on a set A. For all $x \in A$, let $[x]$ denote the set*

$$[x] = \{y \in A \mid yEx\}.$$

The set $[x]$ is called the ***equivalence class*** *(with respect to E) determined by x.*

Theorem 1.2.6 *Let E be an equivalence relation on the set A. Then the following properties hold:*
(1) for all $x \in A$, $[x] \neq \emptyset$,
(2) for all $x, y \in A$, if $y \in [x]$, then $[x] = [y]$,
(3) for all $x, y \in A$, either $[x] = [y]$ or $[x] \cap [y] = \emptyset$,
(4) $A = \cup_{x \in A}[x]$, i.e., A is the union of all equivalence classes with respect to E.

Proof. (1) Let $x \in A$. Since E is reflexive, xEx. Hence $x \in [x]$ and so $[x] \neq \emptyset$.

(2) Let $y \in [x]$. Then yEx and by the symmetric property of E, xEy. Let $u \in [y]$. Then uEy. Since uEy and yEx, the transitivity of E implies that

uEx. Hence $u \in [x]$. Thus $[y] \subseteq [x]$. Now let $u \in [x]$. Then uEx. Since uEx and xEy, uEy by transitivity and so $u \in [y]$. Hence $[x] \subseteq [y]$. Consequently, $[x] = [y]$.

(3) Let $x, y \in A$. Suppose $[x] \cap [y] \neq \emptyset$. Then there exists $u \in [x] \cap [y]$. Thus $u \in [x]$ and $u \in [y]$. By (2), $[x] = [u] = [y]$.

(4) Let $x \in A$. Then $x \in [x] \subseteq \cup_{x \in A}[x]$. Thus $A \subseteq \cup_{x \in A}[x]$. Also, $\cup_{x \in A}[x] \subseteq A$. Hence $A = \cup_{x \in A}[x]$. ■

Definition 1.2.7 *Let A be a set and $\mathcal{P}$ be a collection of nonempty subsets of A. Then $\mathcal{P}$ is called a* **partition** *of A if the following properties hold:*

(1) for all B, $C \in \mathcal{P}$, either $B = C$ or $B \cap C = \emptyset$,

(2) $A = \cup_{B \in \mathcal{P}} B$. ■

The next theorem is an easy consequence of Theorem 1.2.6.

Theorem 1.2.8 *Let E be an equivalence relation on the set A. Then*

$$\mathcal{P} = \{[x] \mid x \in A\}$$

is a partition of A. ■

Given an equivalence relation E on a set A, the set of all equivalence classes forms a partition of A by Theorem 1.2.8. It also follows that corresponding to any partition, we can associate an equivalence relation in a natural way. We state this formally in the next result. Its proof follows easily.

Theorem 1.2.9 *Let $\mathcal{P}$ be a partition of the set A. Define a relation E on A by for all $x, y \in A$, xEy if there exists $B \in \mathcal{P}$ such that x, $y \in B$. Then E is an equivalence relation on A and the equivalence classes are precisely the elements of $\mathcal{P}$.*

The relation E in Theorem 1.2.9 is called the **equivalence relation** on A **induced by the partition** $\mathcal{P}$.

Given a relation R from a set A into a set B and a relation S from B into a set C, there is a relation T from A into C that arises in a natural way as follows: $(a, c) \in T$ for $a \in A$ and $c \in C$ if and only if there exists $b \in B$ such that $(a, b) \in R$ and $(b, c) \in S$. This relation T is called the **composition** of R and S and is denoted by $S \circ R$. More formally we have the following definition.

Definition 1.2.10 *Let R be a relation from a set A into a set B and S be a relation from B into a set C. The* **composition** *of R and S, denoted by $S \circ R$, is the relation from A into C defined by*

$$x(S \circ R)z \text{ if there exists } y \in B \text{ such that } xRy \text{ and } ySz$$

for all $x \in A$, $z \in C$.

It is important to note that composition is associative.

Let R be a relation on a set A. We define recursively relations R^n on A, $n \in \mathbb{N}$, as follows: Let

$$R^1 = R.$$

Suppose R^n has been defined for $n \in \mathbb{N}$. Define

$$R^{n+1} = R \circ R^n.$$

Definition 1.2.11 *Let R be a relation from a set A into a set B. The **inverse** of R, denoted by R^{-1}, is the relation from B into A defined by*

$$xR^{-1}y \text{ if } yRx$$

for all $x \in B$, $y \in A$.

We are now interested in defining a lattice.

Definition 1.2.12 *A relation R on a set S is called a **partial order** on S if it satisfies the following conditions:*

*(1) $(x, x) \in R$ for all $x \in S$ (i.e., R is **reflexive**).*

*(2) $\forall x, y \in S$, if $(x, y) \in R$ and $(y, x) \in R$, then $x = y$ (i.e., R is **antisymmetric**).*

*(3) $\forall x, y, z \in S$, if $(x, y) \in R$ and $(y, z) \in R$, then $(x, z) \in R$ (i.e., R is **transitive**).*

A partial on a set S is usually denoted by $\leq$. We often write $x \leq y$ rather than $(x, y) \in \leq$.

Let R be a partial order on S. Then R is called a **total order** or **linear order** if $\forall x, y \in S$, either $(x, y) \in R$ or $(y, x) \in R$.

Definition 1.2.13 *A set S together with a partial order is called a **partially ordered set (poset)**. A set S together with a total order is called a **totally ordered set** or a **chain**.*

Definition 1.2.14 *Let $(S, \leq)$ be a poset and $\{x, y\} \subseteq S$. Then $z \in S$ is called an **upper bound** of $\{x, y\}$ if $x \leq z$ and $y \leq z$. An element $w \in S$ is called a **lower bound** of $\{x, y\}$ if $w \leq x$ and $w \leq y$.*

Definition 1.2.15 *Let $(S, \leq)$ be a poset and $\{x, y\} \subseteq S$. Then $z \in S$ is called a **least upper bound** of $\{x, y\}$ if z is an upper bound of $\{x, y\}$ and if $z' \in S$ is an upper bound of $\{x, y\}$, then $z \leq z'$. An element $w \in S$ is called a **greatest lower bound** of $\{x, y\}$ if w is a lower bound of $\{x, y\}$ and if $w' \in S$ is a lower bound of $\{x, y\}$, then $w' \leq w$.*

Definition 1.2.16 *Let $(L, \leq)$ be a poset. Then $(L, \leq)$ (or simply L) is called a **lattice** if for every $x, y \in L$, x and y have a least upper bound, denoted $x \vee y$, and greatest lower bound, denoted $x \wedge y$.*

A lattice $(L, \leq)$ is called **distributive** if $x \wedge (y \vee z) = (x \wedge y) \vee (x \wedge z)$ for all $x, y, z \in L$.

1.3 Functions

In this section, we review some of the basic properties of functions.

Definition 1.3.1 *Let A and B be nonempty sets. A relation α from A into B is called a* **function** *(or* **mapping***) from A into B if*

(1) $Dom(\alpha) = A$ and

(2) for all $(x, y), (x', y') \in \alpha$, $x = x'$ implies $y = y'$.

When (2) is satisfied by a relation α, we say that α is **well defined** *or* **single-valued***.*

We use the notation $\alpha : A \to B$ to denote a function α from a set A into a set B. For $(x, y) \in \alpha$, we usually write $\alpha(x) = y$ and say that y is the **image** of x under α and x is a **preimage** of y under α.

If α is a relation from A into B such that (2) of Definition 1.3.1 holds, but (1) is not required, then α is called a **partial function** from A into B.

Definition 1.3.2 *Let α be a function from a set A into a set B. Then*

(1) α is called **one-one** *if for all $x, x' \in A$, $\alpha(x) = \alpha(x')$ implies $x = x'$. (In this case, α is sometimes called* **injective***.)*

(2) α is called **onto** *B (or α* **maps** *A* **onto** *B) if $Im(\alpha) = B$. (In this case, α is sometimes called* **surjective***.)*

(3) If α is both injective and surjective, it is called **bijective***.*

Let A be a nonempty set. The function $\iota : A \to A$ defined by $\iota(x) = x$ for all $x \in A$ is a one-one function of A onto A. ι is called the **identity map** on A.

Let A, B, and C be nonempty sets and $\alpha : A \to B$ and $\beta : B \to C$. By Definition 1.2.10, the **composition** $\circ$ of α and β, written $\beta \circ \alpha$, is the relation from A into C defined as follows:

$$\begin{aligned}\beta \circ \alpha \;=\; & \{(x, z) \mid x \in A,\ z \in C, \text{ there exists } y \in B \\ & \text{such that } \alpha(x) = y \text{ and } \beta(y) = z\}.\end{aligned}$$

Let $\alpha : A \to B$ and $\beta : B \to C$ and $(x, z) \in \beta \circ \alpha$, i.e., $(\beta \circ \alpha)(x) = z$. Then by the definition of composition of functions, there exists $y \in B$ such that $\alpha(x) = y$ and $\beta(y) = z$. Now

$$z = \beta(y) = \beta(\alpha(x)).$$

Hence $(\beta \circ \alpha)(x) = \beta(\alpha(x))$.

We state the following basic results without proof.

Theorem 1.3.3 *Suppose that $\alpha : A \to B$ and $\beta : B \to C$. Then the following properties hold:*

(1) $\beta \circ \alpha : A \to C$, i.e., $\beta \circ \alpha$ is a function from A into C;

(2) If α and β are one-one, then $\beta \circ \alpha$ is one-one;

(3) If α is onto B and β is onto C, then $\beta \circ \alpha$ is onto C. ■

Theorem 1.3.4 *Let $\alpha : A \to B$, $\beta : B \to C$, and $\gamma : C \to D$. Then*

$$\gamma \circ (\beta \circ \alpha) = (\gamma \circ \beta) \circ \alpha.$$

That is, composition of functions is associative. ■

1.4 Fuzzy Subsets

The concept of a fuzzy subset of a set was introduced by L.A. Zadeh [157] in 1965 as a means to represent uncertainty. This notion was originally introduced as a method to consider imprecision and ambiguity occurring in human discourse and thought.

Let X be a set and A be a subset of X. The **characteristic function** of A is the function C_A of X into $\{0, 1\}$ defined by $C_A(x) = 1$ if $x \in A$ and $C_A(x) = 0$ if $x \notin A$. The characteristic function can be used to indicate either membership or nonmembership in a subset of X. This notion can be generalized in a way that introduces the notion of a fuzzy subset of X.

Definition 1.4.1 *A **fuzzy subset** f of X is a function of X into the closed interval $[0, 1]$.*

Let f be a fuzzy subset of a set X. For all $x \in X$, $f(x)$ can be thought of as the degree of membership of x in f. We sometimes use the notation f_A for a fuzzy subset of X, where A is thought of as a fuzzy set and f_A gives the grade of membership of elements of X in A. At times A may be merely a description of a fuzzy subset f of X.

Definition 1.4.2 *Let f be a fuzzy subset of X.*

*(1) Let $t \in [0, 1]$. Define $f_t = \{x \in X \mid f(x) \geq t\}$. We call f_t a **t-cut** or a **level set.***

*(2) The **support** of f is defined to be the set,*

$$Supp(f) = \{x \in X \mid f(x) > 0\}.$$

We let $\mathcal{FP}(X)$ denote the **fuzzy power set** of X, i.e., the set of all fuzzy subsets of X.

Definition 1.4.3 *Let f and g be fuzzy subsets of a set X. Define $\overline{f}$, $f \cap g$, $f \cup g$ as follows:*

$$\begin{aligned} \overline{f}(x) &= 1 - f(x), \\ (f \cap g)(x) &= \min\{f(x), g(x)\}, \\ (f \cup g)(x) &= \max\{f(x), g(x)\}, \end{aligned}$$

*for all $x \in X$. $\overline{f}$ is called the **complement** of f, $f \cap g$ is called the **intersection** of f and g, and $f \cup g$ is called the **union** of f and g.*

We use $\wedge$ to denote min and infimum and $\vee$ to denote max and supremum. Using these symbols, $(f \cap g)(x) = f(x) \wedge g(x)$ and $(f \cup g)(x) = f(x) \vee g(x)$ for all $x \in X$.

The notion of union and intersection can be extended to a family of fuzzy subsets of X. Let $\{f_i\}_{i \in I}$ be a family of fuzzy subsets of X, where I is an index set. Define $\cap_{i \in I} f_i$ and $\cup_{i \in I} f_i$ as follows:

$$\begin{aligned}(\cap_{i \in I} f_i)(x) &= \wedge_{i \in I} f_i(x),\\ (\cup_{i \in I} f_i)(x) &= \vee_{i \in I} f_i(x).\end{aligned}$$

Thus if I is a finite set, say $I = \{1, 2, \ldots, n\}$, then $\cap_{i \in I} f_i = f_1 \cap f_2 \cap \ldots \cap f_n$ and $\cup_{i \in I} f_i = f_1 \cup f_2 \cup \ldots \cup f_n$. In this case, we sometimes write

$$(\cap_{i \in I} f_i)(x) = f_1(x) \wedge f_2(x) \wedge \ldots \wedge f_n(x)$$

and

$$(\cup_{i \in I} f_i)(x) = f_1(x) \vee f_2(x) \vee \ldots \vee f_n(x).$$

If f is a fuzzy subset of X, then f is said to have the **sup property** if for every subset Y of X, there exists $y_0 \in Y$ such that

$$f(y_0) = \{f(y) \mid y \in Y\}.$$

Definition 1.4.4 *Let X, Y, and Z be nonempty sets and let f be a fuzzy subset of $X \times Y$ and g be a fuzzy subset of $Y \times Z$. Define the fuzzy subset $f \circ g$ of $X \times Z$ by*

$$(f \circ g)(x, z) = \vee\{f(x, y) \wedge g(y, z) \mid y \in Y\}$$

$\forall x \in X$ and $\forall z \in Z$.

If f is a fuzzy subset of $X \times X$, we define $f^1 = f$ and $f^{n+1} = f \circ f^n$ for all $n \in \mathbb{N}$.

Further information about fuzzy subsets can be found in [25, 162].

1.5 Semigroups

In this section, we provide some basic results concerning semigroups that are needed later for our presentation of fuzzy codes, fuzzy automata, and fuzzy languages.

Let X be a nonempty set. A function from $X \times X$ into X is called a **binary operation** on X. If $*$ is a binary operation on X, then the pair $(X, *)$ is called a **mathematical system**. If $(X, *)$ is a mathematical system such that $\forall a, b, c \in X$, $(a*b)*c = a*(b*c)$, then $*$ is called **associative** and $(X, *)$ is called a **semigroup**. Let $(X, *)$ be a mathematical system. If there exists $e \in X$ such that $\forall a \in X$, $a * e = a = e * a$, then e is called an **identity** of $(X, *)$ and $(X, *)$ is said to have an identity. If $(X, *)$ has an identity, then it follows easily that the identity is unique. A semigroup $(X, *)$ is called a **monoid** if it has an identity.

Let $(X, *)$ be a semigroup and $\equiv$ be an equivalence relation on X. Then $\equiv$ is called a **right (left) congruence relation** on X if $\forall a, b, c \in X$, $a \equiv b$ implies $a * c \equiv b * c$ $(c * a \equiv c * b)$. A right and left congruence relation $\equiv$ on X is called a **congruence relation**. The number of congruence classes of $\equiv$, i.e., equivalence classes of $\equiv$, is called the **index** of $\equiv$.

Let $(X, *)$ be a semigroup and let S be a nonempty subset of X. Then $(S, *)$ is said to be a **subsemigroup** of $(X, *)$ if $(S, *)$ is a mathematical system. (Here for $(S, *)$, we mean $*$ restricted to $S \times S$.) Let $(X, *)$ be a monoid and let $(S, *)$ be a subsemigroup of $(X, *)$. If e is the identity of $(X, *)$ and $e \in S$, then $(S, *)$ is called a **submonoid** of $(X, *)$. We often write X for $(X, *)$ when the operation $*$ is understood. Let $a \in X$. We define $a^1 = a$ and if a^n is defined for $n \in \mathbb{N}$, we define $a^{n+1} = a^n * a$.

We now state some well-known theorems and definitions.

Theorem 1.5.1 *Let X be a semigroup (monoid). The intersection of any collection of subsemigroups (submonoids) of X is a subsemigroup (submonoid) of X.* ■

Definition 1.5.2 *Let X be a semigroup (monoid) and let S be a subset of X. Let $\langle S \rangle$ denote the intersection of all subsemigroups (submonoids) of X that contain S. Then $\langle S \rangle$ is called the subsemigroup (submonoid) of X **generated** by S. If $\langle S \rangle = X$, then S is called a set of **generators** for X.*

Let $(X, *)$ be a semigroup and let e be an element not in X. Let X' denote $X \cup \{e\}$. Extend $*$ from $X \times X$ into X to $*'$ from $X' \times X'$ into X' as follows: $a *' b = a * b$ if $a, b \in X$; $a *' e = a = e *' a$ if $a \in X$; $e *' e = e$. Then $(X', *')$ is a monoid and X is a subsemigroup of X'. Even if X were a monoid, it is not a submonoid of X' since e is not its identity.

Definition 1.5.3 *Let $(X, *)$ and $(Y, \cdot)$ be semigroups. A function α from X into Y is called a **homomorphism** if $\forall a, b \in X$, $\alpha(a * b) = \alpha(a) \cdot \alpha(b)$. Let α be a homomorphism of X into Y. If α is one-one, then α is called a **monomorphism**. If α is onto Y, then α is called an **epimorphism**. If α is both a monomorphism and epimorphism, then α is called an **isomorphism** and X and Y are said to be **isomorphic**.*

Theorem 1.5.4 *Let X, Y, and Z be semigroups. If α is a homomorphism of X into Y and β is a homomorphism of Y into Z, then $\beta \circ \alpha$ is a homomorphism of X into Z.* ■

Let S be a set. A **free semigroup** on the set S is a semigroup F together with a function $\alpha : S \to F$ such that for any semigroup X and every function $\beta : S \to X$, there exists a unique homomorphism $\gamma : F \to X$ such that $\gamma \circ \alpha = \beta$.

Theorem 1.5.5 *Let F be a free semigroup on a set S together with a function $\alpha : S \to F$. Then α is one-one and $\langle \alpha(S) \rangle = F$.*

Proof. Let $a, b \in S$ be such that $a \neq b$. Let X be a semigroup containing more than one element and consider a function $\beta : S \to X$ such that $\beta(a) \neq \beta(b)$. Now $\gamma \circ \alpha = \beta$. Since $\gamma(\alpha(a)) = \beta(a) \neq \beta(b) = \gamma(\alpha(b))$, it follows that $\alpha(a) \neq \alpha(b)$. Hence α is one-one.

We now show that $\langle \alpha(S) \rangle = F$. Let $\langle \alpha(S) \rangle = A$. Then the function α defines a function $\beta : S \to A$ such that $\iota \circ \beta = \alpha$, where ι is the identity homomorphism $\iota : A \to F$. By the definition of a free semigroup, there exists a homomorphism $\gamma : F \to A$ such that $\gamma \circ \alpha = \beta$. Consider the diagram:

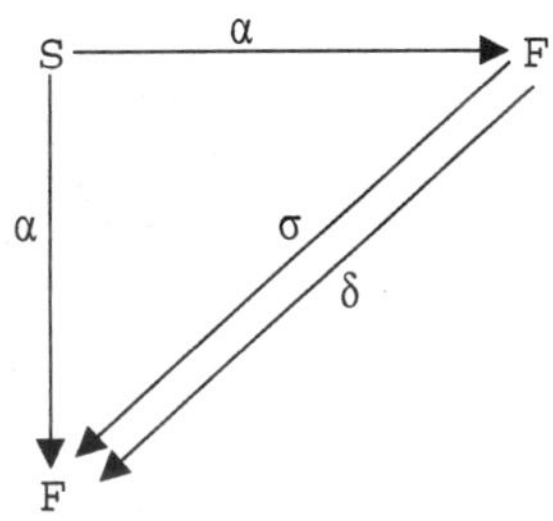

where σ is the identity isomorphism and $\delta = \iota \circ \gamma$. Since $\sigma \circ \alpha = \alpha$, $\delta \circ \alpha = \iota \circ \gamma \circ \alpha = \iota \circ \beta = \alpha$, it follows that $\iota \circ \gamma = \delta = \sigma$ from the uniqueness property defining free semigroups. The homomorphism ι must be an epimorphism. Thus $A = X$ and so $\alpha(S)$ generates F. ∎

Theorem 1.5.6 *Let (F, α) and (F', α') be free semigroups on the same set S. Then there exists a unique isomorphism $\sigma : F \to F'$ such that $\sigma \circ \alpha = \alpha'$.*

Proof. Since (F, α) is a free semigroup on the set S, there exists a homomorphism $\sigma : F \to F'$ such that $\sigma \circ \alpha = \alpha'$. Similarly, there exists a homomorphism $\delta : F' \to F$ such that $\delta \circ \alpha' = \alpha$. Let $\gamma = \delta \circ \sigma$ and ι be the identity isomorphism of F. In the diagram,

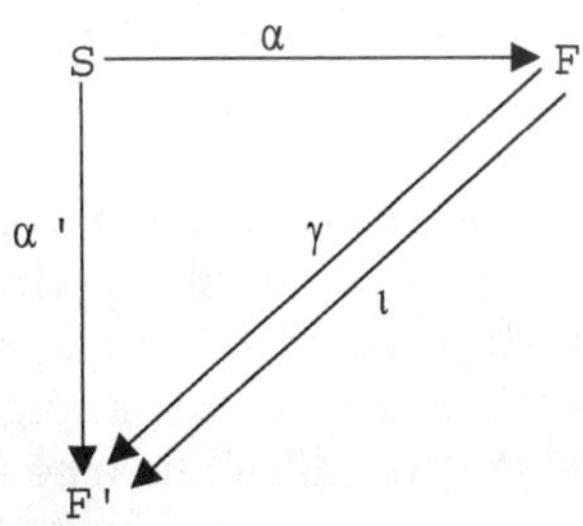

it follows that $\gamma \circ \alpha = \delta \circ \sigma \circ \alpha = \delta \circ \alpha' = \alpha$, $\iota \circ \alpha = \alpha$. From the uniqueness, it follows that $\delta \circ \sigma = \gamma = \iota$. Since ι is an isomorphism, it follows that σ is a monomorphism. Similarly, it follows that $\sigma \circ \delta$ is the identity endomorphism on F'. Thus σ is an epimorphism. Hence σ is an isomorphism. ∎

Theorem 1.5.7 *Let S be a set. Then there exists a free semigroup on S.*

Proof. Let F denote the set of all finite sequences of elements (repetitions allowed) of S. Let $x = (a_1, \ldots, a_m)$, $y = (b_1, \ldots, b_n) \in F$. Define

$$xy = (a_1, \ldots, a_m, b_1, \ldots, b_n).$$

Clearly, for any $x, y, z \in F$, $x(yz) = (xy)z$. Thus under this operation, F is a semigroup.

Define $\alpha : S \to F$ as follows: $\forall a \in S$, define $\alpha(a) = (a)$, the sequence consisting of the element a. We now show that (F, α) is a free semigroup on S.

Let $\beta : S \to X$ be any function from S into a semigroup X. Define $\gamma : F \to X$ by $\gamma(a_1, \ldots, a_m) = \beta(a_1) \ldots \beta(a_m)$ for all $(a_1, \ldots, a_m) \in F$. Clearly, γ is a homomorphism. Let $a \in S$. Then $(\gamma \circ \alpha)(a) = \gamma(\alpha(a)) = \gamma((a)) = \beta(a)$. Hence $\gamma \circ \alpha = \beta$.

Let $\delta : F \to X$ be a homomorphism such that $\delta \circ \alpha = \beta$. We show that $\gamma = \delta$. Let $(a_1, \ldots, a_m) \in F$. Then $\delta(a_1, \ldots, a_m) = \delta((a_1) \ldots (a_m)) = \delta((a_1)) \ldots \delta((a_m)) = \delta(\alpha(a_1)) \ldots \delta(\alpha(a_m)) = \beta(a_1) \ldots \beta(a_m) = \gamma(a_1, \ldots, a_m)$. Thus $\gamma = \delta$. Therefore, (F, α) is a free semigroup on S. ∎

Theorem 1.5.8 *Let X be a semigroup and let S be a set of generators of X. Then every element of X can be written as the product of a finite sequence of elements in S.*

Proof. Let (F, α) be the free semigroup on S as constructed in the proof of Theorem 1.5.7. Then by the definition of a free semigroup, there exists a homomorphism $\gamma : F \to X$ such that $\gamma \circ \alpha = \beta$, where β is the inclusion map $\beta : S \to X$.

We now prove that γ is an epimorphism. Clearly, $\gamma(F)$ is a subsemigroup of X. Since $S = \beta(S) = (\gamma \circ \alpha)(S) = \gamma(\alpha(S)) \subseteq \gamma(F)$ and S generates X, it follows that $\gamma(F) = X$. Thus γ is an epimorphism.

Let $x \in X$. Then there exists $(a_1, \ldots, a_m) \in F$ such that $\gamma(a_1, \ldots, a_m) = x$. Now $x = \gamma(a_1, \ldots, a_m) = \gamma((a_1) \ldots (a_m)) = \gamma((a_1)) \ldots \gamma((a_m)) = \gamma(\alpha(a_1)) \ldots \gamma(\alpha(a_m)) = \beta(a_1) \ldots \beta(a_m) = a_1 \ldots a_m$, where $a_1, \ldots, a_m \in S$. ∎

Let S be a set. Then S determines a unique free semigroup (F, α). Since $\alpha : S \to F$ is injective, we may identify S with its image $\alpha(S)$ in F. We can then consider S to be a subset of F such that S generates F. Also, any function $\beta : S \to X$, where X is a semigroup, extends to a unique homomorphism $\gamma : F \to X$. We call F the **free semigroup generated** by S.

Consider the monoid $\widehat{F} = F \cup \{e\}$, where F is the free semigroup generated by S and $xe = x = ex \ \forall x \in \widehat{F}$. Then every function $\beta : S \to X$, where X is a monoid, extends to a unique proper homomorphism $\gamma : \widehat{F} \to X$. This monoid $\widehat{F}$ is called the **free monoid generated** by the set S.

We now present the approach usually taken in language and automata theory. Let A be a set. Call A an **alphabet** and the elements of A, **letters.** A **word** in A is a nonempty finite sequence $x_1 \dots x_n$, where $x_i \in A$, $i = 1, \dots, n$. Then two words $x_1 \dots x_n$ and $y_1 \dots y_m$ are equal if an only is $n = m$ and $x_i = y_i$, for $i = 1, \dots, n$. Let A^+ denote the set of all words in A.

Then A^+ together with the binary operation of concatenation is a free semigroup on A. The set $A^* = A^+ \cup \{e\}$, where e denotes the empty word, is a free monoid on A. It is easy to see that every word w in A^+ has a unique factorization as a product of elements of A and so A generates A^+.

Proposition 1.5.9 *Let M be a moniod and $S = M \backslash \{e\}$. Then M is free if and only if every element of S has a unique factorization as a product of elements of $A = S \backslash S^2$.*

Proof. If $M = A^*$, then $A = S \backslash S^2$ and the unique factorization property follows from the definition of A^*. Conversely, suppose the unique factorization property holds. Then the canonical injection ϕ of $A = S \backslash S^2$ into M extends to a homomorphism $\hat{\phi} : A^* \to M$. From

$$\hat{\phi}(a_1 a_2 \dots a_n) = \phi(a_1)\phi(a_2)\dots\phi(a_n) = a_1 a_2 \dots a_n,$$

and the fact that every element of S has a unique factorization as a product of elements of A, it follows that ϕ is an isomorphism. Thus M is free. ■

Definition 1.5.10 *A monoid M is called **equidivisible** if for all a, b, c, $d \in M$, $ab = cd$ implies either $a = cu$, $ub = d$ for some $u \in M$ or $av = c$, $b = vd$ for some $v \in M$.*

A general study of equidivisibility can be found in [81].

Proposition 1.5.11 *Let M be a monoid and $S = M \backslash \{e\}$. Then M is free if and only if M is equidivisible and $\cap_{n \in \mathbb{N}} S^n = \emptyset$.*

Proof. That M is free implies equidivisibility and $\cap_{n \in \mathbb{N}} S^n = \emptyset$ is immediate. Conversely, assume M is equidivisible and $\cap_{n \in \mathbb{N}} S^n = \emptyset$. Then the group of units of M is trivial, for if $u \neq e$ is invertible, then $u = u u^{-1} u u^{-1} u \dots \in S^n$ for every $n \in \mathbb{N}$. However, this is impossible since $\cap S^n = \emptyset$. Suppose there exists $k \in \mathbb{N}$ such that $S^k = S^{k+1}$. Then $S^k = \emptyset$. Thus $S = \emptyset$ and M is the free monoid on the empty set. Suppose that the chain of ideals

$$S \supset S^2 \supset \dots \supset S^k \supset S^{k+1} \supset \dots$$

is strictly descending. Then for every $m \in S$ there exists an integer k such that $m \in S^k \backslash S^{k+1}$, else $m \in \bigcap_{n \in \mathbb{N}} S^n$, a contradiction. Thus $m = s_1 s_2 \ldots s_k$ with $s_i \in A = S \backslash S^2$ for all i, $i = 1, 2, \ldots, k$. Hence every $m \in M$, $m \neq e$, has a factorization. We now consider uniqueness. For any two factorizations of $m = a_{i_1} a_{i_2} \ldots a_{i_k} = a_{j_1} a_{j_2} \ldots a_{j_l}$ as a product of elements of A, we let $n = k \wedge l$ and proceed by induction on n. For $n = 1$, $m = a_{i_1} = a_{j_1}$ by the definition of A. For $n > 1$, either $a_{i_1} = a_{j_1} u$, $u a_{i_2} \ldots a_{i_k} = a_{j_2} \ldots a_{j_l}$ for some $u \in M$, or $a_{i_1} v = a_{j_1}$ and $a_{i_2} \ldots a_{i_k} = v a_{j_2} \ldots a_{j_l}$ for some $v \in M$ since M is equidivisible. The definition of A yields either u or $v = e$, $a_{i_1} = a_{j_1}$ and uniqueness follows from the induction hypothesis applied to $a_{i_2} \ldots a_{i_k} = a_{j_2} \ldots a_{j_l}$. By Proposition 1.5.9, M is free. ■

The proof of the next following two corollaries can be found in [75].

Corollary 1.5.12 [79] *Let M be a monoid. Then M is free if and only if M is equidivisible and there exists a homomorphism α from M into the monoid $(\mathbb{N}^0, +, 0)$ such that $\alpha^{-1}(0)$ is the identity of M.* ■

Corollary 1.5.13 *A monoid M is free if and only if M is cancellative, is equidivisible, has a trivial group of units and every $m \in M$, $m \neq e$, has only a finite number of nontrivial left factors.* ■

Example 1.5.14 *Let S be a set and $\mathbb{R}^0 = \mathbb{R}^+ \cup \{0\}$. Let $F(\mathbb{R}^0, S)$ denote the set of all functions defined on intervals $(0, r]$ of $\mathbb{R}^0$ into S. An element of $F(\mathbb{R}^0, S)$ is denoted by (α, r), where α is the function and r is the length of the interval on which α is defined. Define an operation on $F(\mathbb{R}^0, S)$ by juxtaposition of graphs, that is, by*

$$(\alpha, r)(\beta, s) = (\gamma, r + s), \text{ where } \gamma(x) = \begin{cases} \alpha(x) & \text{if } x \in (0, r], \\ \beta(x - r) & \text{if } x \in (r, r + s]. \end{cases}$$

It follows that $F(\mathbb{R}^0, S)$ is cancellative and equidivisible and has a trivial group of units. However, a function has infinitely many left factors. By Corollary 1.5.13, $F(\mathbb{R}^0, S)$ is not a free monoid.

1.6 Codes

Let M be a submonoid of a free monoid A^* on the alphabet A and let M^+ denote the semigroup $M \backslash \{e\}$. It follows from Proposition 1.5.11 that $\bigcap_{n \in \mathbb{N}} (M^+)^n = \emptyset$. Thus every $m \in M^+$ can be written as a product $c_1 c_2 \ldots c_k$, where $c_i \in M^+ \backslash (M^+)^2$ for $i = 1, \ldots, k$ and so $C = M^+ \backslash (M^+)^2$ is a set of generators of M. Let $C' \subseteq M^+$ be any set of generators of M. Then for all $c \in C$, $c = c'_1 c'_2 \ldots c'_n$, where $c'_i \in C' \subseteq M^+$, $i = 1, \ldots, n$ and $n \in \mathbb{N}$. Thus $n = 1$ and $c \in C'$. Hence $C \subseteq C'$. We have thus proved the next result

Proposition 1.6.1 *Let M be submonoid of a free monoid. Then M has a unique minimal set of generators $C = M^+ \backslash (M^+)^2$, where $M^+ = M \backslash \{e\}$. C is called the **base** of M.* ■

Example 1.6.2 *Consider the free monoid A^* on $A = \{a, b\}$. Then $M = \{a^i \mid i \in \mathbb{N}^0, i \neq 1\}$ is a submonoid. M has base $C = \{a^2, a^3\}$. It follows that M is not free over C since $a^6 \in M$ has two decompositions $a^2a^2a^2$ and a^3a^3 as a product of elements of C. Hence by Proposition 1.5.9, M is not free.*

Proposition 1.6.3 *Let M be a submonoid of a free monoid A^*. Then the following conditions are equivalent:*

(1) M is free.

(2) For all $w \in A^*$, if $Mw \cap M = \emptyset$ and $wM \cap M \neq \emptyset$, then $w \in M$.

(3) For all $w \in A^*$, if $Mw \cap M \cap wM \neq \emptyset$, then $w \in M$.

Proof. (1)⇒(2): Suppose that $m_1 w \in M$ and $wm_2 \in M$ for some $m_1, m_2 \in M$. Then $m_1 (wm_2) = (m_1 w) m_2 \in M$. Since M is equidivisible by Proposition 1.5.11, it follows that either $m_1 = m_1 wm$ for some $m \in M$ or $m_1 w = m_1 m'$ for some $m' \in M$. In the first case, we have by cancellation in A^* that $e = wm$ and thus $w = m = e$. In the second case, $w = m'$. Thus $w \in M$.

(2)⇒(3): The implication is immediate.

(3)⇒(1): Suppose that M is not free. Let C denote the base of M. Then there exists $w \in M^+$ with at least two distinct factorizations as a product of elements of C. Choose such a $w \in M^+$ of minimal length in A^*. Now

$$w = c_{i_1} c_{i_2} \ldots c_{i_p} = c_{j_1} c_{j_2} \ldots c_{j_q}, \tag{1.1}$$

where $c_{i_k} \in C$ for $k = 1, \ldots, p$, $c_{j_l} \in C$ for $l = 1, \ldots, q$, and $c_{i_1} \neq c_{j_1}$. By equidivisibility in A^*, we have either $c_{i_1} = c_{j_1} u$ for some $u \in A^*$ or $c_{i_1} v = c_{j_1}$ for some $v \in A^*$. Consider the first case. Then $c_{i_1} = c_{j_1} u$ and (1.1) imply that $uc_{i_2} \ldots c_{i_p} = c_{j_2} \ldots c_{j_p}$. It follows that

$$c_{j_2} c_{j_3} \ldots c_{j_p} c_{i_1} = c_{j_2} c_{j_3} \ldots c_{j_p} c_{j_1} u = uc_{i_2} c_{i_3} \ldots c_{i_p} c_{i_1}$$

By (3), $u \in M$ and $c_{i_1} = c_{j_1} u$ implies $u = e$. Hence $c_{i_1} = c_{j_1}$. Cancelling in (1.1) by c_{i_1} yields $w' = c_{i_2} \ldots c_{i_p} = c_{j_2} \ldots c_{j_q}$ with two distinct factorizations as a product of elements of C and $|w'| < |w|$. However, this contradicts the minimality of the length of w. The argument for the second case is similar. Thus M is free. ■

Definition 1.6.4 *A subset C of a free monoid A^* is called a **code** over A if C is the base of a free submonoid M of A^*. We write $M = C^*$.*

Let C be a subset of A^*, B a set, and $\gamma : B \to C$ a bijection. Then γ extends to a homomorphism $\hat{\gamma} : B^* \to M$ onto the submonoid M of A^*

generated by C. It follows that C is a code if and only if $\hat{\gamma}$ is an isomorphism. In this case, γ is called an **encoding** of B in the alphabet A. However it is not always easy to determine if a subset C of A^* is a code. This can be seen by [75, Example 2.31, p. 107].

The next proposition provides a method for constructing examples of codes.

Proposition 1.6.5 *Let M be a submonoid of a free monoid A^*. Let C be the base of M. Then the following conditions are equivalent:*
(1) For all $w \in A^*$, if $Mw \cap M \neq \emptyset$, then $w \in M$.
(2) $CA^+ \cap C = \emptyset$.

Proof. Clearly, (1) implies (2). Suppose that (2) holds. Suppose further that there exist $m \in M$ and $w \in A^*$ be such that $mw \in M$. We prove that $w \in M$ by induction on the length of m (in A^*). If $m = 1$, then $w \in M$. Suppose $|m| > 0$. Then we factor m and mw as products of elements of C. This gives $c_{i_1}c_{i_2}\dots c_{i_k}w = c_{j_1}c_{j_2}\dots c_{j_l}$. Since $CA^+ \cap C = \emptyset$ and A^* is equidivisible, $c_{i_1} = c_{j_1}, c_{i_2}\dots c_{i_k}w = c_{j_2}\dots c_{j_l}$. Thus $w \in M$ by the induction hypothesis. ■

Definition 1.6.6 *Let C be a code over the alphabet A. Then C is called a* ***prefix*** *(resp.* ***suffix****)* ***code*** *if $CA^+ \cap C = \emptyset$ (resp. $A^+C \cap C = \emptyset$). C is called a* ***biprefix code*** *if it is a prefix and a suffix code.*

By condition (2) in Proposition 1.6.5, it follows that no word of C is a proper left factor of another word of C. Define the relation $\underset{l}{\leq}$ in A^* by $u \underset{l}{\leq} v$ if v is a left factor of u. Then it follows that $\underset{l}{\leq}$ is a partial ordering of A^*. In Figures 1.1(a) and (b), we represent the top part of A^* partially ordered by $\underset{l}{\leq}$ in the cases, where A has two and three elements, respectively.

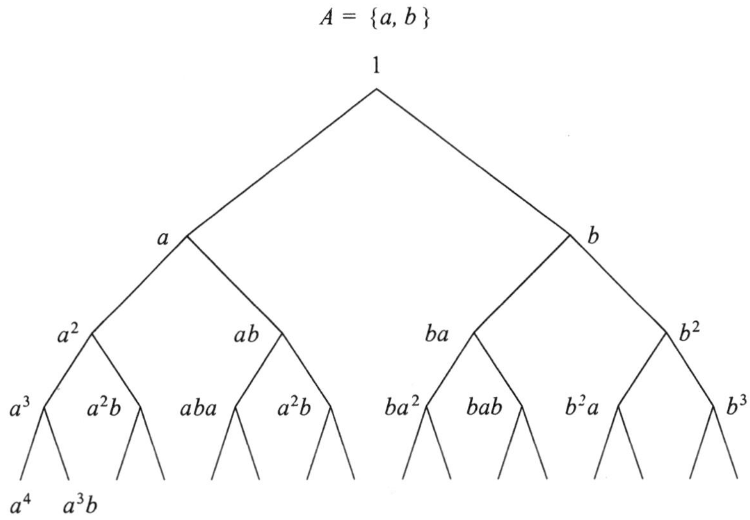

Figure 1.1 (a)[1]

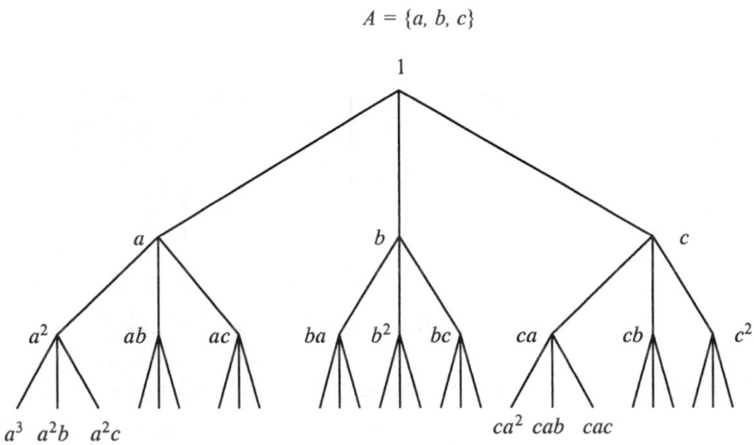

Figure 1.1(b)[1]

A subset C of A^* is a prefix code if and only if for every $c \in C$ and $w \in A^*$, $w \underset{l}{\leq} c$ and $w \neq c$ implies $w \notin C$. Thus to obtain examples of

---

[1]Semigroup and Combinatorial Applications, G. Lallament, copyright 1979, Reprinted by premission of John Wiley & Sons, Inc.

prefix codes, it suffices to select subsets C of A^* that will be end points with respect to $\underset{l}{\leq}$. For example, the falling trees in Figure 1.2(a) and (b)

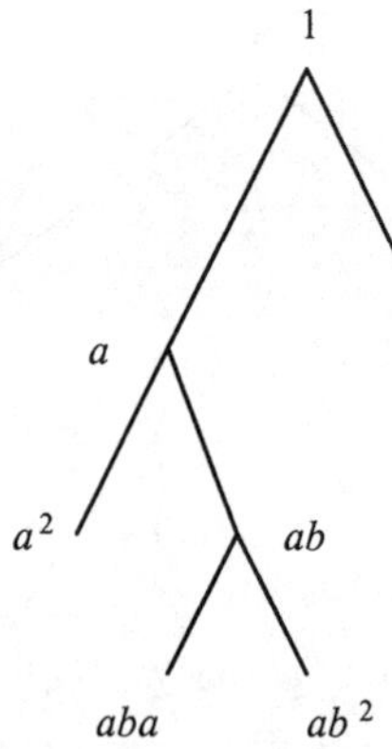

Figure 1.2(a)[2]

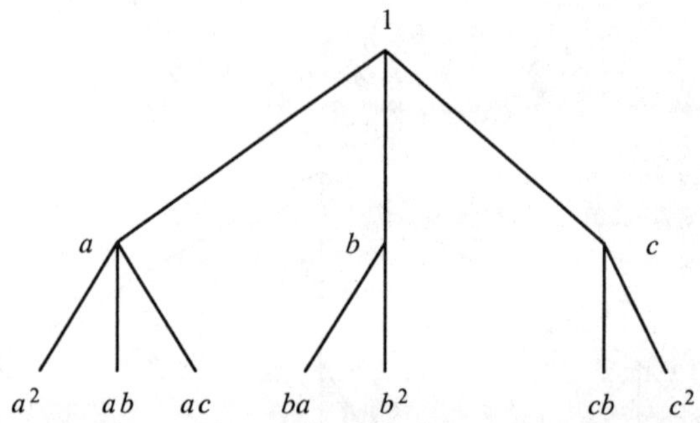

Figure 1.2(b)[2]

yield the prefix codes $C_1 = \{a^2, aba, ab^2, b\}$ over $\{a, b\}$ and $C_2 = \{a^2, ab, ac, ba, b^2, cb, c^2\}$ over $\{a, b, c\}$, respectively. The set $\{a^n b \mid n \in \mathbb{N}\}$ is an example of an infinite prefix code over $\{a, b\}$. It is represented by an infinite falling tree with end points $a^n b$, $n \in \mathbb{N}$.

There is no simple characterization of general codes analogous to condition (2) of Proposition 1.6.5 for prefix codes. We have the following result however. Its proof can be found in [75].

---

[2] Semigroup and Combinatorial Applications, G. Lallament, copyright 1979, Reprinted by premission of John Wiley & Sons, Inc.

Proposition 1.6.7 *Let C be a subset of the free monoid A^*. Define the sequence of subsets D_i recursively as follows: let $D_0 = C$ and*

$$D_i = \{w \in A^+ \mid D_{i-1}w \cap C \neq \emptyset \text{ or } Cw \cap D_{i-1} \neq \emptyset\}.$$

Then C is a code over A if and only if $C \cap D_i = \emptyset$ for all $i \in \mathbb{N}$. ■

If C of Proposition 1.6.7 is finite, then the length of the words in D_i for all $i \in \mathbb{N}$, is bounded by the maximal length of words in C. Therefore, there is only a finite number of distinct D_i, and the proposition gives an algorithm for deciding if C is a code or not.

Example 1.6.8 *(1) Let $C = \{a, a^3b, aba\}$. Then $D_0 = C$, $D_1 = \{a^2b, ba\}$, $D_2 = \{ab\}$, $D_3 = \{a, b\}$. Since $C \cap D_3 \neq \emptyset$, C is not a code.*

(2) Let $C = \{a, a^2b, bab, b^2\}$. Then $D_0 = C$, $D_1 = \{ab\}$, $D_2 = \{b\}$, $D_3 = \{ab, b\} = D_4 = \ldots$. Thus C is a code since $C \cap D_i = \emptyset$ for all $i \in \mathbb{N}$.

We now consider algebraic properties of codes.

A code C over an alphabet A is called a **maximal code** if $C \neq A$, and if for every code $C', C \subseteq C'$ implies $C = C'$. The codes $C = A^n$ $(n \geq 1)$ are examples of maximal codes. This follows since if $w \in A^+$ is not in A^n, then $A^n \cup \{w\}$ cannot be a code since w^n has two distinct decompositions over $A^n \cup \{w\}$. A^n is called the **homogeneous code** of degree n.

Proposition 1.6.9 *If C is a maximal code over the alphabet A, then $C^* \cap A^*wA^* \neq \emptyset$ for all $w \in A^*$.*

Proof. Suppose that $A = \{a\}$. Then $C = \{a^m\}$ for some $m > 1$. Hence the desired result holds. Suppose that A has more than one element. Assume that there exists $w \in A^*$ such that $C^* \cap A^*wA = \emptyset$. We may assume that for every $u \in A^+$ and $v \in A^*$, $w \neq uvu$ for if there exist u, v such that $w = uvu$, taking a letter $a \in A$ distinct from the first letter of u and letting $w' = wa^{|w|}$ it follows that $C^* \cap A^*w'A^* = \emptyset$ and $w' \neq uvu$ for all $u \in A^+$, $v \in A^*$. We now prove that $C' = C \cup \{w\}$ is a code. Suppose that $z = c_{i_1}c_{i_2}\cdots c_{i_m} = c_{j_1}c_{j_2}\cdots c_{j_n}$, where $c_{i_k}, c_{j_l} \in C'$. If all c_{i_k} and c_{j_l} are in C, then $m = n$ and $c_{i_1} = c_{j_1}, \ldots, c_{i_m} = c_{j_m}$. If $c_{i_k} = w$ for some k, $1 \leq k \leq m$, there exists l, $1 \leq l \leq p$, such that $c_{j_l} = w$ else $C^* \cap A^*wA^* \neq \emptyset$. Selecting for c_{i_k} and c_{j_l} the first occurrences of w on both sides of the two factorizations of z above we have that

$$c_{i_1}c_{i_2}\cdots c_{i_{k-1}}wc_{i_{k+1}}\cdots c_{i_m} = c_{j_1}c_{j_2}\cdots c_{j_{l-1}}wc_{j_{l+1}}\cdots c_{j_n},$$

where $c_{i_1}, \ldots, c_{i_{k-1}}, c_{j_1}, \ldots, c_{j_{l-1}} \in C$.

Suppose $|c_{i_1}c_{i_2}\cdots c_{i_{k-1}}| > |c_{j_1}c_{j_2}\cdots c_{j_{l-1}}|$. Then by equidivisability it follows that

$$c_{i_1}c_{i_2}\cdots c_{i_{k-1}} = c_{j_1}c_{j_2}\cdots c_{j_{l-1}}u, \tag{1.2}$$

$$uwc_{i_{k+1}} \cdots c_{i_m} = wc_{j_{l+1}} \cdots c_{j_n}, \tag{1.3}$$

where $u \neq e$. Now w cannot be a left factor of u else (1.2) contradicts $C^* \cap A^*wA^* = \emptyset$. Hence by (1.3), $w = uw'$ and $wc_{i_{k+1}} \cdots c_{i_m} = w'c_{j_{l+1}} \cdots c_{j_n}$ for some $w' \in A^*$. This last equality yields $w = w'v$ for some $v \in A^*$. However, $w = uw' = w'v$ with $u \neq e$ implies $w = xyx$ for some $x \in A^+$ and $y \in A^*$. This is impossible. Hence

$$\left|c_{i_1}c_{i_2} \cdots c_{i_{k-1}}\right| = \left|c_{j_1}c_{j_2} \cdots c_{j_{l-1}}\right|$$

and so $k = l$, $c_{i_1} = c_{j_1}, \ldots, c_{i_{k-1}} = c_{i_{l-1}}$. An argument using induction on the length of z gives $m = n$ and $c_{i_k} = c_{i_l}, \ldots, c_{i_m} = c_{i_n}$. Thus $C' = C \cup \{w\}$ is a code. However, this contradicts the maximality of C. Therefore $C^* \cap A^*wA^* \neq \emptyset$ for all $w \in A^*$. ■

The converse of Proposition 1.6.9 is not true. The code $C = \{a^{|w|+1}bw \mid w \in A^*\}$ over the alphabet $A = \{a, b\}$ is a prefix code such that C^* intersects all left ideals of A^*. However, $C \cup \{a\}$ is also a code. The top of the tree representing C is given in Figure 1.3.

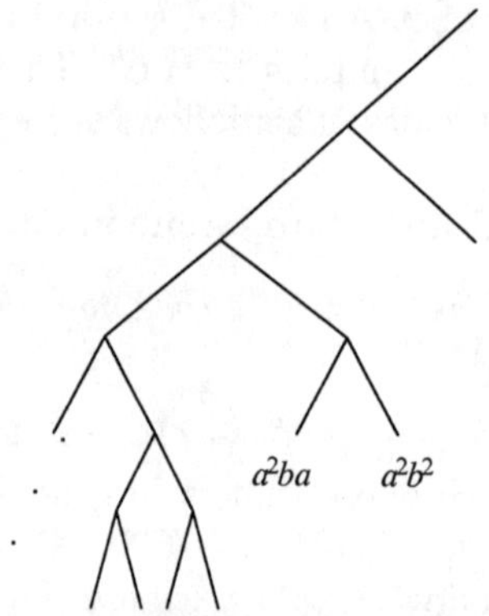

Figure 1.3[3]

Proposition 1.6.10 *Let C be a prefix code over an alphabet A. Then the following conditions are equivalent:*

(1) For every prefix code C' over A, $C \subseteq C'$ implies $C = C'$.

(2) For every $w \in A^*$, w has a left factor in C or is a left factor of some $c \in C$.

(3) For every $w \in A^*$, $wA^* \cap C^* \neq \emptyset$.

(4) $A^* = C^*P$ where P is the set of all proper left factors of words in C.

Proof. (2) ⇔ (4): Immediate.

(1) ⇒ (2): Suppose w is not a left factor of some $c \in C$. Then w has a left factor in C else $C \cup \{w\}$ is a prefix code by Definition 1.6.6.

---

[3]Semigroup and Combinatorial Applications, G. Lallament, copyright 1979, Reprinted by premission of John Wiley & Sons, Inc.

(2) $\Rightarrow$ (3): By induction on $|w|$, every $w \in A^*$ can be written as $w = c_1c_2 \dots c_i w'$, where w' has no left factor in C. Applying (2) to w', there exists $c \in C$ such that $w'w'' = c$ for some $w'' \in A^*$. It follows that $ww'' \in C^*$.

(3) $\Rightarrow$ (1): Suppose there exists a prefix code C' over A such that $C \subset C'$. Then for $w \in C' \backslash C$, $C \cup \{w\}$ is a prefix code. By (3), there exists $w' \in A^*$ such that $ww' = c_1c_2 \dots c_n$, where $c_i \in C$ for $i = 1, 2, \dots, n$. It follows that $w = c_1 \in C$, a contradiction. ∎

If C is a finite code satisfying (3) of Proposition 1.6.10, then C is necessarily a prefix code, [75, Chapter 6, Proposition 4.2]. The converse of Proposition 1.6.9 is also true with C finite, [75, Chapter 6, Proposition 4.1].

Definition 1.6.11 *A prefix code C satisfying any one of* (1), (2), (3), *or* (4) *of Proposition 1.6.10 is called* ***complete****, otherwise C is called* ***incomplete****.*

An example can be found in [75, p. 118] of a complete prefix code that is not a maximal code.

A complete prefix code C over A is represented by a tree, where all the vertices, except the end points, cover exactly $|A|$ vertices. In this case, w has at most one left factor in C for all $w \in A^*$.

Definition 1.6.12 *Let C be a code on the alphabet A. Then C is said to* ***decompose*** *over a code D if C is contained in the submonoid D^* of A^* generated by D. C is called* ***indecomposable*** *if it decomposes only over A or over itself, i.e., $C \subseteq D^*$ implies either $D = C$ or $D = A$.*

An example illustrating Definition 1.6.12 can be found in [75, p. 119].

Let B and D be codes over the alphabets Y and A, respectively. Suppose that there is a bijection $\phi : D \to Y$. Then there is a unique isomorphism $\hat{\phi} : D^* \to Y^*$ extending ϕ. The composite of B over D by ϕ, denoted by $B \underset{\phi}{\otimes} D$, is the code defined by

$$B \underset{\phi}{\otimes} D = \{w \in D^* \mid \phi(w) \in B\}.$$

The operation $\otimes$ is called the **composition of code.** It consists in rewriting a code B over an alphabet Y as a code over A using an encoding ϕ of Y on D. Clearly, a code C decomposes over D in the sense of Definition 1.6.12 if and only if $C = B \underset{\phi}{\otimes} D$ for some code B over an alphabet in bijection with D.

The proof of the following result can be found in [75, p. 119].

Proposition 1.6.13 *Every finite maximal (resp. complete prefix) code is a composite of a finite number of maximal (resp. complete prefix) indecomposable codes.* ∎

The product of the two codes is not, in general, a code. For example, let $C = C_1 \cdot C_2$, where $C_1 = \{a, ba\}$ and $C_2 = \{a, ab\}$. Then C is not a code. However, the product of two prefix codes is a prefix code.

1.7 Finite-State Machines

The theory of machines has had a strong influence on the development of computer systems and their associated languages and software. This theory has also found applications in such areas of science as biology, biochemistry, psychology, and others.

In Sections 1.7–1.11, we review some basic results of automata and language theory. Further details can be found in [43].

Definition 1.7.1 *A six-tuple* $M = (Q, X, Y, f, g, s)$ *is called a* ***finite-state machine*** *if* $Q, X,$ *and* Y *are finite nonempty sets,* $f : Q \times X \to Q$, $g : Q \times X \to Y$, *and* $s \in Q$.

The members of Q are called states. The members of X and Y are called **input** and **output symbols**, respectively. The functions f and g are called the **state transition** and **output functions**, respectively. The state s is called the **initial state.**

Example 1.7.2 *Let* $Q = \{q_0, q_1\}$, $X = \{a, b\}$, *and* $Y = \{0, 1\}$. *Let the functions* $f : Q \times X \to Q$ *and* $g : Q \times X \to Y$ *be as described in Table 1.1.*

	f		g	
$Q \setminus X$	a	b	a	b
q_0	q_0	q_1	0	1
q_1	q_1	q_1	1	0

Table 1.1

Then $M = (Q, X, Y, f, g, q_0)$ *is a finite-state machine. The interpretation of Table 1.1 is as follows:*

$$\begin{aligned} f(q_0, a) &= q_0 & g(q_0, a) &= 0 \\ f(q_0, b) &= q_1 & g(q_0, b) &= 1 \\ f(q_1, a) &= q_1 & g(q_1, a) &= 1 \\ f(q_1, b) &= q_1 & g(q_1, b) &= 0 \end{aligned}$$

The next-state and output functions can also be defined by a transition diagram.

Example 1.7.3 *The transition diagram for the finite-state machine of Example* 1.7.2 *is drawn in Figure* 1.4.

The transition diagram is known to be what is called a ***digraph*** *in graph theory. The vertices are the states. The initial state is indicated by an arrow* *as shown. If the finite-state machine is in state q and inputting x causes output y and a move to state q', a directed edge is drawn from the vertex q to the vertex q' and labelled x/y. The transition diagram of Figure* 1.4 *is obtained.*

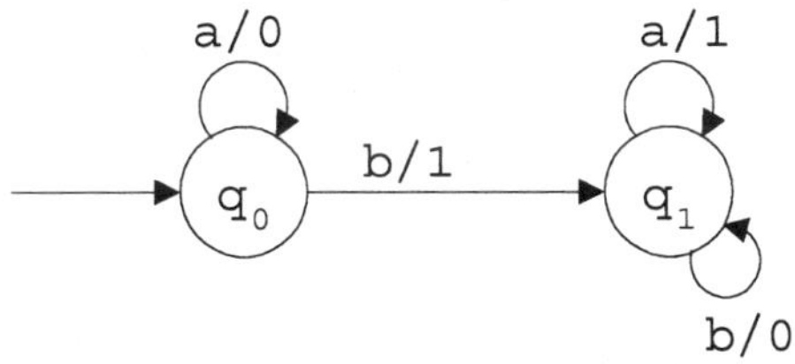

Figure 1.4

If S is a set such that $1 \leq |S| < \infty$, let S^* denote the free monoid generated by S with operation, concatenation. The elements of S^* are called **strings** or **words** over S. The empty string, Λ, is the identity of S^*. Let $S^+ = S^* \backslash \{\Lambda\}$.

Definition 1.7.4 *Let $M = (Q, X, Y, f, g, q)$ be a finite-state machine. An* ***input string*** *for M is a string over X. The string*

$$y_1 \cdots y_n$$

is the ***output string*** *for M corresponding to the input string*

$$x_1 \dots x_n$$

if there exists states $q_0 \dots q_n \in Q$ such that

$$\begin{aligned} q_0 &= q \\ q_i &= f(q_{i-1}, x_i) \qquad \text{for } i = 1, \dots, n_i \\ y_i &= g(q_{i-1}, x_i) \qquad \text{for } i = 1, \dots, n. \end{aligned}$$

1.8 Finite-State Automata

A **finite-state automaton** is a special kind of finite-state machine. It is of special interest due to its relationship to languages as shown in Section 1.10.

Definition 1.8.1 *A* ***finite-state automaton*** $A = (Q, X, Y, f, g, s)$ *is a finite-state machine such that the set of output symbols Y is* $\{0, 1\}$ *and such that the current state determines the last output. Those states for which the last output was* 1 *are called* ***accepting states.***

Example 1.8.2 *Consider the transition diagram of the finite-state machine A defined by the following table. The initial state is q_0. A is a finite-state automation and the set of accepting states is $\{q_1, q_2\}$.*

	f		g	
$Q\backslash X$	a	b	a	b
q_0	q_0	q_1	0	1
q_1	q_0	q_2	0	1
q_2	q_0	q_2	0	1

The transition diagram is shown in Figure 1.5. *If the finite-state machine is in state q_0, the last output was* 0. *If the machine is in either state q_1 or q_2, the last output was* 1. *Thus A is a finite-state automaton. The accepting states are q_1 and q_2.*

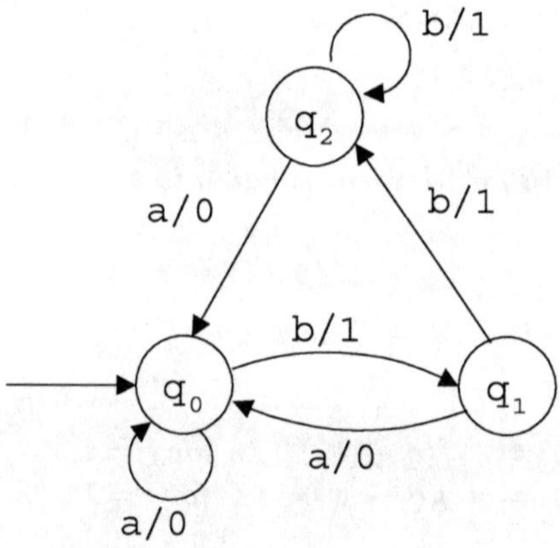

Figure 1.5

From Example 1.8.2, it follows that the finite-state machine defined by a transition diagram is a finite-state automaton if the set of output symbols is $\{0, 1\}$ and if, for each state q, all incoming edges to q have the same output label.

The transition diagram of a finite-state automaton is often drawn with the accepting states in double circles and with the output symbols omitted. If we draw the transition diagram of Figure 1.5 in this way, we obtain the transition diagram of Figure 1.6.

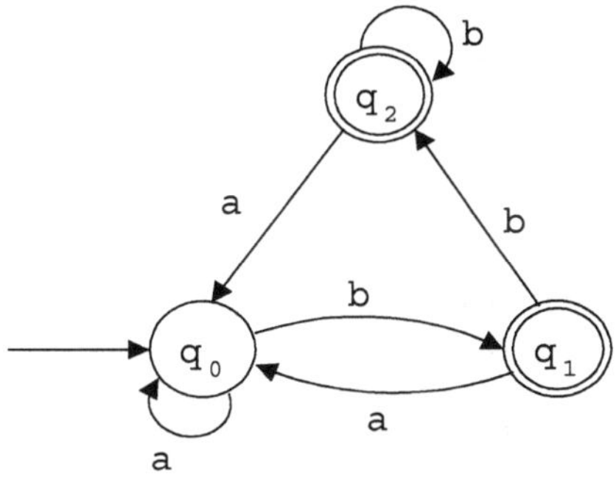

Figure 1.6

Example 1.8.3 *We now draw the transition diagram of the finite-state automaton of Figure 1.7 as a transition diagram of a finite-state machine.*

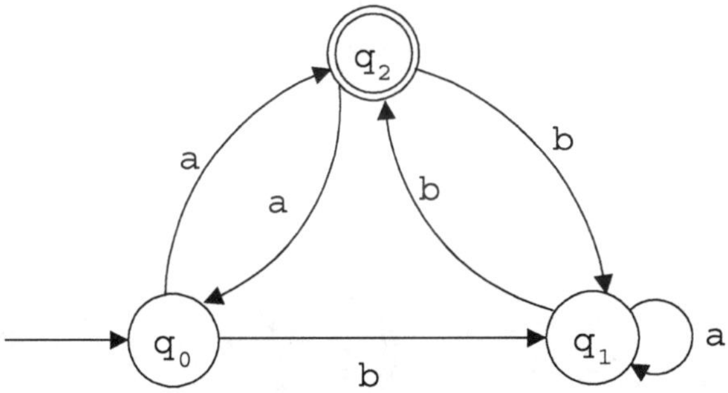

Figure 1.7

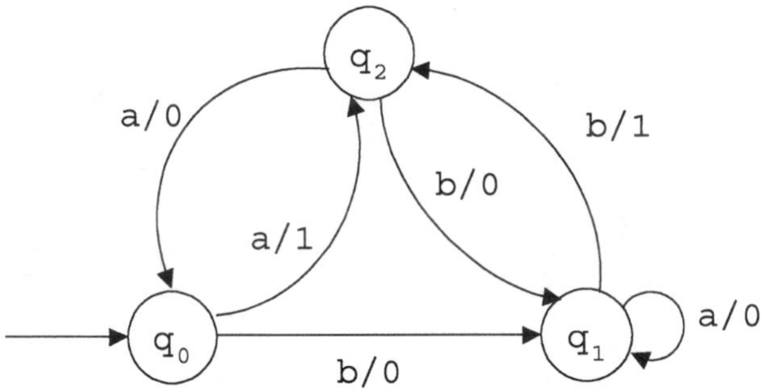

Figure 1.8

Since q_2 is an accepting state, we label all its incoming edges with output 1 as in Figure 1.8. Since the states q_0 and q_1 are not accepting, all their *incoming edges are labelled with output* 0. *The transition diagram of Figure 1.8 is obtained.*

An alternative to Definition 1.8.1 can be obtained by regarding a finite-state automaton A as a 5-tuple $(Q, X, f, \mathcal{A}, g)$, where
(1) a finite set Q of *states,*
(2) a finite set X of *input symbols,*
(3) a *next-state function* f from $Q \times X$ into Q,
(4) a subset $\mathcal{A}$ of Q of *accepting states,*
(5) an *initial state* $q \in Q$.

Example 1.8.4 *The transition diagram of the finite-state automaton* $A = (Q, X, f, \mathcal{A}, q)$, *where* $Q = \{q_0, q_1, q_2\}$, $X = \{a, b\}$, $\mathcal{A} = \{q_2\}$, $q = q_0$, *and* f *is given by the following table,*

$Q \setminus X$	f: a	b
q_0	q_1	q_0
q_1	q_2	q_0
q_2	q_2	q_0

is shown in Figure 1.9.

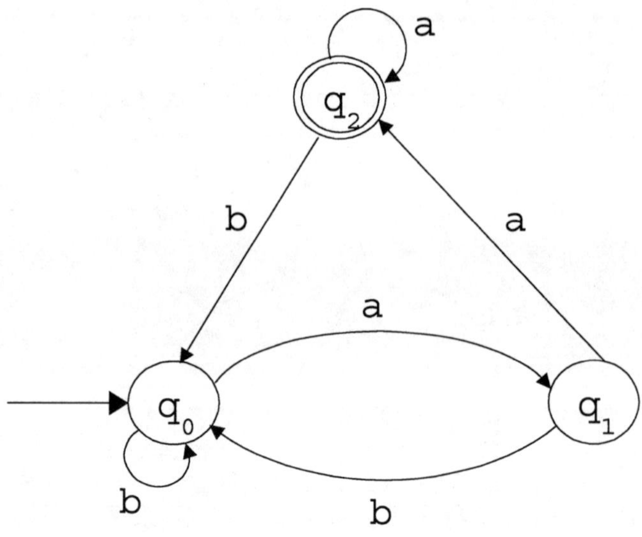

Figure 1.9

If a string is input to a finite-state automaton, we end at either an accepting or a nonaccepting state. The status of this final state determines whether the string is accepted by the finite-state automaton.

Definition 1.8.5 *Let $A = (Q, X, f, \mathcal{A}, q)$ be a finite-state automaton. Let $x = x_1 \cdots x_n$ be a string over X. If there exists states $q_0, \ldots, q_n$ satisfying*

(1) $q_0 = q$;

(2) $f(q_{i-1}, x_i) = q_i$ for $i = 1, \ldots, n$;

(3) $q_n \in \mathcal{A}$;

then x is said to be ***accepted*** *by A. The empty string is said to be* ***accepted*** *if $q \in \mathcal{A}$. Let $Ac(A)$ denote the set of strings accepted by A. Then A is said to* ***accept*** *$Ac(A)$.*

Let $x = x_1 \ldots x_n$ be a string over X. Define states $q_0, \ldots, q_n$ by conditions (1) and (2) above. Then the (directed) path $(q_0, \ldots, q_n)$ is called the **path representing** x in A.

It follows from Definition 1.8.5 that if the path P represents the string x in a finite-state automaton A, then A accepts x if and only if P ends at an accepting state.

Example 1.8.6 *We design a finite-state automaton that accepts precisely those strings over $\{a, b\}$ that contain an odd number of a's.*

We use two states:

E : An even number of a's are in the string.

O : An odd number of a's are in the string.

The initial state is E and the accepting state is O. We obtain the transition diagram shown in Figure 1.10.

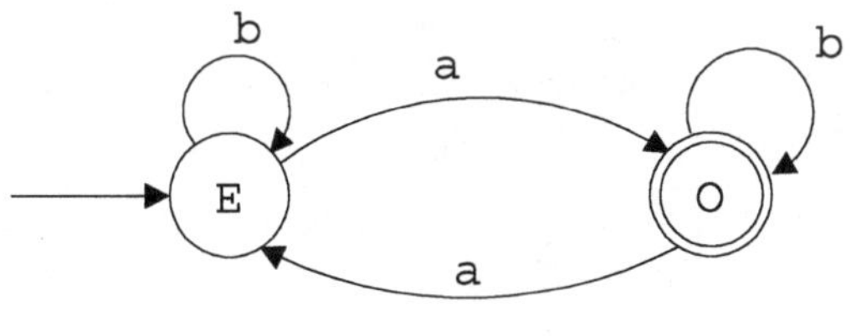

Figure 1.10

A finite-state automaton can be thought of as an algorithm to decide whether or not a given string is accepted.

Finite-state automata A and A' are called **equivalent** if $Ac(A) = Ac(A')$.

1.9 Languages and Grammars

A derivation within a grammar is described as follows: Considering the current string with the starting symbol s, we search the productions for a rule whose left-hand side x is a substring of the current string. Any rule that is matched in this manner can be applied by replacing the substring with the right-hand side y of the rule. A derivation is complete when there are only terminal symbols in the current string.

Definition 1.9.1 *Let A be a finite set. A* ***(formal) language*** *L over A is a subset of A^*, the set of all strings over A.*

Example 1.9.2 *Let $A = \{a, b\}$. The set L of all strings over A containing an odd number of a's is a language over A. As we saw in Example 1.8.6, L is precisely the set of strings over A accepted by the finite-state automaton of Figure 1.10.*

One method to define a language is to give a list of rules that the language must obey.

Definition 1.9.3 *A* ***phrase-structure grammar*** *or* ***type 0 grammar*** *(or, simply,* ***grammar****) G is a 4-tuple $G = (N, T, P, s)$, where*

(1) N is a finite set whose elements are called ***nonterminal symbols,***

(2) T is a finite set whose elements are called ***terminal symbols,*** *where $N \cap T = \emptyset$,*

(3) P is a finite subset of $[(N \cup T)^* \backslash T^*] \times (N \cup T)^*$, called the ***set of productions,***

(4) $s \in N$ and is called the ***starting symbol.***

Example 1.9.4 *Let $N = \{s, S\}$, $T = \{a, b\}$, and $P = \{s \to bs, s \longrightarrow aS, S \longrightarrow bS, S \longrightarrow b\}$. Then $G = (N, T, P, s)$ is a grammar.*

Given a grammar G, a language $L(G)$ can be constructed from G by using the productions to derive the strings that make up $L(G)$. We begin with the starting symbol and then repeatedly use productions until a string of terminal symbols is obtained. The language $L(G)$ is the set of all such strings of terminals.

Definition 1.9.5 *Let $G = (N, T, P, s)$ be a grammar. If $z \longrightarrow w$ is a production and $xzy \in (N \cup T)^*$, then xwy is said to be* ***directly derivable*** *from xzy. In this case, we write*

$$xzy \Longrightarrow xwy.$$

If $z_i \in (N \cup T)^*$ for $i = 1, \ldots, n$, and z_{i+1} is directly derivable from z_i for $= 1, \ldots, n-1$, we say that z_n is ***derivable*** *from z_1 and write*

$$z_1 \Longrightarrow z_n.$$

We call

$$z_1 \Longrightarrow z_2 \Longrightarrow \cdots \Longrightarrow z_n$$

the ***derivation*** *of z_n (from z_1). By convention, any element of $(N \cup T)^*$ is derivable from itself.*

The **language generated** *by* G, written $L(G)$, consists of all strings over T derivable from s.

Example 1.9.6 *Let G be the grammar of Example 1.9.4. The string $abSbb$ is directly derivable from $aSbb$, written*

$$aSbb \Longrightarrow abSbb,$$

where only the production $S \longrightarrow bS$ is used.

The string $bbab$ is derivable from s, written

$$s \Longrightarrow bbab.$$

The derivation is

$$s \Longrightarrow bs \Longrightarrow bbs \Longrightarrow bbaS \Longrightarrow bbab.$$

The only derivations from s are given as follows:

$$\begin{array}{lll} s & \Longrightarrow & bs \\ & \vdots & \\ & \Longrightarrow & b^n s \qquad\qquad n \geq 0 \\ & \Longrightarrow & b^n aS \\ & \vdots & \\ & \Longrightarrow & b^n ab^{m-1} S \\ & \Longrightarrow & b^n ab^m \qquad\quad n \geq 0, \quad m \geq 1. \end{array}$$

Hence $L(G)$ consists of the strings over $\{a, b\}$ containing precisely one a that end with at least one b.

Grammars are classified according to the types of productions that define the grammars.

Definition 1.9.7 *Let G be a grammar.*

(1) If every production is of the form

$$zAw \longrightarrow zvw, \qquad \text{where } z, w \in (N \cup T)^*, \quad A \in N, \quad v \in (N \cup T)^* \backslash \{\Lambda\}, \tag{1.4}$$

then G is called a ***context-sensitive*** *(or* ***type 1****)* ***grammar****.*

(2) If every production is of the form

$$A \rightarrow v, \text{ where } A \in N, \ v \in (N \cup T)^*, \tag{1.5}$$

then G is called a ***context-free*** *(or* ***type 2****)* ***grammar****.*

(3) If every production is of the form

$$A \rightarrow x \text{ or } A \rightarrow xB \text{ or } A \rightarrow \Lambda \quad \text{where } A, B, \in N, \quad x \in T,$$

then G is called a ***regular*** *(or* ***type 3****)* ***grammar****.*

The definition of a context-sensitive grammar comes from the normal form of a more general definition of such grammars.

According to (1.4), in a context-sensitive grammar, A may be replaced by v if A is in the context of x and y. In a context-free grammar, (1.4) states that A may be replaced by v anytime. A regular grammar has simple substitution rules: A nonterminal symbol is replaced by a terminal symbol, or by a terminal symbol followed by a nonterminal symbol, or by the null string.

It is important to note that a regular grammar is a context-free grammar and that a context-free grammar with no productions of the form $A \to \Lambda$ is a context-sensitive grammar.

Some definitions allow x to be replaced by a string of terminals in Definition 1.9.7(3). However, it can be shown that the two definitions produce the same languages.

Let G be a context-free grammar. A derivation $w_0 \Rightarrow w_1 \Rightarrow \ldots \Rightarrow w_n$ is called a **leftmost derivation** if for $i = 0, 1, \ldots, n$, $w_i = xAy$, $w_{i+1} = xzy$, and $A \to z$ is a production, where $x \in T^*$, $A \in N$, $z \in L(G)$, and $y \in (N \cup T)^*$. G is called **ambiguous** if it generates an element of $L(G)$ by two or more distinct left derivations.

Example 1.9.8 *Consider the grammar G defined as follows:*

$$T = \{a, b, c\}, \quad N = \{s, A, B, C, D, E\}$$

with productions

$$\begin{array}{llll} s \to aAB, & s \to aB, & A \to aAC, & A \to aC, \\ B \to Dc, & D \to b, & CD \to CE, & CE \to DE, \\ DE \to DC, & Cc \to Dcc, & & \end{array}$$

and starting symbol s. Then G is context-sensitive. For example, the production $CE \to DE$ ($\Lambda CE \to \Lambda DE$) allows C to be replaced by D if C is followed by E and the production $Cc \to Dcc$ ($\Lambda Cc \to \Lambda Dcc$) allows C to be replaced by Dc if C is followed by c.

DC can be derived from CD since

$$CD \Longrightarrow CE \Longrightarrow DE \Longrightarrow DC.$$

The string $a^3\, b^3 c^3$ is in $L(G)$, since we have

$$\begin{array}{llllll} s & \Longrightarrow aAB & \Longrightarrow & aaACB & \Longrightarrow & aaaCCDc \\ & \Longrightarrow aaaDCCc & \Longrightarrow & aaaDCDcc & \Longrightarrow & aaaDDCcc \\ & \Longrightarrow aaaDDDccc & \Longrightarrow & aaabbbccc. & & \end{array}$$

It follows that

$$L(G) = \{a^n b^n c^n \mid n = 1, 2, \ldots\}.$$

Definition 1.9.9 *A language L is* ***context sensitive*** *(respectively,* ***context-free, regular****) if there is a context-sensitive (respectively, context-free, regular) grammar G with $L = L(G)$.*

Example 1.9.10 *By Example* 1.9.8, *the language*

$$L(G) = \{a^n b^n c^n \mid n = 1, 2, \dots\}$$

is context-sensitive. In [40, p.127], *it is shown that there is no context-free grammar G with $L = L(G)$. Thus L is not context-free.*

Example 1.9.11 *Consider the grammar G defined as follows*

$$T = \{a, b\}, \qquad N = \{s\},$$

with productions

$$s \rightarrow asb, \qquad s \rightarrow ab$$

and starting symbol s. Then G is context-free. The only derivations of s are

$$\begin{aligned} s &\Longrightarrow asb \\ &\vdots \\ &\Longrightarrow a^{n-1}sb^{n-1} \\ &\Longrightarrow a^{n-1}abb^{n-1} = a^n b^n. \end{aligned}$$

Hence $L(G)$ consists of the strings over $\{a, b\}$ of the form $a^n b^n$, $n = 1, 2 \dots$. This language is context-free. In Section 1.11, we show that $L(G)$ is not regular.

It follows from Examples 1.9.10 and 1.9.11 that the set of context-free languages that do not contain the null string is a proper subset of the set of context-sensitive languages and that the set of regular languages is a proper subset of the set of context-free languages. It also follows that there are languages that are not context-sensitive.

Example 1.9.12 *The grammar G defined in Example* 1.9.4 *is regular. Hence the language*

$$L(G) = \{b^n a b^m \mid n = 0, 1, \dots; m = 1, 2, \dots\}$$

it generates is regular.

Grammars G and G' are called **equivalent** if $L(G) = L(G')$.

1.10 Nondeterministic Finite-State Automata

In this section and Section 1.11, we show that regular grammars and finite-state automata are essentially the same in that either is a specification of a regular language. We next illustrate how a finite-state automaton can be converted to a regular grammar.

Example 1.10.1 *We show how to write the regular grammar given by the finite-state automaton of Figure 1.10.*

Let the terminal symbols be the input symbols a, b. Let the nonterminal symbols be the states E and O. The initial state E becomes the starting symbol. The productions correspond to the directed edges. If there is an edge labeled x from S to R, we create the production

$$S \rightarrow xR.$$

Hence we obtain the productions

$$E \rightarrow bE, \quad E \rightarrow aO, \quad O \rightarrow aE, \quad O \rightarrow bO. \tag{1.6}$$

If S is an accepting state, we include the production

$$S \rightarrow \Lambda.$$

In this example, we obtain the additional production

$$O \rightarrow \Lambda. \tag{1.7}$$

Then the grammar $G = (N, T, P, E)$, where $N = \{O, E\}$, $T = \{a, b\}$, and P consists of the productions (1.6) *and* (1.7), *generates the language $L(G)$. $L(G)$ is the same as the set of strings accepted by the finite-state automaton of Figure 1.10.*

The proof of the following result can be found in [43,100].

Theorem 1.10.2 *Let A be a finite-state automaton. Let s be the initial state. Let T denote the set of input symbols and let N denote the set of states. Define productions*

$$S \rightarrow xR$$

if there is an edge labeled x from S to R, and

$$S \rightarrow \Lambda$$

if S is an accepting state. Let G be the grammar

$$G = (N, T, P, s).$$

Then G is regular and the set of strings accepted by A is equal to $L(G)$. ■

We now consider the reverse situation, i.e., given a regular grammar G, we want to construct a finite-state automaton A such that $L(G)$ is precisely the set of strings accepted by A. The procedure of Theorem 1.10.2 cannot simply be reversed as the next example shows.

Example 1.10.3 *Consider the regular grammar defined as follows:*

$$T = \{a, b\}, \quad N = \{s, q\}$$

with productions

$$s \to bs, \; s \to aq, \; q \to bq, \; q \to b$$

and starting symbol s.

Let the nonterminal symbols be the states with s as the initial state. For each production of the form

$$S \to xR,$$

we draw an edge from state S to state R and label it x. The productions

$$s \to bs, \quad s \to aq, \quad q \to bq$$

give the graph shown in Figure 1.11.

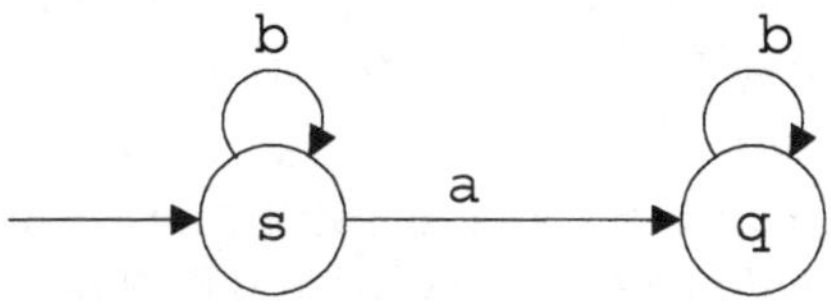

Figure 1.11

The production $q \to b$ is equivalent to the two productions

$$q \to bF, \quad F \to \Lambda,$$

where F is an additional nonterminal symbol. The productions

$$s \to bs, \quad s \to aq, \quad q \to bq, \quad q \to bF$$

give the graph shown in Figure 1.12. From the production

$$F \to \Lambda,$$

it follows that F should be an accepting state.

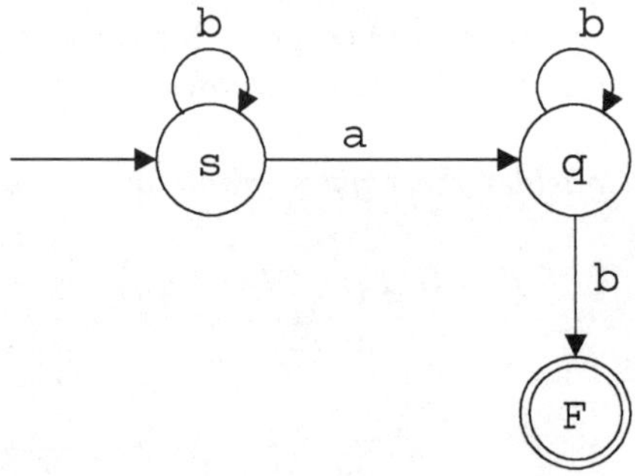

Figure 1.12

However, the graph of Figure 1.12 is not a finite-state automaton for several reasons. Vertex C has no outgoing edge labeled a and vertex F has no outgoing edges at all. Also, vertex C has two outgoing edges labeled b. A diagram like that of Figure 1.12 yields a different kind of automaton called a **nondeterministic finite-state automaton.** The word "nondeterministic" is used since if in Figure 1.12 the automaton is in state C and b is input, a choice of next states exists, i.e., the automaton either remains in state C or goes to state F.

Definition 1.10.4 *A* ***nondeterministic finite-state automaton*** *A is a 5-tuple* $A = (Q, X, f, \mathcal{A}, s)$, *where*
(1) Q *is a finite set of states,*
(2) X *is a finite set of input symbols,*
(3) f *is a next-state function from* $Q \times X$ *into* $\mathcal{P}(Q)$,
(4) $\mathcal{A}$ *is a subset of* Q, *the accepting states,*
(5) $s \in Q$ *is the initial state.*

The main difference between a nondeterministic finite-state automaton and a finite-state automaton is that in a finite-state automaton the next-state function maps a state, input pair to a uniquely defined state, while in a nondeterministic finite-state automaton the next-state function maps a state, input pair to a set of states.

Example 1.10.5 *For the nondeterministic finite-state automaton of Figure 1.12, we have*

$$Q = \{s, q, F\}, \quad X = \{a, b\}, \quad \mathcal{A} = \{F\}.$$

The initial state is s and the next-state function f is given by

	f	
$Q \backslash X$	a	b
s	$\{q\}$	$\{s\}$
q	$\emptyset$	$\{q, F\}$
F	$\emptyset$	$\emptyset$

The transition diagram of a nondeterminate finite-state automaton is drawn similarly to that of a finite-state automaton. An edge from state q to each state in the set $f(q,x)$ is drawn and each is labeled x.

Example 1.10.6 *Consider the nondeterministic finite-state automaton given as follows:*

$$Q = \{s,q,p\},\ X = \{a,b\},\ \mathcal{A} = \{q,p\}$$

with initial state s and next-state function

$Q\backslash X$	f: a	b
s	$\{s,q\}$	$\{p\}$
q	$\emptyset$	$\{q\}$
p	$\{q,p\}$	$\emptyset$

Its transition diagram is shown in Figure 1.13. *It is the transition diagram of the nondeterministic automaton of this example.*

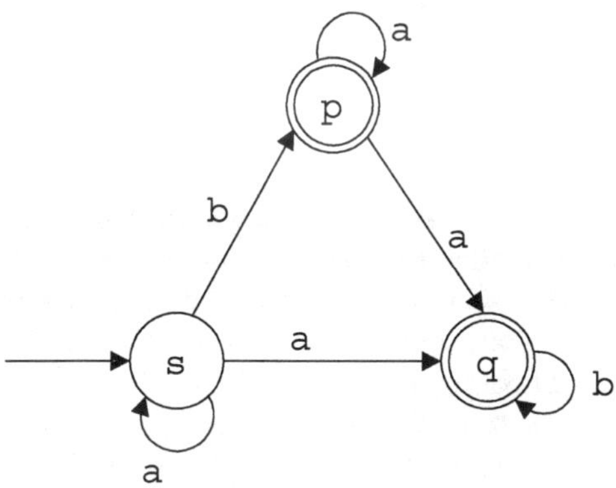

Figure 1.13

A string x is accepted by a nondeterministic finite-state automaton A if there is some path representing x in the transition diagram of A beginning at the initial state and ending in an accepting state.

Definition 1.10.7 *Let $A = (Q,X,f,\mathcal{A},q)$ be a nondeterministic finite-state automaton. The empty string is* **accepted** *by A if and only if $s \in \mathcal{A}$. If $x = x_1 \cdots x_n$ is a nonempty string over X and there exist states $q_0, \ldots, q_n$ such that*

(1) $q_0 = q$;

(2) $q_i \in f(q_{i-1}, x_i)$ *for* $i = 1, \ldots, n$;

(3) $q_n \in \mathcal{A}$;

then x is said to be **accepted** *by A. Let $Ac(A)$ denote the set of strings accepted by A.*

If A and A' are nondeterministic finite-state automata and $Ac(A) = Ac(A')$, then A and A' are said to be ***equivalent****.*

If $x = x_1 \cdots x_n$ is a string over X and there exists states $q_0, \ldots, q_n$ satisfying conditions (1) and (2), then the path $(q_0, \ldots, q_n)$ is called a ***path representing*** *x in A.*

Theorem 1.10.8 *Let $G = (N, T, P, s)$ be a regular grammar. Let $X = T$, $Q = N \cup \{F\}$, where $F \notin N \cup T$,*

$$f(q, x) = \{q' \mid q \to xq' \in P\} \cup \{F \mid q \to x \in P\},$$

and $\mathcal{A} = \{F\} \cup \{q \mid q \to \Lambda \in P\}$.

Then the nondeterministic finite-state automaton $A = (Q, X, f, \mathcal{A}, s)$ accepts precisely the strings $L(G)$. ■

In the next section we show that given a nondeterministic finite-state automaton A, there is a finite-state automaton that is equivalent to A.

1.11 Relationships Between Languages and Automata

In Section 1.10, we showed that if A is a finite-state automaton, then there is a regular grammar G such that $L(G) = Ac(A)$. A partial converse is given by Theorem 1.10.8 where it is shown that if G is a regular grammar, then there is a nondeterministic finite-state automaton A such that $L(G) = Ac(A)$. In this section, we show that if G is a regular grammar, then there is a finite-state automaton A such that $L(G) = Ac(A)$. This result follows from Theorem 1.10.8 once it has been established that any nondeterministic finite-state automaton can be converted to an equivalent finite-state automaton. The method is illustrated by the following example.

Example 1.11.1 *We construct a finite-state automaton equivalent to the nondeterministic finite-state automaton of Figure 1.12.*

The set of input symbols is the same. The states consist of all subsets

$$\emptyset,\ \{s\}, \{q\}, \{F\}, \{s, q\}, \{s, F\}, \{q, F\}, \{s, q, F\}$$

of the original set $Q = \{s, q, F\}$ of states. The initial state is $\{s\}$. The accepting states are all subsets of Q that contain an accepting state of the original nondeterministic finite-state automaton, namely

$$\{F\},\ \{s, F\},\ \{q, F\},\ \{s, q, F\}.$$

Let $X, Y \subseteq Q$. An edge is drawn from X to Y and labeled x if $X = \emptyset = Y$ or if

$$\cup_{q \in X} f(q, x) = Y.$$

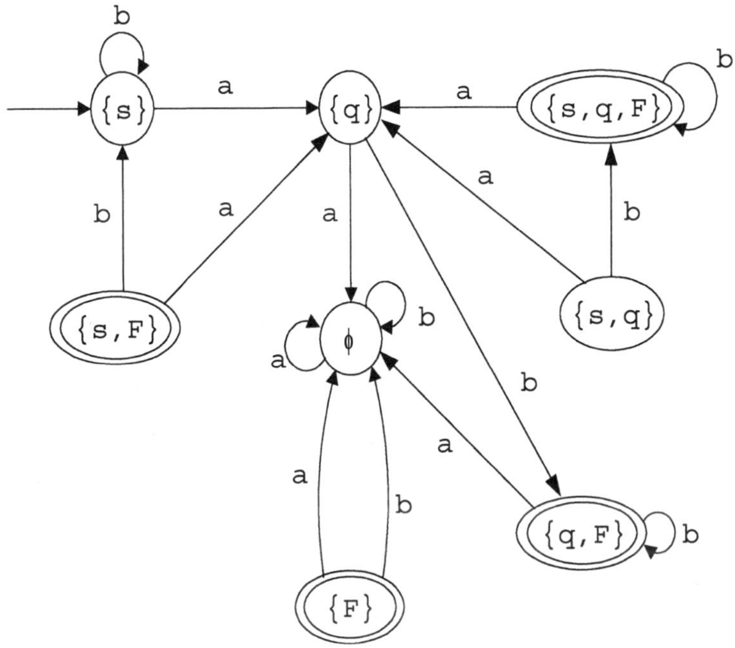

Figure 1.14

The finite-state automaton of Figure 1.14 *is obtained. The states*

$$\{s, F\}, \ \{s, q\}, \quad \{s, q, F\}, \ \{F\}$$

can never be reached and are thus deleted. This yields the simplified, equivalent finite-state automaton of Figure 1.15.

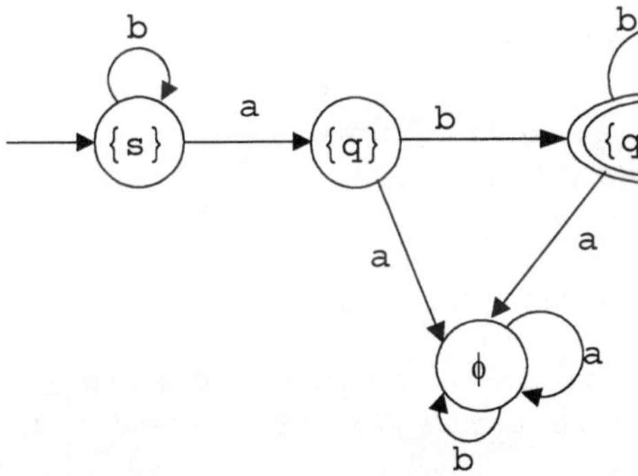

Figure 1.15

Example 1.11.2 *The finite-state automaton equivalent to the nondeterministic finite-state automaton of Example 1.10.6 is given in Figure 1.16.*

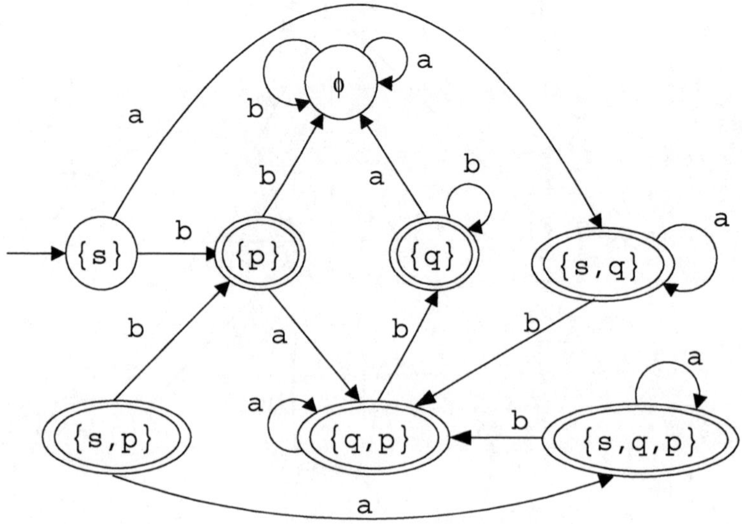

Figure 1.16

The following theorem justifies the method of Examples 1.11.1 and 1.11.2. Its proof can be found in [43, 100].

Theorem 1.11.3 *Let* $A = (Q, X, f, \mathcal{A}, s)$ *be a nondeterministic finite-state automaton. Let*

(1) $Q' = \mathcal{P}(Q)$,

(2) $X' = X$,

(3) $s' = \{s\}$,

(4) $\mathcal{A}' = \{S \subseteq Q \mid S \cap \mathcal{A} \neq \emptyset\}$,

(5) $f'(S, x) = \begin{cases} \emptyset & \text{if } S = \emptyset \\ \cup_{q \in S} f(q, x) & \text{if } S \neq \emptyset. \end{cases}$

Then the finite-state automaton $A' = (Q', X', f', \mathcal{A}', s')$ *is equivalent to* A.

We can now state the following result.

Theorem 1.11.4 *A language* L *is regular if and only if there exists a finite-state automaton that accepts precisely the strings in* L.

Proof. This theorem follows from Theorems 1.10.2, 1.10.8, and 1.11.3. ■

Example 1.11.5 *In this example, we determine a finite-state automaton A that accepts precisely the strings generated by the regular grammar G having productions*

$$s \to bs, s \to aq, q \to bq, q \to b.$$

The starting symbol is s, the set of terminal symbols is $\{a, b\}$, and the set of nonterminal symbols is $\{s, q\}$.

The nondeterministic finite-state automaton A' that accepts $L(G)$ is shown in Figure 1.12. *A finite-state automaton equivalent to A' is shown in Figure* 1.14 *and an equivalent simplified finite-state automaton A is shown in Figure* 1.15. *The finite-state automaton A accepts precisely the strings generated by G.*

Example 1.11.6 *In this example, we show that the language*

$$L = \{a^n b^n \mid n = 1, 2, \dots\}$$

is not regular.

Suppose that L is regular. Then there exists a finite-state automaton A such that $Ac(A) = L$. Assume that A has k states. The string $x = a^k b^k$ is accepted by A. Let P be the path that represents x. Since there are k states, some state q is revisited on the part of the path representing a^k. Thus there is a cycle C, all of whose edges are labeled a, that contains q. We change the path P to obtain a path P' as follows. When we arrive at q in P, we traverse C. After returning to q on C, we continue on P to the end. If the length of C is j, the path P' represents the string $x' = a^{j+k} b^k$. Since P and P' end at the same state q' and q' is an accepting state, x' is accepted by A. This is impossible since x' is not of the form $a^n b^n$. Hence L is not regular.

2
Fuzzy Ideals

2.1 Introduction

We now begin the study of fuzzy semigroup theory.

Let S be a semigroup. Let A and B be subsets of S. Then multiplication of A and B is defined as follows:

$$AB = \{ab \in S \mid a \in A \text{ and } b \in B\}.$$

A nonempty subset A of S is called a **subsemigroup** of S if $AA \subseteq A$. A nonempty subset A of S is called a **left** (**right**) **ideal** of S if $SA \subseteq A$ ($AS \subseteq A$). Further, A is called a **two-sided ideal** of S if it is both a left and a right ideal of S. A nonempty subset A of S is called an **interior ideal** of S if $SAS \subseteq A$, and a **quasi-ideal** of S if $AS \cap SA \subseteq A$. A subsemigroup A of S is called a **bi-ideal** of S if $ASA \subseteq A$. A nonempty subset A is called a **generalized bi-ideal** of S if $ASA \subseteq A$.

Definition 2.1.1 *A semigroup S is called **regular** if for each element a of S, there exists an element $x \in S$ such that $a = axa$.*

Theorem 2.1.2 *A semigroup S is regular if and only if $R \cap L = RL$ for every right ideal R and every left ideal L of S.* ∎

It is natural to extend this property to fuzzy ideals. We show in Theorem 3.1.8 that a semigroup S is regular if and only if

$$f \cap g = f \circ g$$

for every fuzzy right ideal f and every fuzzy left ideal g of S, where $\circ$ is the multiplication of fuzzy subsets. Moreover we study the properties of a semigroup when f and g are other types of fuzzy ideals. In this chapter, we introduce the notion of fuzzy interior ideals, fuzzy quasi-ideals, fuzzy bi-ideals, fuzzy generalized bi-ideals in a semigroup and give some fundamental properties of such fuzzy ideals.

2.2 Ideals in Semigroups

We denote by $L[a]$ ($R[a], J[a], I[a], Q[a], B[a]$) the **principal left (right, two-sided, interior, quasi-, bi-) ideal** of a semigroup S generated by the element $a \in S$, that is,

$$\begin{aligned} L[a] &= \{a\} \cup Sa, \\ R[a] &= \{a\} \cup aS, \\ J[a] &= \{a\} \cup Sa \cup aS \cup Sas, \\ I[a] &= \{a\} \cup \{a^2\} \cup Sas, \\ Q[a] &= \{a\} \cup (aS \cap Sa), \\ B[a] &= \{a\} \cup \{a^2\} \cup aSa. \end{aligned}$$

2.3 Fuzzy Ideals in Semigroups

Let f and g be two fuzzy subsets of S. The **product** $f \circ g$ is defined by

$$(f \circ g)(x) = \begin{cases} \bigvee_{x=yz}\{f(y) \wedge g(z)\}, & \text{if } \exists\, y, z \in S, \text{ such that } x = yz \\ 0 & \text{otherwise.} \end{cases}$$

As is well-known, the operation $\circ$ is associative.
A fuzzy subset f of S is called a **fuzzy subsemigroup** of S if

$$f(ab) \geq f(a) \wedge f(b)$$

for all $a, b \in S$, and is called a **fuzzy left (right)** ***ideal*** of S if

$$f(ab) \geq f(b) \qquad (f(ab) \geq f(a))$$

for all $a, b \in S$. A fuzzy subset f of S is called a **fuzzy two-sided ideal** (or **a fuzzy ideal**) of S if it is both a fuzzy left and a fuzzy right ideal of S. Let A be a subset of S. Recall that we denote by C_A the characteristic function of A

Lemma 2.3.1 *Let A be a nonempty subset of a semigroup S. Then the following properties hold.*

(1) A is a subsemigroup of S if and only if C_A is a fuzzy subsemigroup of S.

(2) A is a left (right, two-sided) ideal of S if and only if C_A is a fuzzy left (right, two-sided) ideal of S. ■

We note that the semigroup S can be considered a fuzzy subset of itself and we write $S = C_S$, i.e., $S(x) = 1$ for all $x \in S$.

Lemma 2.3.2 *Let f be a fuzzy subset of a semigroup S. Then the following properties hold.*

(1) f is a fuzzy subsemigroup of S if and only if $f \circ f \subseteq f$.

(2) f is a fuzzy left ideal of S if and only if $S \circ f \subseteq f$.

(3) f is a fuzzy right ideal of S if and only if $f \circ S \subseteq f$.

(4) f is a fuzzy two-sided ideal of S if and only if $S \circ f \subseteq f$ and $f \circ S \subseteq f$.

Proof. Property (1) follows easily.

We prove that (2) holds. Assume that f is a fuzzy left ideal of S. Let $a \in S$. If $(S \circ f)(a) = 0$, then it is clear that $S \circ f \subseteq f$. Otherwise, there exist elements $x, y \in S$ such that $a = xy$. Then, since f is a fuzzy left ideal of S, we have

$$\begin{aligned} (S \circ f)(a) &= \vee_{a=xy} S(x) \wedge f(y) \\ &\leq \vee_{a=xy} 1 \wedge f(xy) \\ &= \vee(1 \wedge f(a)) \\ &= f(a), \end{aligned}$$

and so $S \circ f \subseteq f$.

Conversely, assume that $S \circ f \subseteq f$. Let x and y be any elements of S. Let $a = xy$. Then we have

$$\begin{aligned} f(xy) &= f(a) \\ &\geq (S \circ f)(a) \\ &= \vee_{a=bc}(S(b) \wedge f(c)) \\ &\geq S(x) \wedge f(y) \\ &= 1 \wedge f(y) \\ &= f(y). \end{aligned}$$

Hence f is a fuzzy left ideal of S. (3) and (4) can be seen in a similar manner. ■

Lemma 2.3.3 *Let S be a semigroup. Then the following properties hold.*

(1) Let f and g be two fuzzy subsemigroups of S. Then $f \cap g$ is also a fuzzy subsemigroup of S.

(2) Let f and g be fuzzy left (right, two-sided) ideal of S. Then $f \cap g$ is also a fuzzy left (right, two-sided) ideal of S.

Proof. (1) Let f and g be any fuzzy subsemigroups of S. Let a and b be any elements of S. Then

$$\begin{aligned}(f \cap g)(ab) &= f(ab) \wedge g(ab) \\ &\geq (f(a) \wedge f(b)) \wedge (g(a) \wedge g(b)) \\ &= (f(a) \wedge g(a)) \wedge (f(b) \wedge g(b)) \\ &\geq (f \cap g)(a) \wedge (f \cap g)(b).\end{aligned}$$

Thus $f \cap g$ is a fuzzy subsemigroup of S. Property (2) can be proved in a similar manner. ■

Lemma 2.3.4 *Let A and B be nonempty subsets of a semigroup S. Then the following properties hold.*
(1) $C_A \cap C_B = C_{A \cap B}$.
(2) $C_A \circ C_B = C_{AB}$.

Proof. (1) Let a be any element of S. Suppose $a \in A \cap B$. Then $a \in A$ and $a \in B$. Thus we have

$$\begin{aligned}(C_A \cap C_B)(a) &= C_A(a) \wedge C_B(a) \\ &= 1 \wedge 1 \\ &= 1 \\ &= C_{A \cap B}(a).\end{aligned}$$

Suppose $a \notin A \cap B$. Then $a \notin A$ or $a \notin B$. Hence we have

$$\begin{aligned}(C_A \cap C_B)(a) &= C_A(a) \wedge C_B(a) \\ &= 0 \\ &= C_{A \cap B}(a).\end{aligned}$$

Therefore we obtain that $C_A \cap C_B = C_{A \cap B}$.

(2) Let a be any element of S. Suppose $a \in AB$. Then $a = xy$ for some $x \in A$ and $y \in B$. Thus we have

$$\begin{aligned}(C_A \circ C_B)(a) &= \vee_{a=uv}\{C_A(u) \wedge C_B(v)\} \\ &\geq C_A(x) \wedge C_B(y) \\ &= 1 \wedge 1 \\ &= 1,\end{aligned}$$

and so

$$(C_A \circ C_B)(a) = 1.$$

Since $a \in AB$, $C_{AB}(a) = 1$. In the case, when $a \notin AB$, we have $a \neq xy$ for all $x \in A$ and $y \in B$. If $a = uv$ for some u, $v \in S$, then we have

$$\begin{aligned}(C_A \circ C_B)(a) &= \vee_{a=uv}\{C_A(u) \wedge C_B(v)\} \\ &= 0 \\ &= C_{AB}(a).\end{aligned}$$

If $a \neq uv$ for all $u,\, v \in S$, then

$$(C_A \circ C_B)(a) = 0 = C_{AB}(a).$$

In any case, we have $C_A \circ C_B = C_{AB}$. ■

Lemma 2.3.5 *If f is a fuzzy right (left) ideal of S, then $f \cup (S \circ f)$ $(f \cup (f \circ S))$ is a fuzzy two-sided ideal of S.*

Proof. Suppose f is a fuzzy right ideal of S. Then $S \circ (f \cup (S \circ f)) = (S \circ f) \cup (S \cup (S \circ f)) = (S \circ f) \cup (S \circ S) \circ f) \subseteq (S \circ f) \cup (S \circ f) = S \circ f \subseteq f \cup (S \circ f)$. Thus $f \cup (S \circ f)$ is a fuzzy left ideal of S. Also $(f \cup (S \circ f)) \circ S = (f \circ S) \cup ((S \circ f) \circ S) = (f \circ S) \cup (S \circ (f \circ S)) \subseteq (f \circ S) \cup (S \circ f) \subseteq f \cup (S \circ f)$. Hence $f \cup (S \circ f)$ is a fuzzy right ideal of S. ■

Proposition 2.3.6 *Let f be a fuzzy subset of S. Then f is a fuzzy subsemigroup (ideal) of S is and only if f_t is a semigroup (ideal) of S $\forall t \in [0,1]$ such that $f_t \neq \emptyset$.* ■

2.4 Fuzzy Bi-ideals in Semigroups

A fuzzy subset f of a semigroup S is called a **fuzzy bi-ideal** of S if

$$f(xyz) \geq f(x) \wedge f(z)$$

for all x, y and z of S.

Lemma 2.4.1 *Let A be a nonempty subset of a semigroup S. Then A is a bi-ideal of S if and only if the characteristic function C_A of A is a fuzzy bi-ideal of S.*

Proof. Assume that A is a bi-ideal of S. It follows from Lemma 2.3.1(1) that C_A is a fuzzy subsemigroup of S. Let x, y and z be any elements of S. If, $z \in A$, then since $C_A(x) = C_A(z) = 1$ and since $xyz \in ASA \subseteq A$, we have

$$C_A(xyz) = 1 = C_A(x) \wedge C_A(z).$$

If $x \in A$ or $z \in A$, then $C_A(x) = 0$ or $C_A(z) = 0$ and so we have

$$C_A(xyz) \geq 0 = C_A(x) \wedge C_A(z).$$

Thus C_A is a fuzzy bi-ideal of S.

Conversely, assume that C_A is a fuzzy bi-ideal of S. Then it follows from Lemma 2.3.1(1) that A is a subsemigroup of S. Let $a = xyz$ be any element of ASA $(x, z \in A)$. Then

$$\begin{aligned} C_A(a) &= C_A(xyz) \\ &\geq C_A(x) \wedge C_A(z) \\ &= 1 \wedge 1 \\ &= 1. \end{aligned}$$

Hence we have $C_A(a) = 1$, and so $a \in A$. Thus $ASA \subseteq A$. Therefore, A is a bi-ideal of S. ■

Lemma 2.4.2 *Let f be a fuzzy subsemigroup of a semigroup S. Then f is a fuzzy bi-ideal of S if and only if $f \circ S \circ f \subseteq f$.*

Proof. Assume that f is a fuzzy bi-ideal of S. Let a be any element of S. In the case when $(f \circ S \circ f)(a) = 0$, it is clear that $f \circ S \circ f \subseteq f$. Otherwise, there exist elements x, y, p and q of S such that $a = xy$ and $x = pq$. Since f is a fuzzy bi-ideal of S, we have

$$f(pqy) \geq f(p) \wedge f(y).$$

Therefore,

$$\begin{aligned}
(f \circ S \circ f)(a) &= \vee_{a=xy}((f \circ S)(x) \wedge f(y)) \\
&= \vee_{a=xy}(((\vee_{x=pq}((f(p) \wedge S(q))) \wedge f(y))) \\
&= \vee_{a=xy}((\vee_{x=pq}(f(p) \wedge 1)) \wedge f(y)) \\
&= \vee_{a=pqy}((f(p) \wedge f(y))) \\
&\leq \vee_{a=pqy}\ f(pqy) \\
&= f(a),
\end{aligned}$$

and so we have $f \circ S \circ f \subseteq f$.

Conversely, assume that $f \circ S \circ f \subseteq f$. Let x, y and z be any elements of S. Set $a = xyz$. Then we have

$$\begin{aligned}
f(xyz) &= f(a) \\
&\geq (f \circ S \circ f)(a) \\
&= \vee_{a=bc}((f \circ S)(b) \wedge f(c)) \\
&\geq (f \circ S)(xy) \wedge f(z) \\
&= (\vee_{xy=pq}(f(p) \wedge S(q))) \wedge f(z) \\
&\geq (f(x) \wedge S(y)) \wedge f(z) \\
&= (f(x) \wedge 1) \wedge f(z) \\
&= f(x) \wedge f(z).
\end{aligned}$$

Thus f is a fuzzy bi-ideal of S. ■

Lemma 2.4.3 *Let f be any fuzzy subset of a semigroup S and g any fuzzy bi-ideal of S. Then the products $f \circ g$ and $g \circ f$ are both fuzzy bi-ideals of S.*

Proof. Since g is a fuzzy bi-ideal of S, by Lemma 2.4.2, we have

$$\begin{aligned}
(f \circ g) \circ (g \circ f) &= f \circ g \circ (f \circ g) \\
&\subseteq f \circ g \circ S \circ g \\
&\subseteq f \circ g.
\end{aligned}$$

Hence it follows from Lemma 2.3.2(1) that $f \circ g$ is a fuzzy subsemigroup of S. Also, we have

$$\begin{aligned}(f \circ g) \circ S \circ (f \circ g) &= f \circ g \circ (S \circ f) \circ g \\ &\subseteq f \circ (g \circ S \circ g) \\ &\subseteq f \circ g.\end{aligned}$$

Thus it follows from Lemma 2.4.2 that $f \circ g$ is a fuzzy bi-ideal of S. Similarly, it can be seen that $g \circ f$ is a fuzzy bi-ideal of S. ∎

Lemma 2.4.4 *Let f and g be two fuzzy bi-ideals of a semigroup S. Then $f \cap g$ is a fuzzy bi-ideal of S.*

Proof. Let a, b and x be any elements of S. Then

$$\begin{aligned}(f \cap g)(ab) &= f(ab) \wedge g(ab) \\ &\geq (f(a) \wedge f(b)) \wedge (g(a) \wedge g(b)) \\ &= (f(a) \wedge g(a)) \wedge (g(a) \wedge g(b)) \\ &= (f \cap g)(a) \wedge (f \cap g)(b),\end{aligned}$$

and

$$\begin{aligned}(f \cap g)(axb) &= f(axb) \wedge g(axb) \\ &\geq (f(a) \wedge f(b)) \wedge (g(a) \wedge g(b)) \\ &= (f(a) \wedge g(a)) \wedge (f(a) \wedge g(a)) \\ &= (f \cap g)(a) \wedge (f \cap g)(b).\end{aligned}$$

Hence $f \cap g$ is a fuzzy bi-ideal of S. ∎

2.5 Fuzzy Interior Ideals in Semigroups

Definition 2.5.1 *A fuzzy subset f of S is called a* ***fuzzy interior ideal*** *of S if $f(xay) \geq f(a)$ for all x, a and y of S.*

Lemma 2.5.2 *Let A a nonempty subset of a semigroup S. Then A is an interior ideal of S if and only if the characteristic function C_A of A is a fuzzy interior ideal of S.*

Proof. First assume that A is an interior ideal of S. Let x, a and y be any elements of S. If $a \in A$, then $xay \in SAS \subseteq A$. This implies that $C_A(xay) = 1 = C_A(a)$. If $a \in A$, then $C_A(xay) \geq 0 = C_A(a)$. Thus C_A is a fuzzy interior ideal of S.

Conversely, assume that C_A is a fuzzy interior ideal of S. Let xay $(x, y \in S, a \in A)$ be any element of SAS. Then $C_A(xay) \geq C_A(a) = 1$. This implies that $C_A(xay) = 1$, and so $xay \in A$. Hence we have $SAS \subseteq A$ and so A is an interior ideal of S. ∎

Lemma 2.5.3 *Let f be a fuzzy subset of a semigroup S. Then f is a fuzzy interior ideal of S if and only if $S \circ f \circ S \subseteq f$.*

Proof. Assume that f is a fuzzy interior ideal of S. Let x be any element of S. If there exist elements y, z, u and v of S such that $x = yz$ and $y = uv$. Then since $f(uvz) \geq f(v)$, we have

$$\begin{aligned}(S \circ f \circ S)(x) &= \vee_{x=yz}\{(S \circ f)(y) \wedge S(z)\} \\ &= \vee_{x=yz}\{(\vee_{y=uv}(S(u) \wedge f(v))) \wedge S(z)\} \\ &\leq \vee_{x=yz}\ \{(\vee_{y=uv}(1 \wedge f(v))) \wedge 1\} \\ &\leq f(x).\end{aligned}$$

In the other case, we have

$$(S \circ f \circ S)(x) = 0 \leq f(x).$$

Therefore, $S \circ f \circ S \subseteq f$.

Conversely, assume that $S \circ f \circ S \subseteq f$. Let x, a and y be any elements of S. Then

$$\begin{aligned}f(xay) &\geq (S \circ f \circ S)(xay) \\ &= \vee_{xay=pq}\{(S \circ f)(p) \wedge S(q)\} \\ &\geq (S \circ f)(xa) \wedge S(y) \\ &= (S \circ f)(xa) \wedge 1 \\ &= \vee_{xa=pq}\{S(p) \wedge f(q)\} \\ &\geq S(x) \wedge f(a) \\ &= 1 \wedge f(a) \\ &= f(a).\end{aligned}$$

Consequently, f is a fuzzy interior ideal of S. ■

As is easily seen, every fuzzy two-sided ideal of a semigroup S is a fuzzy interior ideal of S. The following example shows that the converse of this property does not hold in general.

Example 2.5.4 *Let S be a semigroup of four elements $\{a, b, c, d\}$ with the following multiplication table:*

	a	b	c	d
a	a	a	a	a
b	a	a	a	a
c	a	a	b	a
d	a	a	b	b

Let f be a fuzzy subset of S such that

$$f(a) = 0.7, \qquad f(b) = 0, \qquad f(c) = 0.3, \qquad f(d) = 0.$$

Then f is a fuzzy interior ideal of S which is not a fuzzy two-sided ideal of S. In fact,

$$f(xyz) = f(a) = 0.7 \geq f(y)$$

for every x, y and z of S. Thus f is a fuzzy interior ideal of S. But, since

$$f(dc) = f(b) = 0 \leq 0.3 = f(c),$$

f is not a fuzzy left ideal of S, that is, it is not a fuzzy two-sided ideal of S.

A semigroup S is called **regular** if for each element a of S there exists an element $x \in S$ such that $a = axa$.

Lemma 2.5.5 *Let f be a fuzzy subset of a regular semigroup S. Then the following conditions are equivalent.*
(1) f is a fuzzy ideal of S.
(2) f is a fuzzy interior ideal of S.

Proof. By Lemma 2.5.3 it suffices to prove that (2) implies (1). Assume that (2) holds. Let a and b be any elements of S. Then, since S is regular, there exist elements $x, y \in S$ such that $a = axa$ and $b = byb$. Thus we have

$$f(ab) = f((axa)b) = f((ax)ab) \geq f(a)$$

and

$$f(ab) = f(a(byb)) = f(ab(yb)) \geq f(b).$$

Therefore, f is a fuzzy two-sided ideal of S. ■

A semigroup S is called **left (right) simple** and is called **simple** if it contains no proper two-sided ideal. A semigroup S is called **fuzzy left (right) simple** if every fuzzy left (right) ideal of S is a constant function, and is called a **fuzzy simple** if every two-sided ideal of S is a constant function.

Lemma 2.5.6 *For a semigroup S, the following conditions are equivalent.*
(1) S is a left (right) simple.
(2) S is a fuzzy left (right) simple.

Proof. First assume that S is a left simple. Let f be any fuzzy left ideal of S and a and b any elements of S. Then it follows from [12, p.6] that there exist elements $x, y \in S$ such that $b = xa$ and $a = yb$. Hence since f is a fuzzy left ideal of S,

$$f(a) = f(yb) \leq f(b) = f(xa) \leq f(a),$$

and so $f(a) = f(b)$.

Since a and b are any elements of S, this means that f is a constant function and so S is a fuzzy left simple. Thus we obtain that (1) implies (2).

Conversely, assume that (2) holds. Let A be any left ideal of S. Then, by Lemma 2.3.1(2), C_A is a fuzzy left ideal of S. Thus C_A is a constant function. Let x be any element of S. Then, since A is nonempty, $C_A(x) = 1$, and so $x \in A$. This implies that $S \subseteq A$, that is, $A = S$. Hence S is left simple and (2) implies (1). In the case S is right simple, the proof follows similarly. ■

Lemma 2.5.7 *For a semigroup S, the following conditions are equivalent.*
(1) S is simple.
(2) S is fuzzy simple.
(3) Every fuzzy interior ideal of S is a constant function.

Proof. It can be seen in a similar way as in the proof of Lemma 2.5.6 that (1) and (2) are equivalent. Assume that (2) holds. Let f be any fuzzy interior ideal of S and a and b any elements of S. Then, since S is simple, it follows from [107, Lemma I.3.9] that there exists elements $x, y \in S$ such that $a = xby$. Thus since f is a fuzzy interior ideal of S, we have

$$f(a) = f(xby) \geq f(b).$$

It can be similarly proved that $f(b) \geq f(a)$. Therefore we have $f(a) = f(b)$.
Since a and b are any elements of S, f is a constant function and so (2) implies (3). As is stated above, every fuzzy ideal of S is a fuzzy interior ideal of S. Thus it is clear that (3) implies (2). ■

2.6 Fuzzy Quasi-ideals in Semigroups

Definition 2.6.1 *A fuzzy subset f of a semigroup S is called a **fuzzy quasi-ideal** of S if*

$$(f \circ S) \cap (S \circ f) \subseteq f.$$

As is easily seen, any fuzzy left ideal and any fuzzy right ideal of S is a fuzzy quasi-ideal of S and any fuzzy quasi-ideal of S is a fuzzy bi-ideal of S. The converse of those properties do not hold in general.
In fact, we have the following examples.

Example 2.6.2 *Let $S = \{0, a, b, c\}$ be a semigroup with the following multiplication table:*

	0	a	b	c
0	0	0	0	0
a	0	a	b	0
b	0	0	0	0
c	0	c	0	0

Then $Q = \{0, a\}$ is quasi-ideal of S and is not a left (right, two-sided) ideal of S. Define the fuzzy subset f of S as follows:

$$f(0) = f(a) = 0.7 \quad \text{and} \quad f(b) = f(c) = 0.$$

Then f is a fuzzy quasi-ideal of S and is not a fuzzy left (right, two-sided) ideal of S.

Example 2.6.3 *Let $S = \{0, a, b, c\}$ be a semigroup with the following multiplication table:*

	0	a	b	c
0	0	0	0	0
a	0	0	0	0
b	0	0	0	a
c	0	0	a	b

Then $B = \{0, b\}$ is a bi-ideal of S and is not a quasi-ideal of S. Define the fuzzy subset f of S as follows:

$$f(0) = f(b) = 0.7 \qquad \textit{and} \qquad f(a) = f(c) = 0.$$

Then f is a fuzzy bi-ideal of S and is not a fuzzy quasi-ideal of S.

Lemma 2.6.4 *Let A be a nonempty subset of a semigroup S. Then A is a quasi-ideal of S if and only if the characteristic function C_A is a fuzzy quasi-ideal of S.*

Proof. First assume that A is a quasi-ideal of S. Let a be any element of S. If $a \in A$, then

$$((C_A \circ S) \cap (S \circ C_A))(a) \leq 1 = C_A(a).$$

If $a \notin A$, then $C_A(a) = 0$. On the other hand, assume that

$$((C_A \circ S) \cap (S \circ C_A))(a) = 1.$$

Then

$$\vee_{a=pq}\{C_A(p) \wedge S(q)\} = (C_A \circ S)(a) = 1$$

and

$$\vee_{a=pq}\{S(p) \wedge C_A(q)\} = (S \circ C_A)(a) = 1.$$

This implies that there exist elements b, c, d and e of S with $a = bc = de$ such that

$$C_A(b) = 1 \qquad \text{and} \qquad C_A(e) = 1.$$

Hence

$$a = bc = de \in AS \cap SA \subseteq A,$$

which contradicts that $a \notin A$. Thus we have

$$(C_A \circ S) \cap (S \circ C_A) \subseteq C_A$$

and so C_A is a fuzzy quasi-ideal of S.

Conversely, assume that C_A is a fuzzy quasi-ideal of S. Let a be any element of $AS \cap SA$. Then there exist elements s and t of S and elements b and c of A such that $a = bs = tc$. Thus we have

$$\begin{aligned}(C_A \circ S)(a) &= \vee_{a=pq}\{C_A(p) \wedge S(q)\} \\ &\geq C_A(b) \wedge S(s) \\ &= 1 \wedge 1 \\ &= 1,\end{aligned}$$

and so

$$(C_A \circ S)(a) = 1.$$

Similarly, we have

$$(S \circ C_A)(a) = 1.$$

Hence

$$\begin{aligned}C_A(a) &\geq ((C_A \circ S) \cap (S \circ C_A))(a) \\ &= (C_A \circ S)(a) \wedge (S \circ C_A)(a) \\ &= 1.\end{aligned}$$

Thus $a \in A$ and so $AS \cap SA \subseteq A$. Therefore, A is a quasi-ideal of S. ■

Lemma 2.6.5 *Let f and g be a fuzzy right ideal and a fuzzy left ideal of a semigroup S, respectively. Then $f \cap g$ is a fuzzy quasi-ideal of S.*

Proof.

$$\begin{aligned}((f \cap g) \circ S) \cap (S \circ (f \cap g)) &\subseteq (f \circ S) \cap (S \circ g) \\ &\subseteq f \cap g.\end{aligned}$$

■

Lemma 2.6.6 *Suppose that f or g is a fuzzy quasi-ideal of a semigroup S. Then the product $f \circ g$ is a fuzzy bi-ideal of S.*

Proof. If f is a fuzzy quasi-ideal of S, then we have $f \circ S \circ f \subseteq f$. Thus

$$\begin{aligned}(f \circ g) \circ (f \circ g) &= (f \circ g \circ f) \circ g \\ &\subseteq (f \circ S \circ f) \circ g \\ &\subseteq f \circ g\end{aligned}$$

and

$$\begin{aligned}(f \circ g) \circ S \circ (f \circ g) &= (f \circ (g \circ S) \circ f) \circ g \\ &\subseteq (f \circ (S \circ S) \circ f) \circ g \\ &\subseteq (f \circ S \circ f) \circ g \\ &\subseteq f \circ g.\end{aligned}$$

Consequently, $f \circ g$ is a fuzzy bi-ideal of S. ∎

Lemma 2.6.7 *([17, p.85]) A semigroup S is a group if and only if it contains no proper quasi-ideal.* ∎

Lemma 2.6.8 *For a semigroup, S the following conditions are equivalent.*
(1) S is a group.
(2) Every fuzzy quasi-ideal of S is a constant function.
(3) Every fuzzy bi-ideal of S is a constant function.

Proof. First assume that S is a group with identity e. Let f be any fuzzy bi-ideal of S and a any element of S. Then

$$\begin{aligned}f(a) &= f(eae) \\ &\geq f(e) \wedge f(e) \\ &= f(e) \\ &= f(ee) \\ &= f(aa^{-1})(a^{-1}a)) \\ &= f(a(a^{-1}a^{-1})a) \\ &\geq f(a) \wedge f(a) \\ &= f(a),\end{aligned}$$

and so $f(a) = f(e)$. This implies that f is a constant function and so (1) implies (3). Since every fuzzy quasi-ideal of S is a fuzzy bi-ideal of S, (3) implies (2). Assume that (2) holds. If S is not a group, then it follows from Lemma 2.6.7 that S contains a proper quasi-ideal Q of S. Since Q is nonempty and C_Q is a constant function, we have $C_Q(x) = 1$ for all $x \in S$. This implies that $Q = S$, which contradicts the condition (2). Thus S is a group and so (2) implies (1). ∎

Lemma 2.6.9 *Let f be a fuzzy quasi-ideal of a semigroup S and a any element of A. Then*

$$f(a^n) \leq f(a^{n+1})$$

for every positive integer n.

Proof. For any positive integer n, we have

$$\begin{aligned}(f \circ S)(a^{n+1}) &\leq \vee_{a^{n+1}=xy}\{f(x) \wedge S(y)\} \\ &\leq f(a^n) \wedge S(a) \\ &= f(a^n) \wedge 1 \\ &= f(a^n).\end{aligned}$$

Similarly,

$$(S \circ f)(a^{n+1}) \leq f(a^n).$$

Thus

$$\begin{aligned} f(a^{n+1}) &\leq ((f \circ S) \cap (S \circ f))(a^{n+1}) \\ &= (f \circ S)(a^{n+1}) \wedge (S \circ f)(a^{n+1}) \\ &\leq f(a^n) \wedge f(a^n) \\ &= f(a^n). \end{aligned}$$

■

2.7 Fuzzy Generalized Bi-ideals in Semigroups

Definition 2.7.1 *A nonempty subset A of a semigroup S is called a* ***generalized bi-ideal*** *of S if $ASA \subseteq A$ [74]. A fuzzy subset f of S is called a* ***fuzzy generalized bi-ideal*** *of S if*

$$f(xyz) \geq f(x) \wedge f(z)$$

for all x, y and z of S.

It is clear that every fuzzy bi-ideal of a semigroup S is a fuzzy generalized bi-ideal of S, but the converse of this statement does not hold in general. This is shown by the following example.

Example 2.7.2 *Let $S = \{a, b, c, d\}$ be a semigroup with the following multiplication table:*

	a	b	c	d
a	a	a	a	a
b	a	a	a	a
c	a	a	b	a
d	a	a	b	b

Let f be a fuzzy subset of S such that

$$f(a) = 0.5 \qquad f(b) = 0 \qquad f(c) = 0.2 \qquad f(d) = 0.$$

Then, as is easily seen, f is a fuzzy generalized bi-ideal of S, but not a bi-ideal of S. However, the following theorem shows that the converse of this holds for a regular semigroup.

Lemma 2.7.3 *Every fuzzy generalized bi-ideal of a regular semigroup S is a fuzzy bi-ideal of S.*

Proof. Let f be any fuzzy generalized bi-ideal of S and let a and b be any elements of S. Then, since S is regular, there exists an element $x \in S$ such that $b = bxb$. Thus we have

$$f(ab) = f(a(bxb)) = f(a(bx)b) \geq f(a) \wedge f(b).$$

This implies that f is a fuzzy subsemigroup of S and so f is a fuzzy bi-ideal of S. ■

The following two lemmas are easily seen.

Lemma 2.7.4 *A nonempty subset A of a semigroup S is a generalized bi-ideal of S if and only if C_A is a fuzzy generalized bi-ideal of S.* ■

Lemma 2.7.5 *A fuzzy subset f of a semigroup S is a fuzzy generalized bi-ideal of S if and only if $f \circ S \circ f \subseteq f$.* ■

The following theorem can be seen in a similar way as in the proof of Lemma 2.4.3.

Theorem 2.7.6 *The product of two fuzzy generalized bi-ideals of a semigroup S is a fuzzy generalized bi-ideal of S.* ■

2.8 Fuzzy Ideals Generated by Fuzzy Subsets of Semigroups

We now consider results from [96]. Let S be a semigroup. Clearly the intersection of any collection of fuzzy ideals of S is a fuzzy ideal of S. In fact, the intersection of all fuzzy ideals containing a fuzzy subset f of S is the smallest fuzzy ideal of S containing f. Similar comments hold for various types of fuzzy ideals of S. When S is a monoid, we let e denote the identity of S.

Definition 2.8.1 *Let f be a fuzzy subset of S. The smallest fuzzy left (right, two-sided) ideal of S containing f is called a fuzzy left (right, two-sided) ideal of S generated by f and is denoted by $\langle f \rangle$ $([f], (f))$ (where a fuzzy two-sided ideal means a fuzzy ideal).*

Proposition 2.8.2 *Let S be a monoid. A fuzzy ideal f of S is a fuzzy bi-ideal of S and a fuzzy interior ideal of S both.*

Proof. Clearly, f is a fuzzy subsemigroup of S. For $x, y, z \in S$, $f(xyz) = f((xy)z) \geq f(z)$ and $f(xyz) \geq f(x)$. Thus

$$f(xyz) \geq f(x) \wedge f(z).$$

Hence f is a fuzzy bi-ideal of S. Now for $x, a, y \in S$,

$$f(xay) \geq f(ay) \geq f(a).$$

Thus f is a fuzzy interior ideal of S. ■

Proposition 2.8.3 *Let f be a fuzzy subset of a monoid S. Then f is a fuzzy ideal of S if and only if f is a fuzzy interior ideal of S.*

Proof. By Proposition 2.8.2, the necessity is clear. Now let us show the sufficiency. For $x, y \in S$, $f(xy) = f(xye) \geq f(x)$. Thus f is also a fuzzy ideal of S. ■

Theorem 2.8.4 *Let f be a fuzzy subset of a monoid S. Then $\langle f \rangle = j$, where*

$$j(a) = \vee\{f(x_2) \mid a = x_1x_2, x_1, x_2 \in S\}$$

for all $a \in S$.

Proof. For $a \in S$,

$$j(a) = \vee\{f(x_2) \mid a = x_1x_2, x_1, x_2 \in S\} \geq f(a)\}$$

since $a = ea$ and so $j \supseteq f$. For all $x, y \in S$, we have

$$\begin{aligned} j(xy) &= \vee\{f(x_2) \mid xy = x_1x_2, x_1, x_2 \in S\} \\ &\geq \vee\{f(z_2) \mid xy = (xz_1)z_2, y = z_1z_2, z_1, z_2 \in S\} \\ &\geq \vee\{f(z_2) \mid y = z_1z_2, z_1, z_2 \in S\} = j(y). \end{aligned}$$

Hence j is a fuzzy left ideal of S. If g is a fuzzy left ideal of S and $f \subseteq g$, then for $a \in S$, $f(a) \leq g(a)$, and

$$\begin{aligned} j(a) &= \vee\{f(x_2) \mid a = x_1x_2, x_1, x_2 \in S\} \\ &\leq \vee\{g(x_2) \mid a = x_1x_2, x_1, x_2 \in S\} \\ &\leq \vee\{g(x_1x_2) \mid a = x_1x_2, x_1, x_2 \in S\} = g(a). \end{aligned}$$

Thus it follows that $j \subseteq g$. By Definition 2.8.1, we have that $\langle f \rangle = j$. ■

Theorem 2.8.5 *Let f be a fuzzy subset of a monoid S. Then $[f] = j$, where*

$$j(a) = \vee\{f(x_1) \mid a = x_1x_2, x_1, x_2 \in S\}$$

for all $a \in S$. ■

Theorem 2.8.6 *Let f be a fuzzy subset of a monoid S. Then*
(1) $[\langle f \rangle] = (f)$,
(2) $\langle [f] \rangle = (f)$,
(3) $[\langle f \rangle] = \langle [f] \rangle$.

Proof. First, by Theorem 2.8.5, $[\langle f\rangle]$ is a fuzzy right ideal of S. Second, for any $x, y \in S$,

$$\begin{aligned}[\langle f\rangle](xy) &= \vee\{\langle f\rangle(x_1) \mid xy = x_1x_2, x_1, x_2 \in S\} \\ &= \vee\{\vee\{f(z_2) \mid x_1 = z_1z_2, z_1, z_2 \in S\} \mid xy = x_1x_2, x_1, x_2 \in S\}\end{aligned}$$

and

$$\begin{aligned}[\langle f\rangle](y) &= \vee\{\langle f\rangle(y_1) \mid y = y_1y_2, y_1, y_2 \in S\} \\ &= \vee\{\vee\{f(w_2) \mid y_1 = w_1w_2, w_1, w_2 \in S\} \mid y = y_1y_2, y_1, y_2 \in S\}.\end{aligned}$$

Clearly, $[\langle f\rangle](xy) \geq [\langle f\rangle](y)$, $[\langle f\rangle]$ is also a fuzzy left ideal of S. Thus $[\langle f\rangle]$ is a fuzzy ideal of S. Since $f \subseteq \langle f\rangle \subseteq [\langle f\rangle]$, we have $[\langle f\rangle] \supseteq f$. If g is a fuzzy ideal of S and $g \supseteq f$, then $g \supseteq \langle f\rangle$, since g is a fuzzy left ideal of S, and $g \supseteq [\langle f\rangle]$ since g is a fuzzy right ideal of S. Then by Definition 2.8.1, $[\langle f\rangle]$ is a fuzzy ideal of S generated by f; that is, $[\langle f\rangle] = (f)$. Similarly we can prove $\langle [f]\rangle = (f)$. Hence $[\langle f\rangle] = (f) = \langle [f]\rangle$. ■

By Theorems 2.8.4, 2.8.5, and 2.8.6, respectively, we can establish the following two propositions.

Proposition 2.8.7 *[95] Let S be a monoid and let x_t be a fuzzy singleton of S. Then $\langle x_t\rangle = j$ $([x_t] = j)$, where*

$$j(a) = \begin{cases} t & \text{if there exists } b \in S \text{ such that } a = bx \ \ (a = xb) \\ 0 & \text{otherwise;} \end{cases}$$

for all $a \in S$. ■

Proposition 2.8.8 *[95] Let S be a monoid and let x_t be a fuzzy singleton of S. Then $(x_t) = j$, where*

$$j(a) = \begin{cases} t & \text{if there exist } b, c \in S \text{ such that } a = bxc, \\ 0 & \text{otherwise;} \end{cases}$$

for all $a \in S$. ■

Definition 2.8.9 *Let f be a fuzzy subset of S. The smallest fuzzy interior ideal of S containing f is called a fuzzy interior ideal of S generated by f and is denoted by $\langle f\rangle_I$.*

Theorem 2.8.10 *Let f be a fuzzy subset of a monoid S. Then $\langle f\rangle_I = j$, where*

$$j(a) = \vee\{f(x_2) \mid a = x_1x_2x_3, x_1, x_2, x_3 \in S\}$$

for all $a \in S$.

Proof. For any $x, y \in S$,

$$\begin{aligned} j(xy) &= \vee\{f(x_2) \mid xy = x_1x_2x_3, x_1, x_2, x_3 \in S\} \\ &\geq \vee\{f(z_2) \mid xy = z_1z_2(z_3y), x = z_1z_2z_3, z_1, z_2, z_3 \in S\} \\ &= j(x) \end{aligned}$$

and

$$\begin{aligned} j(xy) &= \vee\{f(x_2) \mid xy = x_1x_2x_3, x_1, x_2, x_3 \in S\} \\ &\geq \vee\{f(w_2) \mid xy = (xw_1)w_2w_3, y = w_1w_2w_3, w_1, w_2, w_3 \in S\} \\ &= j(y). \end{aligned}$$

Thus $j(xy) \geq j(x) \wedge j(y)$ and so j is a fuzzy subsemigroup of S. For $a \in S$.

$$j(a) = \{f(x_2) \mid a = x_1x_2x_3, x_1, x_2, x_3 \in S\} \geq f(a)$$

since $a = eae$ and hence $j \supseteq f$. For any $x, a, y \in S$,

$$\begin{aligned} j(xay) &= \vee\{f(x_2) \mid xay = x_1x_2x_3, x_1, x_2, x_3 \in S\} \\ &\geq \vee\{f(z_2) \mid xay = (xz_1)z_2(z_3y), a = z_1z_2z_3, z_1, z_2, z_3 \in S\} \\ &= j(a). \end{aligned}$$

It follows that j is a fuzzy interior ideal of S. Let g be any fuzzy interior ideal of S such that $g \supseteq f$. Then for all $a \in S$.

$$\begin{aligned} j(a) &= \vee\{f(x_2) \mid a = x_1x_2x_3, x_1, x_2, x_3 \in S\} \\ &\leq \vee\{g(x_2) \mid a = x_1x_2x_3, x_1, x_2, x_3 \in S\} \\ &\leq \vee\{g(x_1x_2x_3) \mid a = x_1x_2x_3, x_1, x_2, x_3 \in S\} \\ &= g(a). \end{aligned}$$

By Definition 2.8.9, we get $\langle f \rangle_I = j$. ■

From Proposition 2.8.3, we have the following result.

Proposition 2.8.11 *Let S be a monoid and let x_t be a fuzzy singleton of S. Then $\langle x_t \rangle_I = j$, where for all $a \in S$,*

$$j(a) = \begin{cases} t & \text{if there exist } x_1, x_2 \in S \text{ such that } a = x_1xx_2, \\ 0 & \text{otherwise.} \end{cases}$$ ■

Set $f = x_t$. Then, from Theorem 2.8.10, we can also prove Proposition 2.8.11.

Definition 2.8.12 *Let f be a fuzzy subset of S. The smallest fuzzy bi-ideal of S containing f is called a fuzzy bi-ideal of S generated by f and is denoted by $\langle f \rangle_B$.*

Definition 2.8.13 *Let f be a fuzzy subset of a monoid S. Then f is called a fuzzy submonoid of S if $f(e) \geq f(x)$ for all $x \in S$.*

Theorem 2.8.14 *Let f be a fuzzy subset of a monoid S. Then $\langle f\rangle_B = j$, where*

$$j(a) = \vee\{f(x_1) \wedge f(x_3) \mid a = x_1x_2x_3, x_1, x_2, x_3 \in S\}$$

for all $a \in S$.

Proof. For any $a \in S$,

$$\begin{aligned} j(a) &= \vee\{f(x_1) \wedge f(x_3) \mid a = x_1x_2x_3, x_1, x_2, x_3 \in S\} \\ &\geq f(e) \wedge f(a) \\ &= f(a), \end{aligned}$$

since $a = eea$ and so $j \supseteq f$. For any $x, y, z \in S$,

$$\begin{aligned} j(x) &= \vee\{f(x_1) \wedge f(x_3) \mid x = x_1x_2x_3, x_1, x_2, x_3 \in S\}, \\ j(z) &= \vee\{f(z_1) \wedge f(z_3) \mid z = z_1z_2z_3, z_1, z_2, z_3 \in S\} \end{aligned}$$

and

$$\begin{aligned} j(xyz) &= \vee\{f(w_1) \wedge f(w_3) \mid xyz = w_1w_2w_3, w_1, w_2, w_3 \in S\} \\ &\geq \vee\{f(x_1) \wedge f(z_3) \mid xyz = x_1(x_2x_3yz_1z_2)z_3, \\ &\quad x = x_1x_2x_3, z = z_1z_2z_3\}. \end{aligned}$$

However,

$$j(x) \wedge j(z) = \vee\{\wedge\{f(x_1) \wedge f(x_3),\ f(z_1) \wedge f(z_3)\} \mid x = x_1x_2x_3, z = z_1z_2z_3\}.$$

Hence $j(xyz) \geq j(x) \wedge j(z)$. Let $y = e$. Then we have

$$j(xz) \geq j(x) \wedge j(z)$$

for all $x, z \in S$. Thus j is a fuzzy subsemigroup of S. Hence j is a fuzzy bi-ideal of S. Let g be a fuzzy bi-ideal of S such that $g \supseteq f$. Then for all $a \in S$,

$$\begin{aligned} j(a) &= \vee\{f(x_1) \wedge f(x_2) \mid a = x_1x_2x_3, x_1, x_2, x_3 \in S\} \\ &\leq \vee\{g(x_1) \wedge g(x_2) \mid a = x_1x_2x_3, x_1, x_2, x_3 \in S\} \\ &\leq \vee\{g(x_1x_2x_3) \mid a = x_1x_2x_3, x_1, x_2, x_3 \in S\} \\ &= g(a) \end{aligned}$$

and so $j \subseteq g$. Hence $\langle f\rangle_B = j$. ∎

Theorem 2.8.15 *Let f be a fuzzy subset of a monoid S. If S is a regular semigroup, then $\langle f\rangle_B = j$, where*

$$j(a) = \vee\{f(x_1) \wedge f(x_3) \mid a = x_1x_2x_3, x_1, x_2, x_3 \in S\}$$

for all $a \in S$.

Proof. From the proof of Theorem 2.8.14, we only need to prove $j(a) \geq f(a)$ for all $a \in S$. Indeed,

$$\begin{aligned} j(a) &= \vee\{f(x_1) \wedge f(x_3) \mid a = x_1x_2x_3, x_1, x_2, x_3 \in S\} \\ &\geq \vee\{f(a) \wedge f(a) \mid a = axa, x \in S\} \\ &= f(a) \end{aligned}$$

for all $a \in S$. ■

Let $f = x_t$. Then by Theorem 2.8.15, we can obtain the following result.

Theorem 2.8.16 *Let S be a monoid and let x_t be a fuzzy singleton of S. If S is a regular semigroup. Then $\langle x_t \rangle_B = j$, where*

$$j(a) = \begin{cases} t & \text{if there exists some } y \in S \text{ such that } a = yxy, \\ 0 & \text{otherwise;} \end{cases}$$

for all $a \in S$.

Inheung Chon has noted the following result.

Theorem 2.8.17 *Let S be a semigroup and let f be a fuzzy subset of S. Then the following properties hold:*

(1) The fuzzy left (right) ideal generated by f is $f \cup C_S f$ ($f \cup fC_S$).

(2) The fuzzy ideal generated by f is $f \cup C_S f \cup fC_S \cup C_S fC_S$. ■

3

Regular Semigroups

In this chapter, we characterize a regular semigroup, a completely regular semigroup, an intra-regular semigroup, and a semisimple semigroup in terms of fuzzy ideals. We also examine types of fuzzy regular subsemigroups.

3.1 Regular Semigroups

Definition 3.1.1 *Let S be a semigroup and $a \in S$. Then a is called* ***regular*** *if there exists an element x in S such that $a = axa$. S is called* ***regular*** *if every element of S is regular.*

The equivalence of (1) and (2) of the following theorem is due to K. Iseki [42] and the equivalence of (1) and (3) is due to O. Steinfeld [136, p.10].

Theorem 3.1.2 *For a semigroup S, the following conditions are equivalent.*

(1) S is regular.

(2) $R \cap L = RL$ for every right ideal R of S and every left ideal L of S.

(3) $A = ASA$ for every quasi-ideal A of S.

Theorem 3.1.3 *For a semigroup S, the following conditions are equivalent.*

(1) S is regular.

(2) $f = f \circ S \circ f$ for every fuzzy quasi-ideal f of S.

(3) $f = f \circ S \circ f$ for every fuzzy generalized bi-ideal f of S.

Proof. First assume that (1) holds. Let f be any fuzzy generalized bi-ideal of S and a any element of S. Then since S is regular, there exists an element $x \in S$ such that $a = axa$. Hence we have

$$\begin{aligned}(f \circ S \circ f)(a) &= \vee_{a=yz}\{(f \circ S)(y) \wedge f(z)\} \\ &\geq (f \circ S)(ax) \wedge f(a) \\ &= (\vee_{ax=pq}\{(f(p) \wedge S(q)\}) \wedge f(a) \\ &\geq (f(a) \wedge S(x)) \wedge f(a) \\ &= (f(a) \wedge 1) \wedge f(a) \\ &= f(a),\end{aligned}$$

and so $f \circ S \circ f \supseteq f$. Since f is a fuzzy generalized bi-ideal of S, $f \circ S \circ f \subseteq f$. Thus $f = f \circ S \circ f$ and so (1) implies (4). It is clear that (4) $\Rightarrow$ (3) $\Rightarrow$ (2). Assume that (2) holds. Let A be any quasi-ideal of S. Then we have

$$ASA \subseteq A(SS) \cap (SS)A \subseteq AS \cap SA \subseteq A.$$

Let a be any element of A. Since C_A is a fuzzy quasi-ideal by Lemma 2.6.4, we have

$$\begin{aligned}\vee_{a=yz}\{(C_A \circ S)(y) \wedge C_A(z)\} &= (C_A \circ S) \circ C_A(a) \\ &= C_A(a) \\ &= 1.\end{aligned}$$

This implies that there exist elements b and c of S such that

$$(C_A \circ S)(b) = 1 \quad \text{and} \quad C_A(c) = 1$$

with $a = bc$. Thus we have

$$\begin{aligned}\vee_{b=pq}\{C_A(p) \wedge S(q)\} &= (C_A \circ S)(b) \\ &= 1.\end{aligned}$$

This implies that there exist elements d and e of S such that

$$C_A(d) = 1 \quad \text{and} \quad S(e) = 1$$

with $b = de$. Thus $d, c \in A$ and $e \in S$ and so $a = bc = (de)c \in ASA$.

Therefore, $A \subseteq ASA$, and so $A = ASA$. Hence it follows from Theorem 3.1.2 that S is regular and so (2) implies (1). ■

Theorem 3.1.4 *Every fuzzy two-sided ideal of a regular semigroup is idempotent.*

Proof. Let f be a fuzzy two-sided ideal of a regular semigroup S. Then by Lemma 2.3.2, we have $f \circ S \circ f \subseteq f$. Hence f is a fuzzy generalized bi-ideal of S. Thus since S is regular, by Theorem 3.1.3, we have

$$f = f \circ S \circ f = f \circ (S \circ f) \subseteq f \circ f \subseteq f \circ S \subseteq f,$$

and so $f \circ f = f$. Therefore f is idempotent. ■

A semigroup S is called **left (right) duo** if every left (right) ideal of S is a two-sided ideal of S. A semigroup S is called **duo** if it is both left and right duo. A semigroup S is called **fuzzy left (right) duo** if every fuzzy left (right) ideal of S is a fuzzy two-sided ideal of S, and is called **fuzzy duo** if it is both fuzzy left and fuzzy right duo.

Theorem 3.1.5 *A regular semigroup S is left (right) duo if and only if it is fuzzy left (right) duo.*

Proof. First assume that S is left duo. Let f be any fuzzy left ideal of S and a, b any elements of S. Then, since the left ideal Sa is a two-sided ideal of S and since S is regular, we have

$$ab \in (aSa)b \subseteq (Sa)S \subseteq Sa.$$

This implies that there exists an element $x \in S$ such that $ab = xa$. Hence since f is a fuzzy left ideal of S, we have

$$f(ab) = f(xa) \geq f(a).$$

Therefore, f is a fuzzy right ideal of S and so f is a fuzzy two-sided ideal. Thus S is a fuzzy left duo. Conversely, assume that S is fuzzy left duo. Let A be any left ideal of S. Then it follows from Lemma 2.3.1 that the characteristic function C_A of A is a fuzzy left ideal of S. Then by the assumption, C_A is a fuzzy two-sided ideal of S. Since A is nonempty, it follows from Lemma 2.3.1 that A is a two-sided ideal of S. Therefore S is left duo. ■

Theorem 3.1.6 *A regular semigroup is duo if and only if it is fuzzy duo.*

Proof. The proof follows from Theorem 3.1.5. ■

Theorem 3.1.7 *For a regular semigroup S, the following conditions are equivalent.*

(1) Every bi-ideal of S is a right (left, two-sided) ideal of S.

(2) Every fuzzy bi-ideal of S is a fuzzy right (left, two-sided) ideal of S.

Proof. First assume that (1) holds. Let f be any fuzzy bi-ideal of S and a, b any elements of S. Since the set aSa is a bi-ideal of S, by the assumption, it is a right ideal of S. Hence since S is regular, we have $ab \in (aSa)S \subseteq aSa$. Thus there exists an element $x \in S$ such that $ab = axa$. Hence since f is a fuzzy bi-ideal of S,

$$f(ab) = f(axa) \geq f(a) \wedge f(a) = f(a).$$

Therefore, f is a fuzzy right ideal of S. Thus (1) implies (2).

Conversely, assume that (2) holds. Let A be any bi-ideal of S. Then it follows from Lemma 2.4.1 that the characteristic function C_A is a fuzzy bi-ideal of S. Hence by the assumption, it is a fuzzy right ideal of S. Thus by Lemma 2.3.1, A is a right ideal of S. Hence (2) implies (1). ■

Theorem 3.1.8 *For a semigroup S the following conditions are equivalent.*

(1) S is regular.

(2) $f \cap g = f \circ g$ for every fuzzy right ideal f and every fuzzy left ideal g of S.

Proof. First assume that (1) holds. Let f be any fuzzy right ideal and g any left ideal of S. Then by Lemma 2.3.2, we have $f \circ g \subseteq f \circ S \subseteq f$ and $f \circ g \subseteq S \circ g \subseteq g$. Thus $f \circ g \subseteq f \cap g$.

Let a be any element of S. Then, since S is regular, there exists an element $x \in S$ such that $a = axa$. Hence we have

$$\begin{aligned}(f \circ g)(a) &= \vee_{a=yz}\{f(y) \wedge g(z)\} \\ &\geq f(ax) \wedge g(a) \\ &\geq f(a) \wedge g(a) \\ &= (f \cap g)(a),\end{aligned}$$

and so $f \circ g \supseteq f \cap g$. Therefore, $f \circ g = f \cap g$ and so (1) implies (2).

Conversely, assume that (2) holds. Let R and L be any right ideal and any left ideal of S, respectively. In order to see that $R \cap L \subseteq RL$ holds, let a be any element of $R \cap L$. Then by Lemma 2.3.1(2), the characteristic functions C_R and C_L of R and L are a fuzzy right ideal and a fuzzy left ideal of S, respectively. Thus we have

$$\begin{aligned}C_{RL}(a) &= (C_R \circ C_L)(a) \\ &= (C_R \cap C_L)(a) \\ &= C_{R \cap L}(a) \\ &= 1,\end{aligned}$$

and so $a \in RL$. Thus $R \cap L \subseteq RL$. Since the inclusion in the other direction always holds, we obtain that $R \cap L = RL$. Hence it follows from Theorem 3.1.2 that S is regular and so (2) implies (1). ■

Now we give some characterizations of a regular semigroup in terms of fuzzy ideals.

Theorem 3.1.9 *For a semigroup S, the following conditions are equivalent.*

(1) S is regular.

(2) $f \cap g = f \circ g \circ f$ for every fuzzy quasi-ideal f and every fuzzy two-sided ideal g of S.

(3) $f \cap g = f \circ g \circ f$ for every fuzzy quasi-ideal f and every fuzzy interior ideal g of S.

(4) $f \cap g = f \circ g \circ f$ for every fuzzy bi-ideal f and every fuzzy two-sided ideal g of S.

(5) $f \cap g = f \circ g \circ f$ *for every fuzzy bi-ideal* f *and every fuzzy interior ideal of* S.

(6) $f \cap g = f \circ g \circ f$ *for every fuzzy generalized bi-ideal* f *and every fuzzy two-sided ideal of* S.

(7) $f \cap g = f \circ g \circ f$ *for every fuzzy generalized bi-ideal* f *and every fuzzy interior ideal of* S.

Proof. First assume that (1) holds. Let f and g be any fuzzy generalized bi-ideal and any fuzzy interior ideal of S, respectively. Then

$$f \circ g \circ f \subseteq f \circ S \circ f \subseteq f$$

and

$$f \circ g \circ f \subseteq S \circ g \circ S \subseteq g.$$

Thus

$$f \circ g \circ f \subseteq f \cap g.$$

Now let a be any element of S. Then, since S is regular, there exists an element $x \in S$ such that $a = axa$ $(= axaxa)$. Thus since g is a fuzzy interior ideal of S, we have

$$\begin{aligned}(f \circ g \circ f)(a) &= \vee_{a=yz}\{f(y) \wedge (g \circ f)(z)\} \\ &\geq f(a) \wedge (g \circ f)(xaxa) \\ &= f(a) \wedge (\vee_{xaxa=pq}\{g(p) \wedge f(q)\}) \\ &\geq f(a) \wedge (g(xax) \wedge f(a)) \\ &\geq f(a) \wedge g(a) \\ &= (f \cap g)(a),\end{aligned}$$

and so $f \circ g \circ f \supseteq f \cap g$. Hence $f \circ g \circ f = f \cap g$ and so (1) implies (7).

It is clear that (7) $\Rightarrow$ (5) $\Rightarrow$ (3) $\Rightarrow$ (2) and (7) $\Rightarrow$ (6) $\Rightarrow$ (4) $\Rightarrow$ (2). Assume that (2) holds. Let f be any fuzzy quasi-ideal of S. Then, since S itself is a fuzzy two-sided ideal of S, we have

$$f = f \cap S = f \circ S \circ f.$$

Thus it follows from Theorem 3.1.3 that S is regular and so (2) implies (1). ∎

Theorem 3.1.10 *For a semigroup* S*, the following conditions are equivalent.*

(1) S *is regular.*

(2) $Q \cap J = QJQ$ *for every quasi-ideal* Q *and every two-sided ideal* J *of* S.

(3) $Q \cap I = QIQ$ *for every quasi-ideal* Q *and every interior ideal* I *of* S.

(4) $B \cap J = BJB$ *for every bi-ideal* B *and every two-sided ideal* J *of* S.

(5) $B \cap I = BIB$ for every bi-ideal B and every interior ideal I of S.

(6) $G \cap J = GJG$ for every generalized bi-ideal G and every two-sided ideal J of S.

(7) $G \cap I = GIG$ for every generalized bi-ideal G and every interior I of S.

Proof. Assume that (1) holds. Let G be any generalized bi-ideal and I any interior ideal of S, respectively. Then $GIG \subseteq GSG \subseteq G$ and $GIG \subseteq SIS \subseteq I$. Thus we have $GIG \subseteq G \cap I$.

Let a be any element of $G \cap I$. By Lemma 2.7.4, C_G is a fuzzy generalized bi-ideal of S. By Lemma 2.5.2, C_I is a fuzzy interior ideal of S. Hence by Lemma 2.3.4, we have

$$\begin{aligned} C_{GIG}(a) &= (C_G \circ C_I \circ C_G)(a) \\ &= (C_G \cap C_I)(A) \\ &= C_{G\cap I}(a) \\ &= 1, \end{aligned}$$

and so $a \in GIG$. Thus we have $G \cap I \subseteq GIG$. Therefore, $G \cap I = GIG$ and so (1) implies (7). It is clear that (7) $\Rightarrow$ (5) $\Rightarrow$ (3) $\Rightarrow$ (2) and (7) $\Rightarrow$ (6) $\Rightarrow$ (4) $\Rightarrow$ (2). Assume that (2) holds. Let Q be any quasi-ideal of S. Now S itself is a two-sided ideal of S,

$$Q = Q \cap S = QSQ \subseteq QS \cap SQ \subseteq Q,$$

and so $Q = QSQ$. Hence it follows from Theorem 3.1.2 that S is regular and so (2) implies (1). ∎

Theorem 3.1.11 *For a semigroup S, the following conditions are equivalent.*

(1) S is regular.

(2) $f \cap g \subseteq f \circ g$ for every fuzzy quasi-ideal f and every fuzzy left ideal g of S.

(3) $f \cap g \subseteq f \circ g$ for every fuzzy bi-ideal f and every fuzzy left ideal g of S.

(4) $f \cap g \subseteq f \circ g$ for every fuzzy generalized bi-ideal f and every fuzzy left ideal g of S.

Proof. Assume that (1) holds. Let f and g be any fuzzy generalized bi-ideal and any fuzzy left ideal of S, respectively. Let a be any element of S. Then, since S is regular, there exists an element $x \in S$ such that $a = axa$. Thus we have

$$\begin{aligned} (f \circ g)(a) &= \vee_{a=yz}\{f(y) \wedge g(z)\} \\ &\geq f(a) \wedge g(xa) \\ &\geq f(a) \wedge g(a) \\ &= (f \cap g)(a), \end{aligned}$$

and so $f \circ g \subseteq f \cap g$. Hence we obtain that (1) implies (4). It is clear that (4) $\Rightarrow$ (3) $\Rightarrow$ (2). Since every fuzzy right ideal of S is a fuzzy quasi-ideal of S, $f \cap g \subseteq f \circ g$ for every fuzzy right ideal f and every fuzzy left ideal g of S. Since inclusion in the other direction always holds, we have $f \cap g = f \circ g$ for every fuzzy right ideal f and every fuzzy left ideal g of S. Thus it follows from Theorem 3.1.8 that S is regular and so (2) implies (1). ■

The following theorem is a special case of Theorem 3.1.11.

Theorem 3.1.12 *For a semigroup S, the following conditions are equivalent.*

(1) S is regular.

(2) $Q \cap L \subseteq QL$ for every quasi-ideal Q and every left ideal L of S.

(3) $B \cap L \subseteq BL$ for every bi-ideal B and every left ideal L of S.

(4) $G \cap L \subseteq GL$ for every generalized bi-ideal G and every left ideal L of S. ■

The left-right dual of Theorem 3.1.11 reads as follows:

Theorem 3.1.13 *For a semigroup S, the following conditions are equivalent.*

(1) S is regular.

(2) $f \cap g \subseteq g \circ f$ for every fuzzy quasi-ideal f, and every fuzzy right ideal g of S.

(3) $f \cap g \subseteq g \circ f$ for every fuzzy bi-ideal f, and every fuzzy right ideal g of S.

(4) $f \cap g \subseteq g \circ f$ for every fuzzy generalized bi-ideal f, and every fuzzy right ideal g of S. ■

The left-right dual of Theorem 3.1.12 reads as follows:

Theorem 3.1.14 *For a semigroup S, the following conditions are equivalent.*

(1) S is regular.

(2) $Q \cap R \subseteq RQ$ for every quasi-ideal Q and every right ideal R of S.

(3) $B \cap R \subseteq RB$ for every bi-ideal B and every right ideal R of S.

(4) $G \cap R \subseteq RG$ for every generalized bi-ideal G and every right ideal R of S. ■

Theorem 3.1.15 *For a semigroup S the following conditions are equivalent.*

(1) S is regular.

(2) $h \cap f \cap g \subseteq h \circ f \circ g$ for every fuzzy right ideal h, every fuzzy quasi-ideal f, and every fuzzy left ideal g of S.

(3) $h \cap f \cap g \subseteq h \circ f \circ g$ for every fuzzy right ideal h, every fuzzy bi-ideal f, and every fuzzy left ideal g of S.

(4) $h \cap f \cap g \subseteq h \circ f \circ g$ for every fuzzy right ideal h, every fuzzy generalized bi-ideal f, and every fuzzy left ideal g of S.

Proof. First assume that (1) holds. Let h, f and g be any fuzzy right ideal, any fuzzy generalized bi-ideal, and any fuzzy left ideal of S, respectively.

Let a be any element of S. Then, since S is regular, there exists an element $x \in S$ such that $a = axa$. Hence we have

$$\begin{aligned}(h \circ f \circ g)(a) &= \vee_{a=yz}\{h(y) \wedge (f \circ g)(z)\} \\ &\geq h(ax) \wedge (f \circ g)(a) \\ &= h(a) \wedge (\vee_{a=pq}\{f(p) \wedge g(q)\}) \\ &\geq h(a) \wedge (f(a) \wedge g(xa)) \\ &\geq h(a) \wedge (f(a) \wedge g(a)) \\ &= (h \cap f \cap g)(a),\end{aligned}$$

and so $h \circ f \circ g \supseteq h \cap f \cap g$. Thus (1) implies (4).

It is clear that (4) $\Rightarrow$ (3) $\Rightarrow$ (2).

Assume that (2) holds. Let h and g be any fuzzy right ideal and any fuzzy left ideal of S, respectively. Since S itself is a fuzzy quasi-ideal of S, by the assumption, we have

$$h \cap g = h \cap S \cap g \subseteq h \circ S \circ g \subseteq h \circ g.$$

Thus it follows that $h \cap g = h \circ g$ for every fuzzy right ideal h and every fuzzy left ideal g of S. Hence it follows from Theorem 3.1.8 that S is regular and so (2) implies (1). ■

The following theorem is the special case of Theorem 3.1.15.

Theorem 3.1.16 *For a semigroup S, the following conditions are equivalent.*

(1) S is regular.

(2) $R \cap Q \cap L = RQL$ for every right ideal R, every quasi-ideal Q and every left ideal L of S.

(3) $R \cap B \cap L = RBL$ for every right ideal R, every bi-ideal B and every left ideal L of S.

(4) $R \cap G \cap L = RGL$ for every right ideal R, every generalized bi-ideal G, and every left ideal L of S. ■

Definition 3.1.17 *A semigroup S is called **right (left) zero** if $xy = y$ ($xy = x$) for all $x, y \in S$.*

Then we have the following result.

Theorem 3.1.18 *For a regular semigroup S, the following conditions are equivalent.*

(1) The set $E(S)$ of all idempotents of S forms a left (right) zero subsemigroup of S.

(2) For every fuzzy left (right) ideal f of S, $f(e) = f(e')$ for all $e, e' \in E(S)$.

Proof. Assume that (1) holds. Let f be any fuzzy left ideal of S and $e, e' \in E(S)$. Then since $ee' = e$ and $e'e = e$, we have

$$f(e) = f(ee') \geq f(e') = f(e'e) \geq f(e),$$

and so $f(e) = f(e')$. Thus (1) implies (2).

Conversely, assume that (2) holds. Since S is regular, $E(S)$ is nonempty. Thus it follows from Lemma 2.3.1 that the characteristic function $C_{L[e']}$ of the principal left ideal $L[e']$ of S is a fuzzy left ideal of S. Thus

$$C_{L[e']}(e) = C_{L[e']}(e') = 1,$$

and so $e \in L[e'] = Se'$. Hence for some $x \in S$, we have

$$e = xe' = x(e'e') = (xe')e' = ee'.$$

Thus $E(S)$ is a left zero subsemigroup of S and so (2) implies (1). The case when S is right zero is proved similarly. ■

Theorem 3.1.19 *For a regular semigroup S, the following conditions are equivalent.*

(1) S is a group.

(2) For every fuzzy bi-ideal f of S, $f(e) = f(e')$ for all $e, e' \in E(S)$.

Proof. First assume that S is a group. Let f be any fuzzy bi-ideal of S. Then it follows from Lemma 2.6.8 that f is a constant function. This implies that $f(e) = f(e')$ for all $e, e' \in E(S)$ and so (1) implies (2). Conversely, assume that (2) holds. Let e and e' be any elements of $E(S)$. Then it follows from Lemma 2.4.1 that the characteristic function $C_{B[e']}$ of the principal bi-ideal $B[e']$ is a fuzzy bi-ideal of $\dot{S}$. Since $e' \in B[e']$, we have

$$C_{B[e']}(e) = C_{B[e']}(e') = 1.$$

Hence $e \in B[e'] = e'Se'$, which implies that $e = e'xe'$ for some $x \in S$. Similarly, we obtain that $e' = eye$ for some $y \in S$. Thus we have $e = e'xe' = e'x(e'e') = (e'xe')e' = ee' = e(eye) = (ee)ye = eye = e'$. Since S is regular, $E(S)$ is nonempty. Therefore, S contains exactly one idempotent. Hence it follows from [12, p.33] that S is a group and so (2) implies (1). ■

We now consider the work in [5].

The quasi-ideal $(X)_q$ of a semigroup S generated by a nonempty subset X of S is of the form $(X)_q = (X \cup SX) \cap (X \cup XS)$. The principle quasi-ideal $(x)_q$ of the semigroup S generated by the element $x \in S$ is of the form $(x)_q = (x \cup Sx) \cap (x \cup xS) = x \cup (Sx \cap xS)$ (cf. [137]).

A semigroup S is called **intra-regular** if for all $x \in S$, there exist $a, b \in S$ such that $x = ax^2b$ (cf. [137]).

Recall that by C_A, we mean the characteristic function of a subset A of the semigroup S. The characteristic function of S is sometimes denoted by

S. A nonempty fuzzy subset f of S is a fuzzy bi-ideal if $f \circ f \subseteq f$ and $f \circ C_S \circ f \subseteq f$.

Lemma 3.1.20 *Let f, g and h be the fuzzy subsets of S. Then the following properties hold:*

(1) $f \circ (g \cup h) = (f \circ g) \cup (f \circ h)$; $(g \cup h) \circ f = (g \circ f) \cup (h \circ f)$;
(2) $f \circ (g \cap h) \subseteq (f \circ g) \cap (f \circ h)$; $(g \cap h) \circ f \subseteq (g \circ f) \cap (h \circ f)$.

Proof. (1) Let $x \in S$. Then

$$(f \circ (g \vee h))(x) = 0 = (f \circ g)(x) \vee (f \circ h)(x)$$

if x is not expressible as $x = yz$; otherwise

$$\begin{aligned}(f \circ (g \vee h))(x) &= \vee_{x=yz}\{f(y) \wedge (g \vee h)(z)\} \\ &= \vee_{x=yz}\{f(y) \wedge (g(z) \vee h(z))\} \\ &= \vee_{x=yz}\{(f(y) \wedge g(z)) \vee (f(y) \wedge h(z))\} \\ &= (\vee_{x=yz}\{f(y) \wedge g(z)\}) \vee (\vee_{x=yz}\{f(y) \wedge h(z)\}) \\ &= (f \circ g)(x) \vee (f \circ h)(x).\end{aligned}$$

Thus $f \circ (g \cup h) = (f \circ g) \cup (f \circ h)$.

(2) $(f \circ (g \cup h))(x) = 0 = (f \circ g)(x) \wedge (f \circ h)(x)$ if x is not expressible as $x = yz$; otherwise

$$\begin{aligned}(f \circ (g \wedge h))(x) &= \vee_{x=yz}\{f(y) \wedge (g \wedge h)(z)\} \\ &= \vee_{x=yz}\{f(y) \wedge (g(z) \wedge h(z))\} \\ &= \vee_{x=yz}\{(f(y) \wedge g(z)) \wedge (f(y) \wedge h(z))\} \\ &\leq (\vee_{x=yz}\{f(y) \wedge g(z)\}) \wedge (\vee_{x=yz}\{f(y) \wedge h(z)\}) \\ &\leq (f \circ g)(x) \wedge (f \circ h)(x).\end{aligned}$$

Thus $f \circ (g \cap h) \subseteq (f \circ g) \cap (f \circ h)$.

Similarly we can show that $(g \cup h) \circ f = (g \circ f) \cup (h \circ f)$ and $(g \cap h) \circ f \subseteq (g \circ f) \cap (h \circ f)$. ■

Lemma 3.1.21 *Let f, g and h be fuzzy subsets of S. If $f \subseteq g$, then $f \circ h \subseteq g \circ h$ and $h \circ f \subseteq h \circ g$.*

Proof. Let $x \in S$, then $(f \circ h)(x) = 0 = (g \circ h)(x)$ if x is not expressible as $x = yz$; otherwise

$$\begin{aligned}(f \circ h)(x) &= \vee_{x=yz}\{f(y) \wedge h(z)\} \\ &\leq \vee_{x=yz}\{g(y) \wedge h(z)\} \text{ (since } f(y) \leq g(y)) \\ &= (g \circ h)(x).\end{aligned}$$

Similarly we can show that $h \circ f \subseteq h \circ g$. ■

Recall that a nonempty subset f of a semigroup S is a fuzzy quasi-ideal of S if $(S \circ f) \cap (f \circ S) \subseteq f$.

Lemma 3.1.22 *If f is a nonempty fuzzy subset of S, then*

$$f \cup (S \circ f)(f \cup (f \circ S))$$

is a fuzzy left (right) ideal of S.

Proof. Let f be a fuzzy nonempty subset of S. Then $S \circ \{f \cup (S \circ f)\} = (S \circ f) \cup (S \circ (S \circ f))$ (Lemma 3.1.20) $= (S \circ f) \cup ((S \circ S) \circ f) \subseteq (S \circ f) \cup (S \circ f) = S \circ f \subseteq f \cup (S \circ f)$. Thus $f \cup (S \circ f)$ is a fuzzy left ideal of S. ∎

Proposition 3.1.23 *Every fuzzy quasi-ideal of S is a fuzzy subsemigroup of S.*

Proof. Let f be a fuzzy quasi-ideal of S. Since $f \subseteq S$, $f \circ f \subseteq S \circ f$ and $f \circ f \subseteq f \circ S$ (Lemma 3.1.21). Hence $f \circ f \subseteq (S \circ f) \cap (f \circ S) \subseteq f$ as f is a fuzzy ideal of S. Thus $f \circ f \subseteq f$. That is, f is a fuzzy subsemigroup of S. ∎

Proposition 3.1.24 *Each one-sided fuzzy ideal of a semigroup S is a fuzzy quasi-ideal of S.*

Proof. Let f be a fuzzy left ideal of S. Then $S \circ f \subseteq f$ and hence $(S \circ f) \cap (f \circ S) \subseteq S \circ f \subseteq f$. Thus f is a quasi-ideal of S. ∎

For any nonempty fuzzy subset f of S, $(f \cup (S \circ f)) \cap (f \cup (f \circ S))$ is a fuzzy quasi-ideal of S.

Proposition 3.1.25 *Every fuzzy quasi-ideal f of a semigroup S is the intersection of the fuzzy left ideal $f \cup (S \circ f)$ and the fuzzy right ideal $f \cup (f \circ S)$ of S.*

Proof. As shown in Lemma 3.1.22, $f \cup (S \circ f)$ (resp. $f \cup (f \circ S)$) is a fuzzy left (resp. right) ideal of S.

In Lemma 2.6.5, we have shown that the intersection of a fuzzy left ideal and a fuzzy right ideal is a fuzzy quasi-ideal. Hence it suffices to show that $f = (f \cup (S \circ f)) \cap (f \cup (f \circ S))$. Since $f \subseteq f \cup (S \circ f)$ and $f \subseteq f \cup (f \circ S)$, it follows that

$$\begin{aligned} f &\subseteq (f \cup (S \circ f)) \cap (f \cup (f \circ S)) \\ &= ((f \cup (S \circ f)) \cap f) \cup ((f \cup (S \circ f)) \cap (f \circ S)) \\ &\subseteq (f \cup ((f \cup (S \circ f)) \cap (f \circ S))) \\ &\qquad \text{(since } (f \cup (S \circ f)) \cap f \subseteq f) \\ &= f \cup ((f \cup (S \circ f)) \cup ((S \circ f) \cap (f \circ S))) \\ &\subseteq f \cup ((f \cap (f \circ S)) \cup f) \\ &\qquad \text{(since } (S \circ f) \cap (f \circ S) \subseteq f) \\ &\subseteq f \cup (f \cup f) \\ &\qquad \text{(since } f \cap (f \circ S) \subseteq f) \\ &= f. \end{aligned}$$

Thus $f = (f \cup (S \circ f)) \cap (f \cup (f \circ S))$. ∎

Propositions 3.1.24 and 3.1.25 yield the following result.

Theorem 3.1.26 *A nonempty fuzzy subset f of a semigroup S is a fuzzy quasi-ideal of S if and only if f is the intersection of a fuzzy right ideal and a fuzzy left ideal of S.*

Proposition 3.1.27 *The intersection of any set of fuzzy quasi-ideals of S is either empty or a fuzzy quasi-ideal of S.*

Proof. Let $\{f_i\}_{i\in I}$ be a family of fuzzy quasi-ideals of S. Suppose $\cap_{i\in I} f_i$ is nonempty. Then

$$\begin{aligned} & (S \circ (\cap_{i\in I} f_i)) \cap ((\cap_{i\in I} f_i) \circ S) \\ \subseteq\ & (\cap_{i\in I}(S \circ f_i)) \cap (\cap_{i\in I}(f_i \circ S)) \\ \subseteq\ & \cap_{i\in I}((S \circ f_i) \cap (f_i \circ S)) \subseteq \cap_{i\in I} f_i \text{ (since } (S \circ f_i) \cap (f_i \circ S) \subseteq f_i). \end{aligned}$$

Thus $\cap_{i\in I} f_i$ is a fuzzy quasi-ideal of S. ■

Proposition 3.1.28 *Every fuzzy quasi-ideal of S is a fuzzy bi-ideal of S.*

Proof. Let f be a fuzzy quasi-ideal of S. Then by Proposition 3.1.23, S is a fuzzy subsemigroup of S i.e., $f \circ f \subseteq f$. Since $f \circ S \subseteq S$,

$$f \circ S \circ f \subseteq S \circ f \tag{3.1}$$

Also, since $S \circ f \subseteq S$,

$$f \circ S \circ f \subseteq f \circ S \tag{3.2}$$

From (3.1) and (3.2), $f \circ S \circ f \subseteq (S \circ f) \cap (f \circ S)$, but $(S \circ f) \cap (f \circ S) \subseteq f$ since f is a fuzzy quasi-ideal. Hence $f \circ S \circ f \subseteq f$. Thus f is a fuzzy bi-ideal of S. ■

The converse of Proposition 3.1.28 is not generally true as shown by the following example.

Example 3.1.29 *Let $S = \{0, 1, 2, 3\}$ be the semigroup defined by the following table.*

•	0	1	2	3
0	0	0	0	0
1	0	0	0	0
2	0	0	0	1
3	0	0	1	2

Define $f : S \to [0, 1]$ by $f(0) = 1$, $f(1) = 0.6$, $f(2) = 0.7$ and $f(3) = 0.5$. Then f is not a fuzzy quasi-ideal of S, but f is a fuzzy bi-ideal of S. We have $(f \circ S)(0) = 1$, $(f \circ S)(1) = 0.7$, $(f \circ S)(2) = 0.5$, $(f \circ S)(3) = 0$; $(S \circ f)(0) = 1$, $(S \circ f)(1) = 0.7$, $(S \circ f)(2) = 0.5$, $(S \circ f)(3) = 0$. Thus $((f \circ S) \cap (S \circ f))(0) = 1$, $((f \circ S) \cap (S \circ f))(1) = 0.7$, $((f \circ S) \wedge (S \circ f))$ $(2) = 0.5$, $((f \circ S) \wedge (S \circ f))(3) = 0$, $\Rightarrow (f \circ S) \cap (S \circ f) \not\subseteq f$. Hence f is *not a fuzzy quasi-ideal of S. But $(f \circ S \circ f)(0) = 1$, $(f \circ S \circ f)(1) = 0.5$, $(f \circ S \circ f)(2) = 0$, $(f \circ S \circ f)(3) = 0$. Thus $f \circ S \circ f \subseteq f$. Also $(f \circ f)(1) = 0.5$, $(f \circ f)(2) = 0.5$, and $(f \circ f)(3) = 0$. $f \circ f \subseteq f$. Hence f is a fuzzy bi-ideal of S.*

The product of two quasi-ideals need not be a quasi-ideal and so the product of two fuzzy quasi-ideals need not be a fuzzy quasi-ideal.

Theorem 3.1.30 *The following conditions on a semigroup S are equivalent.*

(1) S is regular.

(2) For every fuzzy right ideal h and fuzzy left ideal k of S, $h \circ k = h \cap k$.

(3) For every fuzzy right ideal h and fuzzy left ideal k of S,

(a) $h^2 = h \circ h = h$,

(b) $k^2 = k \circ k = k$, and

(c) $h \circ k$ is a fuzzy quasi-ideal of S.

(4) The set of all fuzzy quasi-ideals of S is a regular semigroup under the operation $\circ$.

(5) Every fuzzy quasi-ideal f of S has the form $f = f \circ S \circ f$.

Proof. (1) $\Rightarrow$ (2): The result here follows from Theorem 3.1.8.

(2) $\Rightarrow$ (3): Since h is a fuzzy right ideal of S, $h \cup (S \circ h)$ is a fuzzy two-sided ideal of S by Lemma 2.3.5. Now by the hypothesis, $h = h \cap (h \cup (S \circ h)) = h \circ (h \cup (S \circ h)) = (h \circ h) \cup (h \circ (S \circ h))$ (Lemma 3.1.20) $= (h \circ h) \cup ((h \circ S) \circ h)$ (since $\circ$ is associative) $\subseteq (h \circ h) \cup (h \circ h)$ (since h is a fuzzy right ideal) $= h \circ h$. Thus $h \subseteq h \circ h$. Also, $h \circ h \subseteq h \circ S \subseteq h$. Hence $h = h \circ h = h$. Similarly, $k = k \circ k = k^2$. By (2), $h \circ k = h \cap k$. By Lemma 2.6.5, $h \cap k$ is a fuzzy quasi-ideal of S. Thus $h \circ k$ is a fuzzy quasi-ideal of S.

(3) $\Rightarrow$ (4): Let f be a fuzzy quasi-ideal of S. Then by Lemma 3.1.22, $f \cup (S \circ f)$ is a fuzzy left ideal of S. Thus by (3) (a), $f \subseteq f \cup (S \circ f) = (f \cup (S \circ f)) \circ (f \cup (S \circ f)) = ((f \cup (S \circ f)) \circ f) \cup ((f \cup (S \circ f)) \circ (S \circ f))$ (Lemma 3.1.20) $= ((f \circ f) \cup (S \circ f) \circ f) \cup ((f \circ (S \circ f)) \cup ((S \circ f) \circ (S \circ f)))$ (Lemma 3.1.20) $= ((f \circ f) \cup (S \circ (f \circ f))\} \cup ((f \circ (S \circ f)) \cup (S \circ f)^2) \subseteq ((S \circ f) \cup (S \circ f)) \cup (S \circ (S \circ f) \cup (S \circ f))$ (since $f \circ f \subseteq f$ and $S \circ f$ is a fuzzy left ideal) $\subseteq (S \circ f) \cup ((S \circ f) \cup (S \circ f)) = S \circ f$, i.e., $f \subseteq S \circ f$. Similarly, we can show that $f \subseteq f \circ S$. Thus $f \subseteq (S \circ f) \cap (f \circ S)$. But $(S \circ f) \cap (f \circ S) \subseteq f$ since f is a fuzzy quasi-ideal. Hence

$$(S \circ f) \cap (f \circ S) = f. \tag{3.3}$$

Now

$$h \circ k = (S \circ (h \circ k)) \cap ((h \circ k) \circ S) \tag{3.4}$$

since $h \circ k$ us a fuzzy quasi-ideal, where h is a fuzzy right ideal and k is a fuzzy left ideal of S.

Let f, g be two fuzzy quasi-ideals of S. Then $S \circ f \circ g$ is a fuzzy left ideal and $f \circ g \circ S$ is a fuzzy right ideal of S by Lemma 2.3.2 since $S \circ S \subseteq S$ and $\circ$ is associative. Hence by Proposition 3.1.24, $S \circ f \circ g$ and $f \circ g \circ S$ are fuzzy quasi-ideals. Therefore by (1) above, $S \circ f \circ g = (S \circ f \circ g) \circ S \cap S \circ (S \circ f \circ g) = (S \circ f \circ g) \circ (S \circ f \circ g)$ (by (2)) $= ((S \circ f \circ g) \circ S) \circ (S \circ (S \circ f \circ g))$ (since $S^2 = S$) $= (S \circ f \circ g) \circ (S \circ S) \circ (S \circ f \circ g) = (S \circ f \circ g) \circ S \circ (S \circ f \circ g)$ (since $S \circ S = S$).

Similarly,

$$f \circ g \circ S = (f \circ g \circ S) \circ S \circ (f \circ g \circ S).$$

Now

$$\begin{aligned}
& (S \circ f \circ g) \cap (f \circ g \circ S) \\
= \; & ((S \circ (f \circ g \circ S) \circ (S \circ f \circ g))) \cap ((f \circ g \circ S) \circ S \circ (f \circ g \circ S)) \\
= \; & ((S \circ (f \circ g \circ S) \circ (S \circ f \circ g)))((f \circ g \circ S) \circ (S \circ f \circ g) \circ S) \\
= \; & (f \circ g \circ S) \circ (S \circ f \circ g) \text{ (by (3))} \\
= \; & (S \circ f \circ g) \circ S \circ (S \circ f \circ g) \text{ (since } f \subseteq S) \\
= \; & (f \circ g \circ S) \circ (S \circ g) \text{ (because } S \circ S = S) \\
= \; & f \circ g \circ (S \circ S) \circ g \\
= \; & f \circ (g \circ S \circ g) \\
\subseteq \; & f \circ g \text{ (as } g \circ S \circ g \subseteq g).
\end{aligned}$$

Hence $f \circ g$ is a fuzzy quasi-ideal of S.

Since $\circ$ is an associative, the set of all fuzzy quasi-ideals of S is a semigroup under the operation $\circ$. If f is a fuzzy quasi-ideal of S, then $f = (f \circ S) \cap (S \circ f)$ (by (1) above) $= (f \circ S) \circ (S \circ f)$ (by (3) (c)) $= f \circ (S \circ S) \circ f$ (since $f \circ S \circ f \subseteq f$). Hence $f \circ S \circ f = f$. Thus f is regular. Hence the set of all fuzzy quasi-ideals of S forms a regular semigroup under the operation $\circ$.

(4) $\Rightarrow$ (5): Let f be fuzzy quasi-ideal of S. By assumption, there exists a fuzzy quasi-ideal g of S such that $f = f \circ g \circ f$, and we have $f = f \circ g \circ f \subseteq f \circ S \circ f$ (Lemma 3.1.21) $\subseteq (f \circ S) \cap (S \circ f)$ (since $f \circ S \circ f \subseteq f \circ S \circ S$ $= f \circ S) \subseteq f$. Hence $f = f \circ S \circ f$.

(5) $\Rightarrow$ (1): Let $x \in S$ and $A = \{x\} \cup (Sx \cap xS)$ be the quasi-ideal of S generated by x. Then the characteristic function C_A of A is a quasi-ideal of S. Thus by the assumption, we have $C_A = C_A \circ S \circ C_A$. Hence $1 = C_A(x) = (C_A \circ S \circ C_A)(x) = \vee_{x=yz} \{C_A(y) \wedge (S \circ C_A)(z)\}$. Thus $\exists x, y$ such that $x = yz$ and $C_A(y) = 1$ and $S \circ C_A(z) = 1$. Now $C_A(y) = 1 \Rightarrow y \in A$, i.e., $y = x$ or $y = xs$ where $s \in S$. Also $S \circ C_A(z) = 1 = \vee_{x=ab} \{S(a) \wedge C_B(b)\}$. Hence there exist a, b such that $z = ab$ and $C_A(b) = 1 \Rightarrow b \in A$, i.e., either $b = x$ or $b = s_1 x$, where $s_1 \in S$. Now $x = yz = xcx$, where $c \in S$. Thus x is regular and so S is regular. ■

Theorem 3.1.31 *Let f be a nonempty fuzzy subset of a regular semigroup S. Then the following conditions are equivalent.*

(1) f is a fuzzy quasi-ideal of S.

(2) f may be represented in the form $f = g \circ h$, where g is a fuzzy right ideal and h a fuzzy left ideal of S.

Proof. (1) $\Rightarrow$ (2): Since S is regular, it follows from Theorem 3.1.30 that $f = f \circ S \circ f$ (where f is a fuzzy quasi-ideal of S) $= f \circ (S \circ S) \circ f$ $= (f \circ S) \circ (S \circ f) \Rightarrow f = (f \circ S) \circ (S \circ f)$, where $f \circ S$ is a fuzzy right ideal and $S \circ f$ is a fuzzy left ideal of S.

(2) $\Rightarrow$ (1): Suppose $f = g \circ h$, where g is a fuzzy right ideal and h is a fuzzy left ideal of S. Then by Theorem 3.1.30, $g \circ h$ is a fuzzy quasi-ideal of S. ∎

Theorem 3.1.32 *Let S be a regular semigroup. Then the following assertions hold:*

(1) If f is a fuzzy quasi-ideal of S, then $f^2 = f^3$.

(2) f is a fuzzy quasi-ideal of S if and only if f is a fuzzy bi-ideal of S.

Proof. (1) Let f be a fuzzy quasi-ideal of S. Then $f \circ f \subseteq f$ (since f is a fuzzy subsemigroup) $\Rightarrow f \circ (f \circ f) \subseteq f \circ f \Rightarrow f^3 \subseteq f^2$. Now f^2 is a fuzzy quasi-ideal of S by Theorem 3.1.30. Furthermore, there exists a fuzzy quasi-ideal g of S such that $f^2 = f^2 \circ g \circ f^2 \subseteq f^2 \circ S \circ f^2 = f \circ (f \circ S \circ f) = f \circ f \circ f$ $= f^3$. (Theorem 3.1.30). Thus $f^2 \subseteq f^3$. So $f^2 = f^3$.

(2) By Proposition 3.1.28, every fuzzy quasi-ideal of S is a fuzzy bi-ideal of S. Let f be a fuzzy bi-ideal of S. Then $S \circ f$ (resp. $f \circ S$) is a fuzzy left (resp. right) ideal of S. By Theorem 3.1.30, we have $(S \circ f) \cap (f \circ S) =$ $(S \circ f) \circ (S \circ f) = f \circ (S \circ S) \circ f = f \circ S \circ f \subseteq f$ (since f is a fuzzy bi-ideal of S). Thus f is a fuzzy quasi-ideal of S. ∎

Theorem 3.1.33 *The following conditions are equivalent for a semigroup S.*

(1) For every fuzzy right ideal h and every fuzzy left ideal k of S, $h \circ k = h \cap k \subseteq k \circ h$.

(2) The set of all fuzzy quasi-ideals of S is a band with respect to $\circ$ (that is, each fuzzy quasi-ideal of S is idempotent).

(3) Every fuzzy quasi-ideal of f of S is idempotent (i.e., $f = f^2$).

Proof. (1) $\Rightarrow$ (2): $h \circ k = h \cap k \Rightarrow S$ is regular by Theorem 3.1.30. Hence by the same theorem, the set of all fuzzy quasi-ideals of S is a regular semigroup with respect to $\circ$. Therefore, we have to show that every fuzzy quasi-ideal f of S is idempotent. By Theorem 3.1.30, $f = f \circ S \circ f =$ $(f \circ S \circ f) \circ S \circ (f \circ S \circ f) = (f \circ S) \circ (f \circ S) \circ (S \circ f) \circ (S \circ f)$ (since $S \circ S = S$) $\subseteq (f \circ S) \circ (S \circ f) \circ (f \circ S) \circ (S \circ f)$ (since $h \circ k \subseteq k \circ h$) $= f \circ (S \circ S) \circ f \circ f \circ (S \circ S) \circ f = (f \circ S \circ f) \circ (f \circ S \circ f)$ (since $S \circ S = S$) $= f \circ f$ (since $f \circ S \circ f = f$) $\subseteq f$ (since $f \circ f \subseteq f$). Hence $f = f^2$. Thus the set of all fuzzy quasi-ideals of S is a band with respect to $\circ$.

(2) $\Rightarrow$ (3): Obvious.

(3) $\Rightarrow$ (1): Let h be a fuzzy right ideal and let k be a fuzzy left ideal of S. Then by Lemma 2.6.5, $h \cap k$ is a fuzzy quasi-ideal of S. Hence by (3), $h \cap k = (h \cap k) \circ (h \cap k) \subseteq (h \circ (h \cap k)) \cap (k \circ (h \cap k))$ (Lemma 3.1.20) $\subseteq ((h \circ h) \cap (k \circ h)) \cap ((k \circ h) \cap (k \circ k))$ (Lemma 3.1.20) $= (h \circ h) \cap (k \circ k) \cap (h \circ k) \cap (k \circ h)$ (since h is commutative) $\subseteq h \circ k$. Also from above calculations $h \cap k \subseteq k \circ h$. But $h \circ k \subseteq h \circ S \subseteq h$, $h \circ k \subseteq S \circ h \subseteq k$. Thus $h \circ k \subseteq h \cap k$. Hence by (1) and (2) above, we have $h \circ k = h \cap k \subseteq k \circ h$. ∎

Example 3.1.34 *Let $S = \{a, b, c, d, e\}$ be the semigroup given by the following table:*

$\bullet$	a	b	c	d	e
a	a	a	a	a	a
b	a	a	a	b	c
c	a	b	c	a	a
d	a	a	a	d	e
e	a	d	e	a	a

Then S is a non-commutative regular semigroup, which is not intra-regular. A non-empty fuzzy subset h of S is a fuzzy left ideal of S $\Leftrightarrow$ $h(a) \geq h(x)$ for all $x \in S$, $h(b) = h(d), h(c) = h(e)$. A nonempty fuzzy subset k of S is a fuzzy right ideal of $S \Leftrightarrow k(a) \geq k(x)$ for all $x \in S$, $k(b) = k(c), k(d) = k(e)$, $h \circ h(a) = \vee_{x=yz}\ (h(x) \wedge h(y))$. Note that $h \circ h(a) = h(a)$, since $a = aa$ and $h(a) \geq h(x)$ for all $x \in S$ and $h \circ h(b) = (h(b) \wedge h(d)) \vee (h(c) \wedge h(b)) = h(b) \vee (h(c) \wedge h(b))$ (since $h(b) = h(d)$) $= h(b) \vee h(c) = h(b)$ (if $h(c) \leq h(b)$) $= h(b) \vee h(b) = h(b)$ (if $h(b) \leq h(c)$). Also, $h \circ h(c) = (h(c) \wedge h(c)) \vee (h(b) \wedge h(e)) = h(c)\ \vee (h(b) \wedge h(c))$ (since $h(c) = h(e)$) $= h(c) \vee h(b) = h(c)$ (if $h(b) \leq h(c)$) $= h(c) \vee h(c) = h(c)$ (if $h(c) \leq h(b)$).

Similarly, $h \circ h(d) = (h(d) \wedge h(d)) \vee (h(e) \wedge h(b)) = h(d) \vee (h(e) \wedge h(d))$ (since $h(b) = h(d)$) $= h(d)$ as above. Also, $h \circ h(e) = (h(d) \wedge h(e)) \vee (h(e) \wedge h(c)) = (h(d) \wedge h(e)) \vee h(e)$ (since $h(c) = h(e)$) $= h(e)$ as above. Thus $h^2 = h$. Similarly we can show that $k^2 = k$. Now $k \circ h(a) = \vee_{a=sy}\ \{k(x) \wedge k(y)\}$, $k \circ h(a) = k(a) \wedge h(a)$ (since $a = aa$ and $k(a) \geq k(x)$ and $h(a) \geq h(x)$ for all $x \in S$) $= (k \cap h)(a)$, $k \circ h(b) = (k(b) \wedge h(d)) \vee\ (k(c) \wedge h(b)) = (k(b) \wedge h(b)) \vee (k(b) \wedge h(b))$ $(h(d) = h(b), k(c) = k(b)) = (k \wedge h)(b)$, $k \circ h(c) = (k(c) \wedge h(c)) \vee (k(b) \wedge h(e)) = (k(c) \wedge h(c)) \vee (k(c) \wedge h(c))$ $(h(e) = h(c), k(b) = k(c)) = (k \wedge h)(c)$, $k \circ h(d) = (k(d)\ \wedge h(d)) \vee (k(e) \wedge h(b)) = (k(d) \wedge h(d)) \vee (k(d) \wedge h(d)) = (k \wedge h)(d)$, $k \circ h(e) = (k(d) \wedge h(e)) \vee (k(e) \wedge h(c)) = (k(e) \wedge h(e)) \vee (k(e) \wedge\ h(e)) = (k \wedge h)(e) \Rightarrow k \circ h = k \cap h$.

Let $f(a) = 1$, $f(b) = 4, f(c) = 0.4, f(e) = 0.5$. Then f is a fuzzy quasi-ideal of S. $S \circ f(x) = \vee_{x=yz}\ (S(y) \wedge f(z)) = \vee_{x=yz} f(z)$; $S \circ f\ (a) = 1$; $S \circ f(a) = 1; S \circ f(b) = 0.4$; $S \circ f(c) = 0.5$; $S \circ f(d) = 0.4$; $S \circ f(e) = 0.5$; $f \circ S(x) = \vee_{x=yz} f(y)$; $f \circ S(a) = 1$; $f \circ S(b) = 0.4$; $f \circ S(c) = 0.4$; $f \circ S(d) = 0.5$; $f \circ S(e) = 0.5$. Therefore, $((S \circ f) \cap (f \circ S))(a) = 1 \leq f(a)$; $((S \circ f) \cap (f \circ S))(b) = 0.4 \leq f(b)$; $((S \circ f) \cap (f \circ S))\ (c) = 0.4 \leq f(c)$;

$((S \circ f) \cap (f \circ S))(d) = 0.4 \leq f(d)$; $((S \circ f) \cap (f \circ S))(e) = 0.5 \leq f(e)$. *Thus* f *is a fuzzy quasi-ideal of* S. $f \circ f(a) = \vee_{x=yz}(f(x) \wedge f(y)) = 1$ *since* $a = aa$; $f \circ f(b) = (f(b) \wedge f(d)) \vee (f(c) \wedge f(b)) = 0.4$; $f \circ f(c) = (f(b) \wedge f(e)) \vee (f(c) \wedge f(c)) = 0.4$; $f \circ f(d) = (f(d) \wedge f(d)) \vee \{(f(e) \wedge f(b))\} = 0.4$; $f \circ f\ (e) = (f(d) \wedge f(e)) \vee (f(e) \wedge f(c)) = 0.4$; $\Rightarrow f \circ f \neq f$. *However,*

$$(f \circ f) \circ f(a) = \vee_{x=yz}\{f^2(x) \wedge f(y)\} = 1,$$

$$f^3(b) = (f^2(b) \wedge f(d)) \vee (f^2(c) \wedge f(b)) = 0.4,$$

$$f^3(c) = (f^2(b) \wedge f(e)) \vee (f^2(c) \wedge f(c)) = 0.4,$$

$$f^3(d) = f^2(d) \wedge f(d)) \vee (f^2(e) \wedge f(b)) = 0.4,$$

$$f^3(e) = (f^2(b) \wedge f(e)) \vee (f^2(e) \wedge f(c)) = 0.4.$$

Thus $f \neq f^2 = f^3$.

Example 3.1.35 *Consider the semigroup* $S = \{0, x, 1\}$ *defined by the following table:*

$\bullet$	0	x	1
0	0	0	0
x	0	x	x
1	0	x	1

S is a commutative semigroup that is regular and also intra-regular. If f is a fuzzy quasi-ideal of S, then $f(a) \geq (\vee_{a=bc} f(b)) \wedge (\vee_{a=bc} f(c)) = \vee_{a=bc} f(b)$ *(since S is commutative)* $\Rightarrow f(0) = \vee_{0=bc} f(b) \geq f(a)$ *for all* $a \in S$, $f(x) = f(x) \vee f(1) = f(x) \geq f(1)$, *i.e., every fuzzy quasi-ideal f of this semigroup is such that*

$$f(0) \geq f(x) \geq f(1),$$

$$\begin{aligned} f^2(0) &= \vee_{0=xy}\{f(x) \wedge f(y)\} = f(0) \\ &\qquad \text{since } 0 = 0 \cdot 0 \text{ and } f(0) \geq f(x) \geq f(1), \end{aligned}$$

$$f^2(x) = (f(x) \wedge f(x)) \vee (f(x) \wedge f(1)) = f(x) \text{ since } f(x) \geq f(1),$$

$$f^2(1) = (f(1) \wedge f(1)) = f(1).$$

Thus $f^2 = f$.

Example 3.1.36 *Consider the semigroup $S = \{a, b, c\}$ defined by the following table:*

$\bullet$	a	b	c
a	a	a	a
b	b	b	b
c	c	c	c

Note that S is not commutative but it is regular and intra-regular. In S every nonempty fuzzy subset of S is a fuzzy quasi-ideal of S. If S is a fuzzy quasi-ideal of S, then

$$\begin{aligned} f^2(a) &= \vee_{a=xy}\{f(x) \wedge f(y)\} \\ &= (f(a) \wedge f(a)) \vee (f(a) \wedge f(b)) \vee (f(a) \wedge f(c)) \\ &= f(a), \end{aligned}$$

$$f^2(b) = (f(a) \wedge f(b)) \vee (f(b) \wedge f(b)) \vee (f(b) \wedge f(c)) = f(b),$$

$$f^2(c) = (f(a) \wedge f(c)) \vee (f(b) \wedge f(c)) \vee (f(c) \wedge f(c)) = f(c).$$

Thus $f^2 = f$. Hence every fuzzy quasi-ideal is idempotent.

3.2 Completely Regular Semigroups

In this section, we characterize completely regular semigroups in terms of fuzzy quasi-ideals, fuzzy bi-ideals, and fuzzy generalized bi-ideals.

Definition 3.2.1 *An element a of a semigroup S is called* **completely regular** *if there exists an element $x \in S$ such that $a = axa$ and $ax = xa$.*

Definition 3.2.2 *A semigroup S is called* **completely regular** *if every element of S is completely regular. A semigroup S is called* **left (right) regular** *if for each element a of S, there exists an element $x \in S$ such that $a = xa^2$ $(a = a^2x)$.*

The following characterization of such a semigroup is due to M. Petrich [107, pp.104-105], (see also Clifford and Preston [17,Theorem 4.3]).

Theorem 3.2.3 *For a semigroup S, the following conditions are equivalent.*

(1) S is completely regular.
(2) S is a (disjoint) union of groups.
(3) S is left and right regular, that is, $a \in a^2S \cap Sa^2$ for all $a \in S$.
(4) $a \in a^2Sa^2$ for all $a \in S$.

Theorem 3.2.4 *For a semigroup S, the following conditions are equivalent.*

(1) *S is left (right) regular.*
(2) *For every fuzzy left (right) ideal f of S, $f(a) = f(a^2)$ for all $a \in S$.*

Proof. First assume that (1) holds. Let f be any fuzzy left ideal of S and a any element of S. Then, since S is left regular, there exists an element $x \in S$ such that $a = xa^2$. Hence we have

$$f(a) = f(xa^2) \geq f(a^2) \geq f(a),$$

and so $f(a) = f(a^2)$. Thus (1) implies (2).

Conversely, assume that (2) holds. Let a be any element of S. Then it follows from Lemma 2.3.1 that the characteristic function $C_{L[a^2]}$ of the principal left ideal $L[a^2]$ of S is a fuzzy left ideal of S. Since $a^2 \in L[a^2]$, we have

$$C_{L[a^2]}(a) = C_{L[a^2]}(a^2) = 1.$$

This implies that

$$a \in L[a^2] = \{a^2\} \cup Sa^2.$$

Hence S is left regular and so (2) implies (1). The case when S is regular can be similarly proved. ■

Definition 3.2.5 *A subset P of a semigroup S is called* ***semiprime*** *if $\forall a \in S$, $a^2 \in P$ implies $a \in P$. A fuzzy subset f of a semigroup S is called* ***fuzzy semiprime*** *if*

$$f(a) \geq f(a^2)$$

for all $a \in S$.

The following theorem shows that the concept of fuzzy semiprimality in a semigroup is an extension of semiprimality.

Theorem 3.2.6 *For a nonempty subset A of a semigroup S, the following conditions are equivalent.*

(1) *A is semiprime.*
(2) *The characteristic function C_A of A is a fuzzy semiprime.*

Proof. First assume that (1) holds. Let a be any element of S. If $a^2 \in A$, then, since A is semiprime, $a \in A$. Hence

$$C_A(a) = 1 = C_A(a^2).$$

If $a^2 \in A$, then

$$C_A(a) \geq 0 = C_A(a^2).$$

In any case, we have $C_A(a) \geq C_A(a^2)$ for all $a \in S$ and so C_A is fuzzy semiprime. Thus (1) implies (2).

Conversely, assume that (2) holds. Let $a^2 \in A$ $(a \in S)$. Since C_A is fuzzy semiprime, we have

$$C_A(a) \geq C_A(a^2) = 1,$$

and so $C_A(a) = 1$, that is, $a \in A$. Thus A is semiprime. Hence (2) implies (1). ∎

Theorem 3.2.7 *For any fuzzy subsemigroup f of a semigroup S, the following conditions are equivalent.*

(1) f is fuzzy semiprime.

(2) $f(a) = f(a^2)$ for all $a \in S$.

Proof. It is clear that (2) implies (1). Assume that (1) holds. Let a be any element of S. Then, since f is a fuzzy subsemigroup of S, we have

$$f(a) \geq f(a^2) = f(a) \wedge f(a = f(a),$$

and so $f(a) = f(a^2)$. Thus (1) implies (2). ∎

Now we give characterizations of a completely regular semigroup in terms of fuzzy ideals.

Theorem 3.2.8 *For a semigroup S, the following conditions are equivalent.*

(1) S is completely regular.

(2) Every quasi-ideal of S is semiprime.

(3) Every bi-ideal of S is semiprime.

(4) Every generalized bi-ideal of S is semiprime.

(5) Every fuzzy quasi-ideal of S is semiprime.

(6) Every fuzzy bi-ideal of S is fuzzy semiprime.

(7) Every fuzzy generalized bi-ideal of S is fuzzy semiprime.

(8) $f(a) = f(a^2)$ for every fuzzy quasi-ideal f of S and for all $a \in S$.

(9) $f(a) = f(a^2)$ for every fuzzy bi-ideal f of S and for all $a \in S$.

(10) $f(a) = f(a^2)$ for every fuzzy generalized bi-ideal f of S and for all $a \in S$.

Proof. First assume that (1) holds. Let f be any fuzzy generalized bi-ideal of S and a any element of S. Then, since S is completely regular, there exists an element $x \in S$ such that $a = a^2xa^2$. Thus we have

$$\begin{aligned} f(a) &= f(a^2xa^2) \\ &\geq f(a^2) \wedge f(a^2) \\ &= f(a^2) \\ &= f(a(a^2xa^2)) \\ &= f(a(a^2xa)a) \\ &\geq f(a) \wedge f(a) \\ &= f(a), \end{aligned}$$

and so $f(a) = f(a^2)$. Thus (1) implies (10).

It is clear that (10) $\Rightarrow$ (9) $\Rightarrow$ (8) $\Rightarrow$ (5), (10) $\Rightarrow$ (7) $\Rightarrow$ (6) $\Rightarrow$ (5) and (10) $\Rightarrow$ (4) $\Rightarrow$ (3) $\Rightarrow$ (2).

Assume that (5) holds. Let Q be any quasi-ideal of S and let $a^2 \in Q$ $(a \in S)$. Then, because the characteristic function C_Q of Q is a fuzzy quasi-ideal of S by Lemma 2.6.4, it is fuzzy semiprime. Thus we have

$$C_Q(a) \geq C_Q(a^2) = 1.$$

Hence $a \in Q$ and so Q is semiprime. Thus (5) implies (2). Finally assume that (2) holds. Let a be any element of S. Then because the principal quasi-ideal $Q[a^2]$ generated by a^2 is semiprime and because $a^2 \in Q[a^2]$, we have

$$a \in Q[a^2] = \{a^2\} \cup (a^2S \cap Sa^2).$$

This implies that $a = a^2$ or $a \in a^2S \cap Sa^2$, and so S is completely regular. Thus (2) implies (1). ■

3.3 Intra-regular Semigroups

Recall that a semigroup S is called intra-regular if for each element a of S, there exists elements $x, y \in S$ such that $a = xa^2y$.

The following characterization of such a semigroup is due to A.H. Clifford and G.B. Preston [17, Theorem 4.4].

Theorem 3.3.1 *For a semigroup S, the following conditions are equivalent.*

(1) S is intra-regular.

(2) S is a union of simple semigroups.

(3) Every two-sided ideal of S is semiprime. ■

Theorem 3.3.2 *For a fuzzy subset f of an intra-regular semigroup S, the following conditions are equivalent.*

(1) f is a fuzzy two-sided ideal of S.

(2) f is a fuzzy interior ideal of S.

Proof. It is clear that (1) implies (2). Assume that (2) holds. Let a and b be any elements of S. Then, since S is intra-regular, there exist elements x, y, u and v of S such that $a = xa^2y$ and $b = ub^2v$. Thus f is a fuzzy interior ideal of S and so

$$f(ab) = f((xa^2y(b) = f((xa)a(yb)) \geq f(a)$$

and

$$f(ab) = f(a(ub^2v)) = f((au)b(bv)) \geq f(b).$$

Hence f is a fuzzy two-sided ideal of S and so (2) implies (1). ■

The following is the special case of Theorem 3.3.2.

Theorem 3.3.3 *For a subset A of an intra-regular semigroup S, the following conditions are equivalent.*

(1) A is a two-sided ideal of S.

(2) A is an interior ideal of S. ■

Now we characterize an intra-regular semigroup in terms of fuzzy ideals.

Theorem 3.3.4 *For a semigroup S, the following conditions are equivalent.*

(1) S is intra-regular.

(2) Every fuzzy two-sided ideal of S is fuzzy semiprime.

(3) Every fuzzy interior ideal of S is fuzzy semiprime.

(4) $f(a) = f(a^2)$ for all fuzzy two-sided ideal f of S and for all $a \in S$.

(5) $f(a) = f(a^2)$ for all fuzzy interior ideal f of S and for all $a \in S$.

Proof. First assume that (1) holds. Let f be any fuzzy interior ideal of S and a any element of S. Since S is intra-regular, there exist elements $x, y \in S$ such that $a = xa^2y$. Hence

$$\begin{aligned} f(a) &= f(xa^2y) \\ &\geq f(a^2) \\ &= f(a(xa^2y)) \\ &= f((ax)a(ay)) \\ &\geq f(a), \end{aligned}$$

and so $f(a) = f(a^2)$. Thus (1) implies (5). It is clear that (5) implies (4).

Assume that (4) holds. Then it follows from Lemma 2.3.1 that the characteristic function $C_{J[a^2]}$ of the principal two-sided ideal $J[a^2]$ generated by a^2 is a fuzzy two-sided ideal of S. Since $a^2 \in J[a^2]$, we have

$$C_{J[a^2]}(a) = C_{J[a^2]}(a^2) = 1.$$

This implies that

$$a \in J[a^2] = \{a^2\} \cup Sa^2 \cup a^2S \cup Sa^2S.$$

Hence it is easily seen that S is intra-regular and so (4) implies (1). It is clear that (2) and (4) are equivalent and that (3) and (5) are equivalent. ■

Theorem 3.3.5 *Let S be an intra-regular semigroup. Then for any fuzzy interior ideal f of S and for any elements a and b of S, $f(ab) = f(ba)$.*

Proof. Let f be any fuzzy interior ideal of S, and a, b any elements of S. Then by Theorem 3.3.4, we have $f(ab) = f((ab)^2) = f(a(ba)b) \geq f(ba)$ $= f((ba)^2) = f(b(ab)a) \geq f(ab)$. Thus $f(ab) = f(ba)$. ■

Theorem 3.3.6 *Let f be a fuzzy semiprime fuzzy interior ideal of a semigroup S. Then $f(a^n) \geq f(a^{n+1})$ for all positive integers n.*

Proof. Let n be any positive integer. Then

$$f(a^n) \geq f(a^{2n}) \geq f(a^{4n}) = f(a^{3n-2}a^{n+1}a) \geq f(a^{n+1}).$$

■

Definition 3.3.7 *A semigroup S is called* ***archimedean*** *if for all $a, b \in S$, there exists a positive integer n such that $a^n \in SbS$.*

Theorem 3.3.8 *Let S be an archimedean semigroup. Then every fuzzy semiprime fuzzy interior ideal of S is a constant function.*

Proof. Let f be any fuzzy semiprime fuzzy interior ideal of S and a, b any elements of S. Then, since S is archimedean, there exist elements $x, y \in S$ such that $a^n = xby$ for some positive integer n. Thus we have

$$f(a) \geq f(a^n) = f(xby) \geq f(b).$$

Similarly, we have $f(b) \geq (a)$, and so $f(a) = f(b)$. Since a, b any elements of S, f is a constant function. ■

Theorem 3.3.9 *An archimedean semigroup contains no proper semiprime interior (two-sided) ideal.* ■

Theorem 3.3.10 *A semigroup S is simple if and only if it is intra-regular and archimedean.*

Proof. First assume that S is simple. Let f be any fuzzy interior ideal of S. Then it follows from Lemma 2.5.7 that f is a constant function. Hence for any element a of S, $f(a) = f(a^2)$. Thus it follows from Theorem 3.3.4 that S is intra-regular. Hence by M.Petrich [107, Lemma I, 3.9] there exist elements $x, y \in S$ such that $a^1 = a = xby$. Therefore, S is archimedean.

Conversely, assume that S is intra-regular and archimedean. Let f be any fuzzy interior ideal of S. Then, since S is intra-regular, f is fuzzy semiprime. Since S is archimedean, f is a constant function. Thus it follows from Lemma 2.5.7 that S is simple. ■

Theorem 3.3.11 *Let S be a semigroup S. Then the following conditions are equivalent.*

(1) S is intra-regular.

(2) $L \cap R \subseteq LR$ for every left ideal L and every right ideal R of S.

(3) $f \cap g \subseteq f \circ g$ for every fuzzy left ideal f and every fuzzy right ideal g of S.

Proof. First assume that (1) holds. Let f and g be any fuzzy left ideal and any fuzzy right ideal of S, respectively. Let a be any element of S.

Then, since S is intra-regular, there exist elements $x, y \in S$ such that $a = xa^2y$. Hence we have

$$\begin{aligned}(f \circ g)(a) &= \vee_{a=pq}\{f(p) \wedge g(q)\} \\ &\geq f(xa) \wedge g(ay) \\ &\geq f(a) \wedge g(a) \\ &= (f \cap g)(a),\end{aligned}$$

and so $f \cap g \subseteq f \circ g$. Thus (1) implies (3).

Next assume that (3) holds. Let L and R be any left ideal and any right ideal of S, respectively. Let a be any element $L \cap R$. Then $a \in L$ and $a \in R$. Since by Lemma 2.3.1 C_L and C_R is a fuzzy left ideal and a fuzzy right ideal of S, respectively, by Lemma 2.3.4 we have

$$\begin{aligned}C_{LR}(a) &= (C_L \circ C_R)(a) \\ &\geq (C_L \cap C_R)(a) \\ &= C_{L \cap R}(a) \\ &= 1,\end{aligned}$$

and so $a \in LR$. Thus we obtain that $L \cap R \subseteq LR$. Hence (3) implies (2).

Finally, assume that (2) holds. Let a be any element of S. Then

$$\begin{aligned}a \in L[a] \cap R[a] &\subseteq L[a]R[a] \\ &= (\{a\} \cup Sa)(\{a\} \cup aS) \\ &= \{a^2\} \cup a^2S \cup Sa^2 \cup Sa^2S \\ &\subseteq Sa^2S,\end{aligned}$$

and so S is intra-regular. Thus (2) implies (1). ■

The following is due to Steinfeld [137, Corollary 9.10].

Theorem 3.3.12 *A semigroup S is regular and intra-regular if and only if every quasi-ideal of S is idempotent.* ■

Now we characterize a regular and intra-regular semigroup in terms of fuzzy ideals.

Theorem 3.3.13 *For a semigroup S, the following conditions are equivalent.*

(1) S is regular and intra-regular.

(2) Every fuzzy quasi-ideal of S is idempotent.

(3) Every fuzzy bi-ideal of S is idempotent.

(4) $f \cap g \subseteq f \circ g$ for all fuzzy quasi-ideals f and g of S.

(5) $f \cap g \subseteq f \circ g$ for every fuzzy quasi-ideal f and every fuzzy bi-ideal g of S.

(6) $f \cap g \subseteq f \circ g$ for every fuzzy quasi-ideal f and every fuzzy quasi-ideal g of S.

(7) $f \cap g \subseteq f \circ g$ for every fuzzy bi-ideal f and every fuzzy quasi-ideal g of S.

(8) $f \cap g \subseteq f \circ g$ *for every fuzzy bi-ideals* f *and* g *of* S.

(9) $f \cap g \subseteq f \circ g$ *for every fuzzy bi-ideal* f *and every fuzzy generalized bi-ideal* g *of* S.

(10) $f \cap g \subseteq f \circ g$ *for every fuzzy generalized bi-ideal* f *and every quasi-ideal* g *of* S.

(11) $f \cap g \subseteq f \circ g$ *for every fuzzy generalized bi-ideal* f *and every fuzzy bi-ideal* g *of* S.

(12) $f \cap g \subseteq f \circ g$ *for every fuzzy generalized bi-ideals* f *and* g *of* S.

Proof. Assume that (1) holds. Let f and g be any fuzzy generalized bi-ideals of S, and a any element of S. Then, since S is regular, there exists an element $x \in S$ such that $a = axa$. Since S is intra-regular, there exist elements $y, z \in S$ such that $a = ya^2z$. Thus

$$a = axa = axaxa = ax(ya^2z)xa = (axya)(azxa).$$

Hence

$$\begin{aligned}(f \circ g)(a) &= \vee_{a=pq}\{f(p) \wedge g(q)\}\\ &\geq f(a(xy)a) \wedge g(a(zx)a)\\ &\geq f(a) \wedge g(a)\\ &= (f \cap g)(a),\end{aligned}$$

and so $f \cap g \subseteq f \circ g$. Thus (1) implies (12).

It is clear that (12) $\Rightarrow$ (11) $\Rightarrow$ (10) $\Rightarrow$ (4) $\Rightarrow$ (2), (12) $\Rightarrow$ (9) $\Rightarrow$ (8) $\Rightarrow$ (7) $\Rightarrow$ (4), (12) $\Rightarrow$ (6) $\Rightarrow$ (5) $\Rightarrow$ (4) and (8) $\Rightarrow$ (3) $\Rightarrow$ (2). Assume that (2) holds. Let Q be any quasi-ideal of S and a any element of Q. Since the characteristic function C_Q of Q is a fuzzy quasi-ideal of S by Lemma 2.6.4, we have by Lemma 2.3.4 that

$$C_{QQ}(a) = (C_Q \circ C_Q)(a) = C_Q(a) = 1,$$

and so $a \in Q^2$. Thus $Q \subseteq Q^2$.

Since the converse inclusion always holds, we obtain that $Q = Q^2$ and so (2) implies (1) by Theorem 3.3.12. ■

Theorem 3.3.14 *For a semigroup* S, *the following conditions are equivalent.*

(1) S *is regular and intra-regular.*

(2) $f \cap g \subseteq (f \circ g) \cap (g \circ f)$ *for every fuzzy right ideal* f *and every fuzzy left ideal* g *of* S.

(3) $f \cap g \subseteq (f \circ g) \cap (g \circ f)$ *for every fuzzy right ideal* f *and every fuzzy quasi-ideal* g *of* S.

(4) $f \cap g \subseteq (f \circ g) \cap (g \circ f)$ *for every fuzzy right ideal* f *and every bi-ideal* g *of* S.

(5) $f \cap g \subseteq (f \circ g) \cap (g \circ f)$ *for every fuzzy right ideal* f *and every fuzzy generalized bi-ideal* g *of* S.

(6) $f \cap g \subseteq (f \circ g) \cap (g \circ f)$ *for every fuzzy left ideal* f *and every fuzzy quasi-ideal* g *of* S.

(7) $f \cap g \subseteq (f \circ g) \cap (g \circ f)$ *for every fuzzy left ideal* f *and every fuzzy bi-ideal* g *of* S.

(8) $f \cap g \subseteq (f \circ g) \cap (g \circ f)$ *for every fuzzy left ideal* f *and every fuzzy generalized bi-ideal* g *of* S.

(9) $f \cap g \subseteq (f \circ g) \cap (g \circ f)$ *for all fuzzy quasi-ideals* f *and* g *of* S.

(10) $f \cap g \subseteq (f \circ g) \cap (g \circ f)$ *for every fuzzy quasi-ideal* f *and every fuzzy bi-ideal* g *of* S.

(11) $f \cap g \subseteq (f \circ g) \cap (g \circ f)$ *for every fuzzy quasi-ideal* f *and every fuzzy generalized bi-ideal* g *of* S.

(12) $f \cap g \subseteq (f \circ g) \cap (g \circ f)$ *for all fuzzy bi-ideals* f *and* g *of* S.

(13) $f \cap g \subseteq (f \circ g) \cap (g \circ f)$ *for every fuzzy bi-ideal* f *and every fuzzy generalized bi-ideal* g *of* S.

(14) $f \cap g \subseteq (f \circ g) \cap (g \circ f)$ *for every fuzzy generalized bi-ideals* f *and* g *of* S.

Proof. Assume that (1) holds. Let f and g be any fuzzy generalized bi-ideals of S. Then it follows from Theorem 3.1.13 that $f \cap g \subseteq f \circ g$. Also, we have

$$f \cap g = g \cap f \subseteq g \circ f.$$

Therefore,

$$f \cap g \subseteq (f \circ g) \cap (g \circ f),$$

and (1) so implies (14).

It is clear that (14) $\Rightarrow$ (13) $\Rightarrow$ (12) $\Rightarrow$ (9) $\Rightarrow$ (6) $\Rightarrow$ (2), (14) $\Rightarrow$ (11) $\Rightarrow$ (10) $\Rightarrow$ (9), (14) $\Rightarrow$ (8) $\Rightarrow$ (7) $\Rightarrow$ (6) and (14) $\Rightarrow$ (5) $\Rightarrow$ (4) $\Rightarrow$ (3) $\Rightarrow$ (2). Assume that (2) holds. Let f and g be any fuzzy right ideal and any fuzzy left ideal of S, respectively. Then we have

$$f \cap g \subseteq (f \circ g) \cap (g \circ f) \subseteq g \circ f.$$

Thus it follows from Theorem 3.3.10 that S is intra-regular. On the other hand,

$$f \cap g \subseteq (f \circ g) \cap (g \circ f) \subseteq f \circ g.$$

Since the inclusion $f \circ g \subseteq f \cap g$ always holds, we have $f \cap g = f \circ g$. Hence it follows from Theorem 3.1.8 that S is regular. Therefore, (2) implies (1). ∎

Theorem 3.3.15 *For a semigroup* S, *the following conditions are equivalent.*

(1) S *is regular and intra-regular.*

(2) $f \cap g \subseteq f \circ g \circ f$ *for every fuzzy quasi-ideal* f *and every fuzzy left ideal* g *of* S.

(3) $f \cap g \subseteq f \circ g \circ f$ *for every fuzzy quasi-ideal* f *and every fuzzy right ideal* g *of* S.

(4) $f \cap g \subseteq f \circ g \circ f$ *for all fuzzy quasi-ideals* f *and* g *of* S.

(5) $f \cap g \subseteq f \circ g \circ f$ *for every fuzzy quasi-ideal* f *and every fuzzy bi-ideal* g *of* S.

(6) $f \cap g \subseteq f \circ g \circ f$ *for every fuzzy quasi-ideal* f *and every fuzzy generalized bi-ideal* g *of* S.

(7) $f \cap g \subseteq f \circ g \circ f$ *for every fuzzy bi-ideal* f *and every fuzzy left ideal* g *of* S.

(8) $f \cap g \subseteq f \circ g \circ f$ *for every fuzzy bi-ideal* f *and every fuzzy right ideal* g *of* S.

(9) $f \cap g \subseteq f \circ g \circ f$ *for every fuzzy bi-ideal* f *and every fuzzy quasi-ideal* g *of* S.

(10) $f \cap g \subseteq f \circ g \circ f$ *for all fuzzy bi-ideals* f *and* g *of* S.

(11) $f \cap g \subseteq f \circ g \circ f$ *for every fuzzy bi-ideal* f *and every fuzzy generalized bi-ideal* g *of* S.

(12) $f \cap g \subseteq f \circ g \circ f$ *for every fuzzy generalized bi-ideal* f *and every fuzzy left ideal* g *of* S.

(13) $f \cap g \subseteq f \circ g \circ f$ *for every fuzzy generalized bi-ideal* f *and every fuzzy right ideal* g *of* S.

(14) $f \cap g \subseteq f \circ g \circ f$ *for every fuzzy generalized bi-ideal* f *and every fuzzy quasi-ideal* g *of* S.

(15) $f \cap g \subseteq f \circ g \circ f$ *for every fuzzy generalized bi-ideal* f *and every fuzzy bi-ideal* g *of* S.

(16) $f \cap g \subseteq f \circ g \circ f$ *for all fuzzy generalized bi-ideals* f *and* g *of* S.

Proof. First assume that (1) holds. Let f and g be any fuzzy generalized bi-ideals of S and a any element of S. Then, since S is regular, there exists an element $x \in S$ such that $a = axa$. Since S is intra-regular, there exist elements $y, z \in S$ such that $a = yaaz$. Thus we have

$$\begin{aligned} a &= axa \\ &= axaxaxa \\ &= ax(yaaz)x(yaaz)xa \\ &= (axya)(azxya)(azxa). \end{aligned}$$

Since f and g are fuzzy generalized bi-ideals of S, we have

$$\begin{aligned} (f \circ g \circ f)(a) &= \vee_{a=pq}\{f(p) \wedge (g \circ f)(q)\} \\ &\geq f(axya) \wedge (g \circ f)((azxyz)(azxa) \\ &\geq f(a) \wedge (\vee_{(azxya)(azxa)=uv}\{g(u) \wedge f(v)\}) \\ &\geq f(a) \wedge (g(azxya) \wedge f(azxa)) \\ &\geq f(a) \wedge (g(a) \wedge f(a)) \\ &= (f \cap g)(a), \end{aligned}$$

and so $f \cap g \subseteq f \circ g \circ f$. Thus we obtain that (1) implies (16).

It is clear that (11) $\Rightarrow$ (6) $\Rightarrow$ (5) $\Rightarrow$ (4) $\Rightarrow$ (3), (16) $\Rightarrow$ (15) $\Rightarrow$ (14) $\Rightarrow$ (13) $\Rightarrow$ (8) $\Rightarrow$ (3), (14) $\Rightarrow$ (12) $\Rightarrow$ (7) $\Rightarrow$ (2) and (16) $\Rightarrow$ (11) $\Rightarrow$ (10) $\Rightarrow$ (9) $\Rightarrow$ (8). Assume that (3) holds. Let f be any quasi-ideal of S. Then, since S itself is a fuzzy right ideal of S, we have

$$f = f \cap S = f \circ S \circ f.$$

Hence it follows from Theorem 3.1.3 that S is regular. Let f and g be any fuzzy left ideal and any fuzzy right ideal of S, respectively. Then, since f is a fuzzy quasi-ideal of S, we have

$$f \cap g \subseteq f \circ g \circ f \subseteq f \circ (g \circ S) \subseteq f \circ g.$$

Therefore, it follows from Theorem 3.3.14 that S is intra-regular. Thus (3) implies (1). In a similar way, we can prove that (2) implies (1). ∎

Theorem 3.3.16 *For a semigroup S, the following conditions are equivalent.*

(1) S is regular and intra-regular.

(2) $f \cap g \cap h \subseteq f \circ g \circ h$ for every fuzzy quasi-ideal f, every fuzzy right ideal g, and every fuzzy left ideal h of S.

(3) $f \cap g \cap h \subseteq f \circ g \circ h$ for every fuzzy bi-ideal f, every fuzzy right ideal g, and every fuzzy left ideal h of S.

(4) $f \cap g \cap h \subseteq f \circ g \circ h$ for every fuzzy generalized bi-ideal f, every fuzzy right ideal g, and every fuzzy left ideal h of S.

Proof. Assume that (1) holds. Let f be any fuzzy generalized bi-ideal, g any fuzzy left ideal, and h any fuzzy right ideal of S. Let a be any element of S. Since S is regular, there exists an element $x \in S$ such that $a = axa$. Since S is intra-regular, there exist elements $y, z \in S$ such that

$$a = ya^2z.$$

Hence we have

$$a = (axya)(azxya)(azxa).$$

Thus we have

$$\begin{aligned}(f \circ g \circ h)(a) &= \vee_{a=pq}\{f(p) \wedge (g \circ h)(q)\} \\ &\geq f(axya) \wedge (g \circ h)(azxyaazxa) \\ &\geq f(a) \wedge (\vee_{azxyaazxa=uv}\{g(u) \wedge h(v)\}) \\ &\geq f(a) \wedge (g(azxya) \wedge h(azxa)) \\ &\geq f(a) \wedge g(a) \wedge h(a) \\ &= (f \cap g \cap h)(a),\end{aligned}$$

and so

$$f \cap g \cap h \subseteq f \circ g \circ h.$$

Hence (1) implies (4). It is clear that (4) $\Rightarrow$ (3) $\Rightarrow$ (2). Assume that (2) holds. Let f and g be any fuzzy left ideal and any fuzzy right ideal of S, respectively. Then since f is a quasi-ideal of S and S itself is a fuzzy left ideal of S, by the assumption, we have

$$\begin{aligned} f \cap g &= f \cap g \cap S \\ &\subseteq f \circ g \circ S \\ &= f \circ (g \circ S) \\ &\subseteq f \circ g. \end{aligned}$$

Thus it follows from Theorem 3.3.11 that S is intra-regular. On the other hand, since g is a fuzzy quasi-ideal of S and S itself is a fuzzy right ideal of S, we have

$$\begin{aligned} g \cap f &= g \cap S \cap f \\ &= g \circ S \circ f \\ &= (g \circ S) \circ f \\ &\subseteq g \circ f. \end{aligned}$$

The inclusion $g \circ f \subseteq g \cap f$ always holds. Therefore,

$$g \circ f = g \cap f.$$

Hence it follows from Theorem 3.1.8 that S is regular. Thus (2) implies (1). ∎

3.4 Semisimple Semigroups

In Section 3.3, we proved that every fuzzy quasi-ideal (bi-ideal) of a semigroup S is idempotent if and only if S is regular and intra-regular. In this section, we study a semigroup whose fuzzy left (right, two-sided, interior) ideals are all idempotent.

Definition 3.4.1 *A semigroup S is called* ***left (right) quasi-regular*** *if every left (right) ideal of S is idempotent, and is called* ***quasi-regular*** *if every left ideal and every right ideal of S is idempotent,* [12].

As is easily seen, a semigroup S is left quasi-regular if and only if $a \in SaSa$, that is, there exist elements $x, y \in S$ such that $a = xaya$.

Theorem 3.4.2 *A semigroup S is left (right) quasi-regular if and only if every fuzzy left (right) ideal is idempotent.*

Proof. Let S be a left quasi-regular semigroup and f any fuzzy left ideal of S. Let a be any element of S. Then there exist elements $x, y \in S$ such that $a = xaya$. Thus

$$\begin{array}{rcl}(f \circ f)(a) & = & \vee_{a=pq}\{f(p) \wedge f(q)\} \\ & \geq & f(xa) \wedge f(ya) \\ & \geq & f(a) \wedge f(a) \\ & = & f(a),\end{array}$$

and so $f \circ f \supseteq f$. Since f is a fuzzy left ideal of S, $f \circ f \subseteq f$. Hence $f = f \circ f$ and so f is idempotent.

Conversely, assume that every fuzzy left ideal of S is idempotent. Let a be any element of S. Then, since $C_{L[a]}$ is a fuzzy left ideal of S,

$$\begin{array}{rcl}C_{L[a]L[a]}(a) & = & (C_{L[a]} \circ C_{L[a]})(a) \\ & = & C_{L[a]}(a) \\ & = & 1,\end{array}$$

and so

$$\begin{array}{rcl}a \in L[a]L[a] & = & (\{a\} \cup Sa)(\{a\} \cup Sa) \\ & = & \{a^2\} \cup aSa \cup Sa^2 \cup SaSa \\ & \subseteq & SaSa.\end{array}$$

Thus S is left quasi-regular. The case when S is right quasi-regular can be similarly proved. ∎

Theorem 3.4.3 *For a semigroup S, the following conditions are equivalent.*

(1) S is quasi-regular.

(2) $f = (f \circ S)^2 \cap (S \circ f)^2$ for every fuzzy quasi-ideal f of S.

Proof. Assume that (1) holds. Let f be any fuzzy quasi-ideal of S. Since S is right quasi-regular, it follows from Theorem 3.4.2 that the fuzzy right ideal $f \circ S$ is idempotent. Similarly, the fuzzy left ideal $S \circ f$ is idempotent. Hence

$$(f \circ S)^2 \cap (S \circ f)^2 = (f \circ S) \cap (S \circ f) \subseteq f.$$

In order to prove the converse inclusion holds, let a be any element of S. Then, since S is right quasi-regular, there exist elements $x, y \in S$ such that $a = axay$. Thus we have

$$\begin{array}{rcl}(f \circ S)^2(a) & = & \vee_{a=pq}\{(f \circ S)(p) \wedge (f \circ S)(q)\} \\ & \geq & (f \circ S)(ax) \wedge (f \circ S)(ay) \\ & \geq & (\vee_{ax=uv}\{f(u) \wedge S(v)\}) \wedge (\vee_{ay=uv}\{f(u) \wedge S(v)\}) \\ & \geq & (f(a) \wedge S(x)) \wedge (f(a) \wedge S(y)) \\ & = & f(a),\end{array}$$

and so we have $(f \circ S)^2 \supseteq f$. It can be similarly proved that $(S \circ f)^2 \supseteq f$. Hence

$$(f \circ S)^2 \cap (S \circ f)^2 \supseteq f.$$

Thus

$$f = (f \circ S)^2 \cap (S \circ f)^2,$$

and so (1) implies (2). Conversely, assume that (2) holds. Let f be any fuzzy right ideal of S. Then, since f is a fuzzy quasi-ideal of S, we have

$$f = (f \circ S)^2 \cap (S \circ f)^2 \subseteq (f \circ S)^2 \subseteq f^2 \subseteq f \circ S \subseteq f,$$

and so $f = f^2$. Therefore, it follows from Theorem 3.4.2 that S is right quasi-regular. It can be similarly proved that S is left quasi-regular. Thus (2) implies (1). ∎

Theorem 3.4.4 *For a semigroup S, the following conditions are equivalent.*

(1) S is both intra-regular and left quasi-regular.

(2) $g \cap h \cap f \subseteq g \circ h \circ f$ for every fuzzy quasi-ideal f, every fuzzy left ideal g, and every fuzzy right ideal h of S.

(3) $g \cap h \cap f \subseteq g \circ h \circ f$ for every fuzzy bi-ideal f, every fuzzy left ideal g, and every fuzzy right ideal h of S.

(4) $g \cap h \cap f \subseteq g \circ h \circ f$ for every fuzzy generalized bi-ideal f, every fuzzy left ideal g, and every fuzzy right ideal h of S.

Proof. Assume that (1) holds. Let f be any generalized bi-ideal, g any fuzzy left ideal, and h any fuzzy right ideal of S. Let a be any element of S. Then, since S is intra-regular, there exist elements $x, y \in S$ such that $a = xa^2y$. Since S is left quasi-regular, there exist elements $u, v \in S$ such that $a = uava$. Hence

$$a = uava = u(xaay)va = ((ux)a)((a(yv))a).$$

Therefore,

$$\begin{aligned}
(g \circ h \circ f)(a) &= \vee_{a=pq}\{g(p) \wedge (h \circ f)(q)\} \\
&\geq g((ux)a) \wedge (h \circ f)((a(yv)a) \\
&\geq g(a) \wedge (\vee_{a=pq}\{h(p) \wedge f(q)\}) \\
&\geq g(a) \wedge (h(a(yv)) \wedge f(a)) \\
&\geq g(a) \wedge (h(a) \wedge f(a)) \\
&= (g \cap h \cap f)(a),
\end{aligned}$$

and so

$$f \circ h \circ f \supseteq g \cap h \cap f.$$

Thus (1) implies (4). It is clear that (4) $\Rightarrow$ (3) $\Rightarrow$ (2). Assume that (2) holds. Let g be any fuzzy left ideal and f any fuzzy right ideal of S. Then since the fuzzy left ideal g is a fuzzy quasi-ideal of S and since S itself is a fuzzy right ideal of S, we have

$$g = g \cap S \cap g \subseteq g \circ S \circ g \subseteq g \circ g \subseteq g \circ S \subseteq g,$$

and so $g = g \circ g$. Therefore, S is left quasi-regular. On the other hand, since the fuzzy right ideal f is a fuzzy quasi-ideal of S and since S itself is a fuzzy right ideal of S, we have

$$g \cap f = g \cap S \cap f = g \circ S \circ f \subseteq g \circ f.$$

Hence it follows from Theorem 3.3.13 that S is intra-regular and so (2) implies (1). ■

Theorem 3.4.5 *For a semigroup S, the following conditions are equivalent.*

(1) S is both intra-regular and left quasi-regular.

(2) $L \cap R \cap Q \subseteq LRQ$ for every left ideal L, every right ideal R, and every quasi-ideal Q of S.

(3) $L \cap R \cap B \subseteq LRB$ for every left ideal L, every right ideal R, and every bi-ideal B of S.

(4) $L \cap R \cap G \subseteq LRG$ for every left ideal L, every right ideal R, and every generalized bi-ideal G of S.

Proof. Assume that (1) holds. To see (3), let L be any left ideal, R any right ideal, and G generalized bi-ideal of S. Let $a \in L \cap R \cap G$. Then

$$\begin{aligned} C_{LRG}(a) &= (C_L \circ C_R \circ C_G)(a) \\ &\geq (C_L \cap C_R \cap C_G)(a) \\ &= C_L(a) \wedge C_R(a) \wedge C_G(a) \\ &= 1 \wedge 1 \wedge 1 \\ &= 1. \end{aligned}$$

This implies that $a \in LRG$, and so $L \cap R \cap G \subseteq LRG$. Thus (1) implies (4). It is clear that (4) $\Rightarrow$ (3) $\Rightarrow$ (2) $\Rightarrow$ (1). ■

A semigroup S is called **semisimple** (Clifford and Preston [17, p.76]) if every two-sided ideal of S is idempotent. It is clear that S is semisimple if and only if $a \in (SaS)(SaS)$ for every $a \in S$, that is, there exist elements x, y and z in S such that $a = xayaz$.

Theorem 3.4.6 *Every fuzzy interior ideal of a semisimple semigroup S is a fuzzy two-sided ideal of S.*

Proof. Let f be a fuzzy interior ideal of S. Let a and b be any elements of S. Then, since S is semisimple, there exist elements $x, y, z \in S$ such that $a = xayaz$. Thus we have

$$f(ab) = f((xayaz)b) = f((xay)a(zb)) \geq f(a).$$

Hence f is a fuzzy right ideal of S. Similarly, it can be proved that f is a fuzzy left ideal of S. Thus f is a fuzzy two-sided ideal of S. ■

The following theorem is a special case of Theorem 3.4.6.

Theorem 3.4.7 *Every interior ideal of a semisimple semigroup S is a two-sided ideal of $\dot{S}$.*

Theorem 3.4.8 *For a semigroup S, the following conditions are equivalent.*

(1) S is semisimple.

(2) Every fuzzy two-sided ideal of S is idempotent.

(3) Every fuzzy interior ideal of S is idempotent.

(4) $f \cap g = f \circ g$ for every fuzzy two-sided ideals f and g of S.

(5) $f \cap g = f \circ g$ for every fuzzy two-sided ideal f and every fuzzy interior ideal g of S.

(6) $f \cap g = f \circ g$ for every fuzzy interior ideal f and every fuzzy two-sided ideal g of S.

(7) $f \cap g = f \circ g$ for all fuzzy interior ideals f and g of S.

(8) The set of all fuzzy two-sided ideals of S is a semilattice under the multiplication of fuzzy subsets, that is, $f \circ g = g \circ f$ and $f \circ f = f$ for all fuzzy two-sided ideals f and g of S.

(9) The set of all fuzzy interior ideals of S is a semilattice under the multiplication of fuzzy subsets, that is, $f \circ g = g \circ f$ and $f \circ f = f$ for all fuzzy interior ideals f and g of S.

Proof. Assume that (1) holds. Let f and g be fuzzy interior ideals of S. Note that S itself is a fuzzy interior ideal of S. Since f is a fuzzy two-sided ideal of S by Theorem 3.4.6, we have $f \circ g \subseteq f \circ S \subseteq f$ and $f \circ g \subseteq S \circ g \subseteq g$. Thus $f \circ g \subseteq f \cap g$.

Let a be any element of S. Since S is semisimple, there exist elements x, y, z and w in S such that $a = (xay)(zaw)$. Hence

$$\begin{aligned} (f \circ g)(a) &= \vee_{a=pq}\{f(p) \wedge g(q)\} \\ &\geq f(xay) \wedge g(yaz) \\ &= f(a) \wedge g(a) \\ &= (f \cap g)(a), \end{aligned}$$

and so $f \circ g \supseteq f \cap g$. Thus $f \cap g = f \circ g$ and so (1) implies (7).

It is clear that (7) $\Rightarrow$ (6) $\Rightarrow$ (4), (7) $\Rightarrow$ (5) $\Rightarrow$ (4) $\Rightarrow$ (2), (7) $\Rightarrow$ (3) $\Rightarrow$ (2), and (7) $\Rightarrow$ (9) $\Rightarrow$ (8) $\Rightarrow$ (2).

Assume that (2) holds. Let a be any element of S. Then it follows from Lemma 2.5.2 that the characteristic function $C_{J[a]}$ of the principal two-sided ideal $J[a]$ generated by a is a fuzzy two-sided ideal of S. Hence by Lemma 2.3.4, we have

$$C_{J[a]J[a]}(a) = (C_{J[a]} \circ C_{J[a]})(a) = C_{L[a]}(a) = 1,$$

and so

$$\begin{aligned} a \in J[a]J[a] &= (\{a\} \cup aS \cup Sa \cup SaS)(\{a\} \cup aS \cup Sa \cup SaS) \\ &= \{a^2\} \cup a^2S \cup aSaS \cup aSa \cup aSaS \cup aSSaS \cup SaaS \\ &\quad \cup SaSa \cup SaSaS \cup SaSa \cup SaSaS \cup SaSSa \cup SaSSaS \\ &\subseteq (SaS)(SaS). \end{aligned}$$

Thus S is semisimple and so (2) implies (1). ■

The following theorem can be easily seen.

Theorem 3.4.9 *For a semigroup S, the following conditions are equivalent.*

(1) S is semisimple.

(2) Every interior ideal of S is idempotent.

(3) $A \cap B = AB$ for all two-sided ideals A and B of S.

(4) $A \cap B = AB$ for every two-sided ideal A and every interior ideal B of S.

(5) $A \cap B = AB$ for every interior ideal A and every two-sided ideal B of S.

(6) $A \cap B = AB$ for all interior ideals A and B of S.

(7) The set of all two-sided ideals of S is a semilattice.

(8) The set of all interior ideals of S is a semilattice.

(9) $(J[a])^2 = J[a]$ for every element $a \in S$.

(10) $(I[a])^2 = I[a]$ for every element $a \in S$.

3.5 On Fuzzy Regular Subsemigroups of a Semigroup

We introduce the concepts of fuzzy regular subsemigroups, fuzzy weakly regular subsemigroups, fuzzy completely regular subsemigroups, and fuzzy weakly completely regular subsemigroups. In this section, S and T denote semigroups. The results of Sections 3.5 through 3.7 are mainly from [130].

Let $x \in S$. Then we write

$$R_x = \{x' \mid x' \in S, xx'x = x\}.$$

Proposition 3.5.1 *The following conditions are equivalent for all $x \in S$.*

(1) x is a regular element.

(2) $R_x \neq \emptyset$.

(3) $x \in xSx$.

Let $C_x = \{y \mid y \in S, yx = xy\}$.

Proposition 3.5.2 *The following conditions are equivalent for all $x \in S$.*

(1) x is a completely regular element.

(2) $R_x \cap C_x = \emptyset$.

(3) $x \in x^2Sx^2$

(4) $x \in x^2S \cap Sx^2$.

Proposition 3.5.3 *Let α be a homomorphism from S onto T. Then*

(1) $\alpha(R_x) \subseteq R_{\alpha(x)} \forall x \in S$;
(2) $\alpha^{-1}(R_y) \supseteq R_{\alpha^{-1}(y)} \forall y \in T$.

Proposition 3.5.4 *Let α be a homomorphism from S onto T. Then*
(1) *$\alpha(f)$ is a fuzzy subsemigroup of T if f is a fuzzy subsemigroup of S;*
(2) *$\alpha^{-1}(g)$is a fuzzy subsemigroup of S if g is a fuzzy subsemigroup of T.*

Definition 3.5.5 *If f is a fuzzy subsemigroup of S and $\forall x \in S$, there exists $x' \in R_x$ such that*

$$f(x') \geqslant f(x) \tag{3.5}$$

provided $f(x) \neq 0$, then f is called a ***fuzzy regular subsemigroup*** *of S.*

Proposition 3.5.6 *f is a fuzzy regular subsemigroup of S if and only if $\forall t \in (0,1]$, f_t is a regular subsemigroup of S provided $f_t \neq \emptyset$.*

Proof. By Proposition 2.3.6, it is sufficient to show that f satisfies (3.5) if and only if each element of f_t is a regular element of f_t. In fact, if f satisfies (3.5), then $\exists x' \in R_x$ such that $f(x') \geqslant f(x)$, and furthermore, $\forall x \in f_t$, $f(x') \geqslant f(x) \geqslant t \Rightarrow x' \in f_t$, i.e., f_t is a regular subsemigroup of S. Conversely, suppose $\forall t \in (0,1]$, f_t is a regular subsemigroup of S provided $f_t \neq \emptyset$, but (3.5) does not hold, i.e., $\exists x \in S$ such that $f(x) \neq 0$ and $\forall x' \in R_x$, $f(x') < f(x)$. Set $t = f(x)$. Clearly, $x \in f_t$ and $\forall x' \in R_x$, $x' \notin f_t$. This contradicts the fact that f_t is regular. Thus (3.5) does hold. ■

Proposition 3.5.7 *If A is a nonempty subset of S, then A is a regular subsemigroup of S if and only if C_A, the characteristic function of A, is a fuzzy regular subsemigroup of S.*

Proof. If A is regular subsemigroup of S, then $C_A(xy) \geqslant C_A(x) \wedge C_A(y)$ and $x, y \in A$ implies $xy \in A$. $\forall x \in A$, if $C_A(x) \neq 0$, i.e., $C_A(x) = 1$, then $x \in A$, and furthermore from the regularity of A we know that $\exists x' \in R_x$ such that $x' \in A$, i.e., $C_A(x') = 1$. Therefore, $A(x') \geqslant A(x)$, i.e., (3.5) holds, and C_A is a fuzzy regular subsemigroup of S.

Conversely, if C_A is a fuzzy regular subsemigroup of S, then $\forall x, y \in A$, $C_A(x) = C_A(y) = 1$. Hence $C_A(xy) \geqslant C_A(x) \wedge C_A(y) = 1$. Thus $C_A(xy) = 1$. Hence $xy \in A$. In addition, $C_A(x') \geqslant C_A(x) = 1 \Rightarrow C_A(x') = 1 \Rightarrow x' \in A$. Thus A is a regular subsemigroup of S. ■

Proposition 3.5.8 *Let α be a semigroup surjection homomorphism from S onto T.*

(1) *If f is fuzzy regular subsemigroup of S, then $\alpha(f)$ is a fuzzy regular subsemigroup of T.*

(2) *If g is a fuzzy regular subsemigroup of T, then $\alpha^{-1}(g)$ is a fuzzy regular subsemigroup of S.*

Proof. (1) From Propositions 3.5.4 and 3.5.6, it is sufficient to show that (3.5) holds, or equivalently, $\forall t \in (0,1]$, $(\alpha(f))_t$ is a regular subsemigroup of T provided $(\alpha(f))_t \neq \emptyset$. In fact, if $\exists t_0 \in (0,1]$ such that $(\alpha(f))_{t_0}$ is nonempty and is not regular, then $\exists x \in (\alpha(f))_{t_0}$ such that $\forall x' \in R_x$, $x' \notin (\alpha(f))_{t_0}$, i.e., $(\alpha(f))(x) \geqslant t_0 \neq 0$, and $\forall x' \in R_x$, $(\alpha(f))(x') < t_0$, $\vee_{y\in\alpha^{-1}(x')} f(y) \geqslant t_0$ and $\vee_{z\in\alpha^{-1}(x')} f(z) < t_0$. Let $0 < t < t_0$, $\vee_{y\in\alpha^{-1}(x)} f(y) > t$, and $\vee_{z\in\alpha^{-1}(x')} f(z) < t$. Now $\exists y_0 \in S$ with $\alpha(y_0) = x$ and $f(y_0) > t$, i.e., $y_0 \in f_t, f_t \neq \emptyset$. For any z, x' with $\alpha(z) = x'$, we have $f(z) < t$, i.e., $z \notin f_t$. Clearly, $\forall y_0' \in R_{y_0}$, $\alpha(y_0') = x'$, and so $\forall y' \in R_{y_0}$, $y_0' \notin f_t$, i.e., f_t is not regular. This contradicts the fact that f is a fuzzy regular subsemigroup of S.

(2) If $g \in \mathcal{FP}(T)$ is a fuzzy regular semigroup of T, then $\forall y \in T$, $g(y) \neq 0$ implies $\exists y' \in R_y$ such that $g(y') \geqslant g(y)$. Since α is surjective, $\forall y \in T$, $\exists x \in S$ with $\alpha(x) = y$, and $\forall x \in S$, $g(\alpha(x)) \neq 0$ implies $\exists (\alpha(x))' \in R_{\alpha(x)}$ with $g([\alpha(x)]') \geqslant g(\alpha(x))$. From Proposition 3.5.3, $\forall \alpha(x)' \in R_{\alpha(x)}$, $\exists x \in R_x$ such that $[\alpha(x)]' = \alpha(x')$. Hence

$$\begin{aligned} & g([\alpha(x)]') \geqslant g(\alpha(x)) \\ \Rightarrow\ & g(\alpha(x')) \geqslant g(\alpha(x)) \\ \Rightarrow\ & (\alpha^{-1}(g))(x') \geqslant (\alpha^{-1}(g))(x), \end{aligned}$$

i.e., (3.5) holds, and $\alpha^{-1}(g)$ is a fuzzy regular subsemigroup of S. ■

Proposition 3.5.9 *If f is a fuzzy regular subsemigroup of S, then $f \circ f = f$.*

Proof. From Lemma 2.3.2(1), we have that $f \circ f \subseteq f$. Now $\forall x \in S$, if $f(x) = 0$, then $f \circ f(x) \leqslant f(x) \Rightarrow f \circ f\ (x) = f(x)$, and if $f(x) \neq 0$, then $\exists x' \in R_x$ with $f(x') \geqslant f(x)$ because f is fuzzy regular. Hence, $f \circ f(x) = \vee_{yz=x}\{f(y) \wedge f(z)\} \geqslant (f(xx') \wedge f(x) \geqslant (f(x') \wedge f(x)) = f(x)$, i.e., $f \circ f \supseteq f$. Thus $f \circ f = f$. ■

Let $S' = S \cup \{1\}$ and $x \cdot 1 = 1 \cdot x = x\ \forall x \in S'$. Then S' is a semigroup with identity. $\forall$ fuzzy subsets f of S, we define the fuzzy subset f' of S as follows: $f'(x) = 1$ if $x = 1$ and $f'(x) = f(x)$ if $x \in S$.

Clearly, $\forall t \in [0,1], 1 \in f_t'$.

Proposition 3.5.10 *f is a fuzzy regular subsemigroup of S if and only if $\forall x \in X$, if $f(x) \neq 0$, then $\exists t \in (0,1]$ and an idempotent element $e \in f_t$ such that $xf_t' = ef_t$, where $t = f(x)$.*

Proof. If f is a fuzzy regular subsemigroup of S, then $\forall t \in (0,1]$, f_t is a regular subsemigroup of S provided $f_t \neq \emptyset$. Now, $\forall x \in S$, if $f(x_0) = t_0 \neq 0$, then $x \in f_{t_0}$, and f_{t_0} is nonempty and a regular subsemigroup of S. From [9] , $\exists e \in f_t$ such that $xf_t' = ef_t$.

Conversely, suppose that $\forall x \in S$, there is an idempotent element $e \in f_t'$ such that $xf_t' = ef_t$ when $f(x) = t \neq 0$. We show that f is a fuzzy regular subsemigroup of S. From Proposition 3.5.6, it is sufficient to show that $\forall t \in (0,1]$, f_t is a regular subsemigroup of S provided $f_t \neq \emptyset$. In fact, if $f_t \neq \phi$, then $\forall x \in f_t$, $f(x) \geqslant t$. We set $f(x) = t$ and have $t_0 \geqslant t$. By assumption, there is an idempotent element $e \in f_{t_0}$ such that $xf'_{t_0} = ef_{t_0}$. Therefore, $\exists y \in f_{t_0}$ with $x = ey$ and $\exists z \in f'_{\lambda_{t0}}$ with $xz = e$. Now $ex = e^2y = ey = x$, i.e., $xzx = x$, $z \in R_x$, and $\mu(z) \geqslant \lambda_0 > \lambda$, i.e., $z \in f_t$. Consequently, $\forall x \in f_t$, $\exists z \in R_x$ with $z \in f_t$, i.e., f_t is a regular subsemigroup of S. ■

3.6 Fuzzy Weakly Regular Subsemigroups

Definition 3.6.1 *If f is a fuzzy subsemigroup of S and $\forall x \in S$, $R_x \neq \emptyset$,*

$$\vee_{x' \in R_x} f(x') \geqslant f(x) \tag{3.6}$$

*provided $f(x) \neq 0$, then f is called a **fuzzy weakly regular subsemigroup** of S.*

Proposition 3.6.2 *f is a fuzzy weakly regular subsemigroup of S if and only if $\forall t \in [0,1)$, $f_{t^>}$ is a regular subsemigroup of F if $f_{t^>} \neq \emptyset$, where $f_{t^>}$ is the t-strong cut,*

$$f_{t^>} = \{x \in S \mid f(x) > t\}.$$

Proof. Since f is a fuzzy subsemigroup of S if and only if $\forall t \in [0,1)$, $f_{t^>}$ is a subsemigroup of S provided $f_{t^>} \neq \emptyset$, it is sufficient to consider regularity. In fact, if f satisfies (3.6), then $\forall x \in f_{t^>}$ such that $f(x) > t$, $f_{t^>}$ is a regular subsemigroup of S. Conversely, suppose $\forall t \in [0,1)$, $f_{t^>}$ is regular provided $f_{t^>} \neq \emptyset$, and (3.6) does not hold, i.e., $\exists x_0 \in S$ such that $f(x_0) \neq 0$, $\vee_{x'_0 \in R_{x_0}} f(x'_0) < f(x_0)$. Let $t_0 = (\vee_{x'_0 \in R_{x_0}} f(x'_0), f(x_0))$. Then $x_0 \in f_{t_0^>}$, and $\forall x'_0 \in R_{x_0}$, $f(x'_0) < t_0$, $x'_0 \notin f_{t_0^>}$. This contradicts the regularity of $f_{t^>}$. ■

Proposition 3.6.3 *Let A be a subsemigroup of S. Then A is a weakly regular subsemigroup of S if and only if C_A is a fuzzy weakly subsemigroup of S.*

Proposition 3.6.4 *If α is a semigroup homomorphism from S onto T, then the images and inverse images of fuzzy weakly regular subsemigroups under α are also fuzzy weakly regular.*

Proof. It is sufficient to show that (3.6) holds. Let f be a fuzzy weakly regular subsemigroup of S. To prove that $\alpha(f)$ is a fuzzy weakly regular subsemigroup of T, from Proposition 3.6.2, we need only show that $\forall t \in [0,1)$ $(\alpha(f))_{t^>}$ is a regular subsemigroup of T. Indeed, if $\exists t_0 \in [0,1)$ such that $(\alpha(f))_{t_0^>}$ is nonempty and is not regular, then $\exists y \in (\alpha(f))_{t_0^>}$ such that $\forall y' \in R_y$, $y' \notin (\alpha(f))_{t_0^>}$, i.e., $(\alpha(f))(y) > t_0$, and $\forall y' \in R_y$, $(\alpha(f))(y') \leqslant t_0$.

Conversely, $\vee_{z\in\alpha^{-1}(y)} f(z) > t_0$, and $\forall y' \in R_y$, $\vee_{z\in\alpha^{-1}(y')} f(z) \leqslant t_0$. This implies that $\exists x_0 \in S$ such that $\alpha(x_0) = y$ and $f(x_0) > t_0$, i.e., $x_0 \in f_{t_0^>}$, $f_{t_0^>} \neq \emptyset$. Also, $\forall y' \in R_y$, for any z with $\alpha(z) = y'$, it follows that $f(z) \leqslant t_0$, i.e., $z \notin f_{t_0^>}$. Clearly, $\forall x_0' \in R_{x_0}$, $\alpha(x_0') = y'$. Thus $\forall x_0' \notin f_{t_0^>}$, and $f_{t^>}$ is not regular. This contradicts the fuzzy weak regularity of f.

On the other hand, let g be a fuzzy weakly regular subsemigroup of T. Then $\forall y \in T$, $\vee_{y'\in R_y} g(y') \geqslant g(y)$ when $g(y) \neq 0$. Since α is a surjective homomorphism, $\forall y \in T$, $\exists x \in S$ with $\alpha(x) = y$. Now, $\forall x \in S$, it holds that $\vee_{[\alpha(x)]'\in R_{\alpha(x)}} g(\alpha(x)') \geqslant g(\alpha(x))$ when $g(\alpha(x)) \neq 0$. Thus from Proposition 3.5.3, we obtain $\vee_{x'\in R_x} [\alpha^{-1}(g)](x') = \vee_{\alpha(x')\in\alpha(R_x)} g(\alpha(x)') \geqslant \vee_{[\alpha(x)]'\in R_{\alpha(x)}} g(\alpha(x)') \geqslant g(\alpha(x)) = [\alpha^{-1}(g)](x)$, i.e., (3.5) holds, and $\alpha^{-1}(g)$ is a fuzzy weakly regular subsemigroup of S. ■

Proposition 3.6.5 *(1) f is a fuzzy regular subsemigroup of S implies f is a fuzzy weakly regular subsemigroup of S.*

(2) If f has the sup property, then f is a fuzzy regular subsemigroup if and only if f is a fuzzy weakly regular subsemigroup.

Proposition 3.6.6 *Let f be an fuzzy subsemigroup of S. Then f is fuzzy weakly regular if and only if $\forall x \in S$,*

$$\vee_{x'\in R_x} f(x'xx') \geqslant f(x) \tag{3.7}$$

when $f(x) \neq 0$.

Proof. If (3.6) holds, then $\vee_{x'\in R_x} f(x'xx') \geqslant \vee_{x'\in R_x} f(x') \wedge f(x) = (\vee_{x\in R_x} f(x')) \wedge f(x)) \geqslant (f(x) \wedge f(x)) = f(x)$, i.e., (3.7) holds. Conversely, if (3.7) holds, then $\forall x' \in R_x$, $x'xx' \in R_x$, and $\vee_{x'\in R_x} f(x') \geqslant \vee_{x'\in R_x} f(x'xx') \geqslant f(x)$, i.e., (3.6) holds. ■

Definition 3.6.7 *Let f be a fuzzy subset of S. Then $\forall x, y \in S$, the fuzzy subset xfy of S is defined by $\forall z \in S$, $(xfy)(z) = 0$ if there does not exist $s \in S$ with $xsy = z$ and $xfy(z) = \vee_{xsy=z,s\in S} f(s)$ otherwise.*

Proposition 3.6.8 *Let f be a fuzzy subsemigroup. Then f is fuzzy weakly regular if and only if $\forall x \in S$, $(xfx)(x) \geqslant f(x)$ when $f(x) \neq 0$.*

Proof. The proof is immediate from Definitions 3.6.1 and 3.6.7. ■

Corollary 3.6.9 *Let A be a subsemigroup of S. Then A is regular if and only if $\forall x \in A$, $x \in xAx$.*

3.7 Fuzzy Completely Regular and Weakly Completely Regular Subsemigroups

Definition 3.7.1 *If f is a fuzzy subsemigroup, and $\forall x \in S$, $\exists x^* \in R_x \cap C_x$ such that*

$$f(x^*) \geqslant f(x) \tag{3.8}$$

when $f(x) \neq 0$, then f is called a ***fuzzy completely regular subsemigroup*** *of S. If a fuzzy subsemigroup f of S satisfies $\forall x \in S$, $R_x \cap C_x \neq \emptyset$ and*

$$\vee_{x* \in R_x \cap Cx} f(x^*) \geqslant f(x) \tag{3.9}$$

when $f(x) \neq 0$, then f is called a ***fuzzy weakly completely regular subsemigroup*** *of S.*

Proposition 3.7.2 (1) *Fuzzy complete regularity implies fuzzy weakly complete regularity.*

(2) *If f has the sup property, then f is fuzzy completely regular if and only if f is fuzzy weakly completely regular.*

Proposition 3.7.3 *A is completely regular subsemigroup of S if and only if C_A is a fuzzy completely regular (or fuzzy weakly completely regular) subsemigroup of S.*

Proposition 3.7.4 *f is a fuzzy completely regular (or fuzzy weakly completely regular) subsemigroup of S if and only if $\forall t \in (0,1]$ (respectively $[0,1)$), f_t (respectively $f_{t>}$) is a completely regular subsemigroup of S if f_t ($f_{t>}$) $\neq \emptyset$.*

Proposition 3.7.5 *If $\alpha : S \to T$ is a surjective semigroup homomorphism, then the images and inverse images of fuzzy completely regular (or fuzzy weakly completely regular) subsemigroups under α are also fuzzy completely regular (or fuzzy weakly completely regular) subsemigroups.*

Proposition 3.7.6 *Let f be a fuzzy subsemigroup of S. Then f is fuzzy weakly completely regular if and only if $\forall x \in S$, $(x^2 f x^2)(x) \geqslant f(x)$ when $f(x) \neq 0$.*

Proof. If f is fuzzy weakly completely regular, then we have

$$\begin{aligned} f(x) &\leqslant \vee_{x^* \in R_x \cap C_x} f(x^*) \\ &= \vee_{xx^*x=x} f(x^*) \\ &= \vee_{xx^*xx^*xx^*x=x} f(x^*) \\ &= \vee f(x^*) \\ &\leqslant \vee f((x^{*3})) \\ &\leqslant \vee f(z) \\ &= (x^2 f x^2)(x) \end{aligned}$$

when $f(x) \neq 0$. Conversely, if $\forall x$, $(x^2 f x^2)(x) \geqslant f(x)$ provided $f(x) \neq 0$, then

$$\vee_{x^2 z x^2 = x} f(z) \geqslant f(x).$$

Suppose the f is not fuzzy weakly completely regular. Then $\exists x_0 \in S$, $f(x_0) \neq 0$, and $\vee_{x_0^* \in R_{x_0} \cap C_{x_0}} f(x_0^*) < f(x_0)$. Let $t \in (0, 1]$ with

$$\vee_{x_0^* \in R_{x_0} \cap C_{x_0}} f(x_0^*) < t < f(x_0).$$

Then $\forall x_0^* \in R_{x_0} \cap C_{x_0}$, $f(x_0^*) < t < f(x)$. Now, for any z satisfying the equality $x_0^2 z x_0^2 = x_0$, we write $x_0' = x_0 z x_0$. Clearly, $x_0' \in R_{x_0} \cap C_{x_0}$, and $f(x_0') < t < f(x_0)$, $f(x_0') = f(x_0 z x_0) \geqslant (f(z) \wedge f(x_0), f(z) < t < f(x_0)$. From this we deduce that

$$\vee_{x_0^2 z x_0^2 = x_0} f(z) \leqslant t < f(x_0).$$

This contradicts the assumption. ∎

Corollary 3.7.7 *Let A be a subsemigroup of S. Then A is completely regular if and only if $\forall x \in A$, $x \in x^2 A x^2$.*

Definition 3.7.8 *Let f be a fuzzy subset of S. Define the fuzzy subsets xf and fx of S as follows: $\forall y \in S$*

$$(xf)(y) = \begin{cases} 0 & \text{if there does not exist } z \in S \text{ with } xz = y, \\ \vee_{xz=y} f(z) & \text{otherwise,} \end{cases}$$

$$(fx)(y) = \begin{cases} 0 & \text{if there does not exist } w \in S \text{ with } wx = y, \\ \vee_{wx=y} f(w) & \text{otherwise.} \end{cases}$$

Proposition 3.7.9 *Let f be a fuzzy subsemigroup of S. Then f is fuzzy weakly completely regular if and only if $\forall x \in S$, $(x^2 f \cap f x^2)(x) \geq f(x)$ when $f(x) \neq 0$.*

Proof. If f is fuzzy weakly completely regular, then $\vee_{x^* \in R_x \cap C_x} f(x^*) \geqslant f(x)$ and

$$\begin{aligned} (x^2 f)(x) &= \vee_{x^2 z = x} f(z) \\ &\geqslant \vee_{x^2 z = x} f(z) \\ &= \vee_{xzx = x} f(z) \\ &= \vee_{z \in R_x \cap C_x} f(z) \\ &\geqslant f(x) \\ &\neq 0. \end{aligned}$$

Similarly, $(fx^2)(x) \geqslant f(x)$. Hence $(x^2 f \cap f x^2)(x) \geqslant f(x)$.

Conversely, if $\forall x, (x^2 f \cap f x^2)(x) \geqslant f(x)$ when $f(x) \neq 0$ and f is not fuzzy weakly completely regular, then $\exists x_0 \in S$ such that $f(x_0) \neq 0$ and

$\vee_{x_0^* \in R_{x_0} \cap C_{x_0}} f(x_0^*) < f(x_0)$. Let $t_0 \in (\vee_{x_0^* \in R_{x_0} \cap C_{x_0}} f(x_0^*), f(x_0))$. Then $\forall x_0^* \in R_{x_0} \cap C_{x_0}$, $f(x_0^*) < t < f(x_0)$. Now, if for any z and w with $x_0^2 z = x_0$ and $wx_0^2 = x_0$, we let $x_0' = zx_0w$, then it is easy to show that $x_0' \in R_{x_0} \cap C_{x_0}$. Therefore, $f(x_0') < t < f(x_0)$ and $f(x_0') = f(zx_0w) \geqslant f(z) \wedge f(x_0) \wedge f(w)$. Now, we have $f(z) < t$ and $f(w) < t$, and so

$$\begin{aligned}(x_0^2 f \cap f x_0^2)(x_0) &= (\vee_{x_0^2 z = x_0, z \in S} f(z)) \wedge (\vee_{wx_0^2 = x_0,\ w \in S} f(w)) \\ &= \vee_{x_0^2 z = x_0, wx_0^2 = x_0, z, w \in S} (f(z) \wedge f(w)) \\ &\leqslant \vee_{x_0^2 z = x_0, wx_0^2 = x_0, z, w \in S} [t \wedge f(z)(\text{or } f(w))] \\ &\leqslant t \\ &< f(x_0).\end{aligned}$$

This contradicts the assumption. ∎

Corollary 3.7.10 *Let A be a subsemigroup of S. Then A is completely regular if and only if $\forall x \in A$, $x \in xA^2 \cap Ax^2$.*

3.8 Weakly Regular Semigroups

In [3], the authors characterized regular semigroups, semigroups that are semilattices of left groups, semigroups that are both regular and intra-regular and weakly regular semigroups via fuzzy bi-ideals. Many of their results have been presented above. Thus in this section, we consider only the results in [3] involving weakly regular semigroups. A semigroup S is called **weakly regular** if for every $x \in S, x \in (xS)^2$.

Lemma 3.8.1 *A monoid is weakly regular if and only if $J \cap I = JI$ for all right ideals J and all two-sided ideals I of S.*

Theorem 3.8.2 *Let S be a monoid. Then the following conditions are equivalent.*

(1) S is weakly regular.

(2) $B \cap I \cap J \subseteq BIJ$ for every bi-ideal B, every two-sided ideal I, and every right ideal J of S.

(3) $Q \cap I \cap J \subseteq QIJ$ for every quasi-ideal Q, every two-sided ideal I, and every right ideal J of S. ∎

Theorem 3.8.3 *Let S be a monoid. Then the following conditions are equivalent.*

(1) S is weakly regular.

(2) $B \cap I \subseteq BI$ for every bi-deal B and every two-sided ideal I of S.

(3) $Q \cap I \subseteq QI$ for every quasi-ideal Q and every two-sided ideal I of S.

∎

Theorem 3.8.4 *[2] Let S be a monoid. Then S is weakly regular if and only if $f \cap g \subseteq f \circ g$ for every fuzzy right ideal f and every fuzzy two-sided ideal g of S.* ∎

Theorem 3.8.5 *Let S be a monoid. Then the following conditions are equivalent.*

(1) S is weakly regular.

(2) $f \cap g \cap h \subseteq f \circ g \circ h$ for every bi-ideal f, every fuzzy ideal g, and every fuzzy right ideal h of S.

(3) $f \cap g \cap h \subseteq f \circ g \circ h$ for every quasi-ideal f, every fuzzy ideal g, and every fuzzy right ideal h of S.

Proof. (1) $\Rightarrow$ (2) : Let $x \in S$. Then $x \in (xS)^2$. Thus $x = xsxt$ for some $s, t \in S$. Hence $(f \circ g \circ h)(x) = \vee\{f(x) \wedge (g \circ h)(sxt) \mid x = sxt, s, t \in S\} \geq f(x) \wedge (g \circ h(sxt) = f(x) \wedge (\vee\{g(p) \wedge h(v) \mid sxt = pv\}) \geq f(x) \wedge (g(xsx) \wedge h(xt^2))$ (since $sxt = s(xsxt)t = (sxs)(xt^2)) \geq f(x) \wedge (g(x) \wedge h(x)) = (f \cap g \cap h)(x)$.

(2) $\Rightarrow$ (3) : This implication follows since every fuzzy quasi-ideal of S is a fuzzy bi-ideal of S.

(3) $\Rightarrow$ (1) : Let f be a fuzzy right ideal of S and let $h = C_S$. Then $f \cap g \cap h = f \cap g$ and $f \circ g \circ h = f \circ g$. Hence $f \cap g \subseteq f \circ g$ for every fuzzy right ideal f and every fuzzy ideal g of S. Since $f \circ g \subseteq f \cap g$ always holds, it follows that $f \circ g = f \cap g$. Thus by Theorem 3.8.4, (1) holds. ■

The proof of the next result is similar to that of Theorem 3.8.5.

Theorem 3.8.6 *Let S be a monoid. Then the following conditions are equivalent.*

(1) S is weakly regular.

(2) $f \cap g \subseteq f \circ g$ for every bi-ideal f and every fuzzy two-sided ideal g of S.

(3) $f \cap g \subseteq f \circ g$ for every quasi-ideal f and every fuzzy two-sided ideal g of S. ■

We close this chapter with some comments concerning [95]. In this paper, the authors introduce the notion of a fuzzy regular point. They give a pointwise depiction of fuzzy regular subsemigroups, fuzzy completely regular subsemigroups, fuzzy intra-regular subsemigroups, and fuzzy left (right) regular subsemigroups. They also introduce the concepts of a weakly left (right, intra-) regular subsemigroup.

4
Semilattices of Groups

In this chapter, we characterize a semigroup that is a semilattice of left (right) simple semigroups, a semigroup that is a semilattice a left (right) groups, and a semigroup that is a semilattice of groups in terms of fuzzy ideals. We study fuzzy normal semigroups. We close the chapter with a look at Green's relations and fuzzy ideals.

4.1 A Semilattice of Left (Right) Simple Semigroups

A commutative idempotent semigroup is called a **semilattice**. Let S be a semigroup and Y a semilattice. If ψ be a homomophism from S onto Y, then each subset

$$i\psi^{-1} = \{a \in S \mid \psi(a) = i\}$$

is a subsemigroup of S.

We denote $i\psi^{-1}$ by S_i. Then S is the disjoint union of the subsemigroups S_i $(i \in Y)$, and $S_iS_j \subseteq S_{ij}$, where ij is the product of i and j in the semilattice Y. Thus we can define the following: A semigroup S is called a **semilattice of semigroups of type** T if S is a disjoint union of a family of subsemigroup S_i $(i \in Y)$ of type T such that for each S_i and S_j $(i, j \in Y)$, the products S_iS_j and S_jS_i are both contained in the same subsemigroup S_k $(k \in Y)$. In this section, we study left (right) simple semigroups of type T.

The following characterizations of a semigroup that is a semilattice of left simple semigroups are due to T. Saito [111].

Theorem 4.1.1 *For a semigroup S, the following conditions are equivalent.*

(1) S is a semilattice of left simple semigroups.

(2) S is left regular and every left ideal of S is two-sided.

(3) $AB = A \cap B$ for every left ideals A and B of S. ■

Theorem 4.1.2 *For a left regular semigroup S, the following conditions are equivalent.*

(1) Every left ideal of S is a two-sided ideal of S.

(2) Every fuzzy left ideal of S is a fuzzy two-sided ideal of S.

Proof. Assume that (1) holds. Let f be any fuzzy left ideal of S, and a and b any elements of S. Then, since the left ideal Sa is a two-sided and since S is left regular, we have

$$ab \in (Sa^2)b \subseteq (Sa)bS \subseteq Sa.$$

This implies that there exists an element $x \in S$ such that $ab = xa$. Thus since f is a fuzzy left ideal of S, we have

$$f(ab) = f(xa) \geq f(a).$$

Hence f is a fuzzy right ideal of S, that is, f is a fuzzy two-sided ideal of S. Thus (1) implies (2).

Assume that (2) holds. Let A be any left ideal of S. Then by Lemma 2.3.1 the characteristic function C_A is a fuzzy left ideal of S. Then by the assumption, C_A is a fuzzy right ideal of S. Thus it follows from Lemma 2.3.1 that A is a right ideal of S and so A is a two-sided ideal. Hence (2) implies (1). ■

Now we give a characterization of a semigroup that is a semilattice of left simple semigroups by fuzzy ideals.

Theorem 4.1.3 *For a semigroup S, the following conditions are equivalent.*

(1) S is a semilattice of left simple semigroups.

(2) S is left regular, and every fuzzy left ideal of S is a fuzzy two-sided ideal of S.

(3) $f \circ g = f \cap g$ for every fuzzy left ideals f and g of S.

(4) The set of all fuzzy left ideals of S is a semilattice under the multiplication of fuzzy subsets.

(5) The set of all left ideals of S is a semilattice under the multiplication of subsets.

Proof. It follows from Theorems 4.1.1 and 4.1.2 that (1) and (2) are equivalent. Assume that (2) holds. To see that (3) holds, let f and g be fuzzy left ideals of S and a any element of S. Then, since S is left regular, there exists an element $x \in S$ such that $a = xa^2$. Since by the assumption that f is also a fuzzy right ideal of S, we have

$$\begin{aligned}(f \circ g)(a) &= \vee_{a=yz}(f(y) \wedge g(z)) \\ &\geq f(xa) \wedge g(a) \\ &\geq f(a) \wedge g(a) \\ &= (f \cap g)(a),\end{aligned}$$

and so $f \circ g \supseteq f \cap g$. On the other hand, by the assumption that g is a fuzzy right ideal of S,

$$\begin{aligned}(f \circ g)(a) &= \vee_{a=yz}(f(y) \wedge g(z)) \\ &\leq \vee_{a=yz}(f(yz) \wedge g(yz)) \\ &= \vee_{a=yz}(f(a) \wedge f(a)) \\ &= f(a) \wedge g(a) \\ &= (f \cap g)(a),\end{aligned}$$

and so $f \circ g \subseteq f \cap g$. Thus $f \circ g = f \cap g$ and so (2) implies (3).

It is clear that (3) implies (4).

Assume that (4) holds. To prove (5) holds, let A and B be any left ideals of S and a be any element of BA. Then it follows from Lemma 2.3.1 that the characteristic functions C_A and C_B are fuzzy left ideals of S. Thus by Lemma 2.3.4,

$$\begin{aligned}C_{AB}(a) &= (C_A \circ C_B)(a) \\ &= (C_B \circ C_A)(a) \\ &= C_{BA}(a) \\ &= 1,\end{aligned}$$

and so $a \in AB$. Thus $BA \subseteq AB$. Similarly, we have $AB \subseteq BA$. Hence $AB = BA$.

To see that any left ideal A of S is idempotent, let a be any element of A. Then, since C_A is a fuzzy left ideal, we have

$$C_{A^2}(a) = (C_A \circ C_A)(a) = C_A(a) = 1,$$

and so $a \in A^2$. Thus $A \subseteq A^2$ and so $A = A^2$. Therefore, (4) implies (5).

Finally assume that (5) holds. Let A be any left ideal of S. Then, since S itself is a left ideal, by the assumption we have

$$AS = SA \subseteq A.$$

Thus A is a right ideal of S, and so A is a two-sided ideal of S. Let a be any element of S. Then, since the left ideal Sa is a left ideal of S, we have

$SaS = (Sa)S \subseteq Sa$. Hence

$$\begin{aligned} a \in L[a] &= L[a]L[a] \\ &= (\{a\} \cup Sa)(\{a\} \cup Sa) \\ &= \{a^2\} \cup aSa \cup Sa^2 \cup SaSa \\ &\subseteq \{a^2\} \cup (aS)aSa \cup Sa^2 \cup SaSa \\ &\subseteq \{a^2\} \cup SaSa \cup Sa^2 \\ &\subseteq \{a^2\} \cup Sa^2. \end{aligned}$$

This implies that S is left regular. Thus it follows from Theorem 4.1.1 that S is a semilattice of left simple semigroups. Consequently, (5) implies (1). ∎

The left-right dual of Theorem 4.1.3 reads as follows:

Theorem 4.1.4 *For a semigroup S, the following conditions are equivalent.*

(1) S is a semilattice of right simple semigroups.

(2) S is right regular and every fuzzy right ideal of S is a fuzzy two-sided ideal of S.

(3) $f \circ g = f \cap g$ for every fuzzy right ideals f and g of S.

(4) The set of all fuzzy right ideals of S is a semilattice under the multiplication of fuzzy subsets.

(5) The set of all right ideals of S is a semilattice under the multiplication of subsets. ∎

Theorem 4.1.5 *For a semigroup S the following conditions are equivalent.*

(1) S is a semilattice of left simple semigroups.

(2) For every fuzzy left ideal f of S, $f(a) = f(a^2)$ and $f(ab) = f(ba)$ for all $a, b \in S$.

Proof. First assume that (1) holds. Let f be any fuzzy left ideal of S and a any element of S. Hence it follows from Theorem 4.1.1 that S is left regular. Then it follows from Theorem 3.2.4 that $f(a) = f(a^2)$. Let a and b be any elements of S. Hence it follows from Theorem 4.1.3 that the fuzzy left ideal f is a fuzzy two-sided ideal of S. Thus

$$f(ab) = f((ab)^2) = f(a(ba)b) \geq f(ba).$$

Similarly, we have $f(ba) \geq f(ab)$. Therefore, $f(ab) = f(ba)$, and (1) implies (2).

Assume that (2) holds. Let f be any fuzzy left ideal of S. Since $f(a) = f(a^2)$ for all $a \in S$, it follows from Theorem 3.2.4 that S is left regular. Let a and b be any elements of S. Then, since f is a fuzzy left ideal of S,

$$f(ab) = f(ba) \geq f(a).$$

Hence f is a fuzzy right ideal, that is, f is a fuzzy two-sided ideal of S. Thus it follows from Theorem 4.1.3 that S is a semilattice of left simple semigroups. Hence (1) holds. ■

The left-right dual of Theorem 4.1.5 reads as follows:

Theorem 4.1.6 *For a semigroup S, the following conditions are equivalent.*

(1) S is a semilattice of right simple semigroups.

(2) For every fuzzy right ideal f of S, $f(a) = f(a^2)$ and $f(ab) = f(ba)$ for all $a, b \in S$. ■

4.2 A Semilattice of Left (Right) Groups

A semigroup S is called **right (left) simple** if it contains no proper right (left) ideal. An element a of a semigroup S is said to be **left (right) cancellable** if, for any $x, y \in S$, $ax = ay$ ($xa = ya$) implies $x = y$. A semigroup S is called **left (right) cancellative** if every element of S is left (right) cancellable. A semigroup S is called a **left group** if it is left simple and right cancellative (Clifford and Preston [17, p.37]), that is, for every $a \in S$, there exists a unique element $x \in S$ such that $xa^2 = a$ (M. Petrich [107, Theorem IV.3.9]). Dually, a semigroup S is called a right group if it is right simple and left cancellative.

The following is due to M. Petrich [107, Theorem IV.3.10].

Theorem 4.2.1 *For a semigroup S, the following conditions are equivalent.*

(1) S is a semilattice of left groups.

(2) S is regular and $aS \subseteq Sa$ for every $a \in S$. ■

Theorem 4.2.2 *Let S be a semigroup that is a semilattice of left groups. Then the following properties hold.*

(1) Every fuzzy (generalized) bi-ideal of S is a fuzzy right ideal of S.

(2) Every fuzzy interior ideal of S is a fuzzy left ideal of S.

Proof. (1) Let f be any fuzzy generalized bi-ideal of S, and a and b any elements of S. Then it follows from Theorem 4.2.1 that there exist elements $x, y \in S$ such that $a = axa$ and $ab = ya$. Thus

$$ab = (axa)b = (ax)(ab) = (ax)(ya) = a(xy)a.$$

Since f is a fuzzy generalized bi-ideal of S,

$$f(ab) = f(a(xy)a) \geq f(a) \wedge f(a) = f(a).$$

Hence f is a fuzzy right ideal of S.

(2) Let g be any interior ideal of S, and a and b any elements of S. Then it follows from Theorem 4.2.1 that there exists an element $z \in S$ such that $b = bzb$. Since g is a fuzzy interior ideal of S, we have

$$g(ab) = g(a(bzb)) = g(ab(zb)) \geq g(b).$$

Thus g is a fuzzy left ideal of S. ■

The following result can be easily seen.

Theorem 4.2.3 *For a semigroup S that is a semilattice of left groups, the following properties hold.*

(1) Every (generalized) bi-ideal of S is a right ideal of S.

(2) Every interior ideal of S is a left ideal of S. ■

Now we characterize a semigroup that is a semilattice of left groups in terms of fuzzy ideals.

Theorem 4.2.4 *For a semigroup S, the following conditions are equivalent.*

(1) S is a semilattice of left groups.

(2) $f \cap g = f \circ g$ for every fuzzy quasi-ideal f and every fuzzy left ideal g of S.

(3) $f \cap g = f \circ g$ for every fuzzy quasi-ideal f and every fuzzy two-sided ideal g of S.

(4) $f \cap g = f \circ g$ for every fuzzy quasi-ideal f and every fuzzy interior ideal g of S.

(5) $f \cap g = f \circ g$ for every fuzzy bi-ideal f and every fuzzy left ideal g of S.

(6) $f \cap g = f \circ g$ for every fuzzy bi-ideal f and every fuzzy two-sided ideal g of S.

(7) $f \cap g = f \circ g$ for every fuzzy bi-ideal f and every fuzzy interior ideal g of S.

(8) $f \cap g = f \circ g$ for every fuzzy generalized bi-ideal f and every fuzzy left ideal g of S.

(9) $f \cap g = f \circ g$ for every fuzzy generalized bi-ideal f and every fuzzy two-sided ideal g of S.

(10) $f \cap g = f \circ g$ for every fuzzy generalized bi-ideal f and every fuzzy interior ideal g of S.

(11) $f \cap g = f \circ g$ for every fuzzy one-sided ideal f and every fuzzy two-sided ideal g of S.

(12) $f \cap g = f \circ g$ for every fuzzy one-sided ideal f and every fuzzy interior ideal g of S.

(13) S is regular and every fuzzy left ideal of S is a fuzzy two-sided ideal of S.

Proof. First assume that (1) holds. Let f and g any generalized bi-ideal and any fuzzy interior ideal of S , respectively. Let a be any element of S.

Then, since S is regular by Theorem 4.1.1, there exists an element $x \in S$ such that

$$a = axa(= axaxa).$$

Since g is a fuzzy interior ideal of S,

$$g(xa(xa)) \geq g(a).$$

Thus we have

$$\begin{aligned}(f \circ g)(a) &= \vee_{a=uv}(f(u) \wedge g(v)) \\ &\geq f(a) \wedge g(xa(xa)) \\ &\geq f(a) \wedge g(a) \\ &= (f \cap g)(a),\end{aligned}$$

and so $f \circ g \supseteq f \cap g$.

On the other hand, it follows from Theorem 4.1.2 that f is a fuzzy right ideal and g is a fuzzy left ideal of S. Thus we have

$$\begin{aligned}(f \circ g)(a) &= \vee_{a=uv}(f(u) \wedge g(v)) \\ &\leq \vee_{a=uv}(f(uv) \wedge g(uv)) \\ &= \vee_{a=uv}(f(a) \wedge g(a)) \\ &= f(a) \wedge g(a) \\ &= (f \cap g)(a),\end{aligned}$$

and so $f \circ g \subseteq f \cap g$. Therefore, we obtain that $f \circ g = f \cap g$, and (1) implies (10). It is clear that (10) $\Rightarrow$ (9) $\Rightarrow$ (8) $\Rightarrow$ (5) $\Rightarrow$ (2), (10) $\Rightarrow$ (7) $\Rightarrow$ (6) $\Rightarrow$ (5) $\Rightarrow$ (2), (7) $\Rightarrow$ (4) $\Rightarrow$ (3) $\Rightarrow$ (2) and (4) $\Rightarrow$ (12) $\Rightarrow$ (11). Assume that (2) holds. Then it follows from Theorem 3.1.11 that S is regular. Let Q be any quasi-ideal of S. Then the characteristic function C_Q of Q is a fuzzy quasi-ideal of S. Since S itself is a fuzzy left ideal of S, by the assumption, we have

$$C_Q = C_Q \cap S = C_Q \circ S.$$

Hence C_Q is a fuzzy right ideal of S. Hence it follows from Lemma 2.3.1 that Q is a right ideal of S. Thus any quasi-ideal of S is a right ideal of S. Let a be any element of S. Then the quasi-ideal Sa of S is a right ideal. Since S is regular,

$$aS \subseteq (aSa)S = ((aS)a)S \subseteq (Sa)S \subseteq Sa.$$

Hence it follows from Theorem 4.2.1 that S is a semilattice of left groups, and (2) implies (1).

Assume that (11) holds. Let f and g be any fuzzy right ideal and any fuzzy left ideal of S, respectively. Then S itself is a fuzzy two-sided ideal of S and so by assumption, we have

$$g = g \cap S = g \circ S.$$

Thus f is a fuzzy right ideal, that is, g is a fuzzy two-sided ideal of S. Hence by the assumption, we have $f \cap g \subseteq f \circ g$. The converse inclusion always holds. Thus $f \cap g = f \circ g$, and it follows from Theorem 3.1.8 that S is regular. Let a be any element of S. Then, as is similarly proved in the proof of (2) $\Rightarrow$ (1), we obtain that $aS \subseteq Sa$. Thus it follows from Theorem 4.2.1 that S is a semilattice of left groups, and (11) implies (1).

Next assume that (1) holds. Then it follows from Theorem 4.2.1 that S is regular. Let f be any fuzzy left ideal of S. Then it follows from Theorem 4.2.2(1) that the fuzzy bi-ideal f is a right ideal of S, that is, f is a fuzzy two-sided ideal of S. Thus (1) implies (13).

Conversely, assume that (13) holds. Then, since S is regular, it follows from Theorem 3.1.5 that every left ideal of S is a right ideal of S. Let a be any element of S. Then, as is proved in (2) $\Rightarrow$ (1), we have $aS \subseteq Sa$. Thus (13) implies (1). ∎

The left-right dual of Theorem 4.2.4 reads as follows:

Theorem 4.2.5 *For a semigroup S, the following conditions are equivalent.*

(1) S is a semilattice of right groups.

(2) $f \cap g = g \circ f$ for every fuzzy quasi-ideal f and every fuzzy right ideal g of S.

(3) $f \cap g = g \circ f$ for every fuzzy quasi-ideal f and every fuzzy two-sided ideal g of S.

(4) $f \cap g = g \circ f$ for every fuzzy quasi-ideal f and every fuzzy interior ideal g of S.

(5) $f \cap g = g \circ f$ for every fuzzy bi-ideal f and every fuzzy right ideal g of S.

(6) $f \cap g = g \circ f$ for every fuzzy bi-ideal f and every fuzzy two-sided ideal g of S.

(7) $f \cap g = g \circ f$ for every fuzzy bi-ideal f and every fuzzy interior ideal g of S.

(8) $f \cap g = g \circ f$ for every fuzzy generalized bi-ideal f and every fuzzy right ideal g of S.

(9) $f \cap g = g \circ f$ for every fuzzy generalized bi-ideal f and every fuzzy two-sided ideal g of S.

(10) $f \cap g = g \circ f$ for every fuzzy generalized bi-ideal f and every fuzzy interior ideal g of S.

(11) $f \cap g = g \circ f$ for every fuzzy one-sided ideal f and every fuzzy two-sided ideal g of S.

(12) $f \cap g = g \circ f$ for every fuzzy one-sided ideal f and every fuzzy interior ideal g of S.

(13) S is regular and every right ideal of S is a fuzzy two-sided ideal of S. ∎

We now give, using fuzzy ideals, another characterization of a semigroup that is a semilattice of left groups.

Theorem 4.2.6 *For a semigroup S, the following conditions are equivalent.*

(1) S is a semilattice of left groups.

(2) $f \cap g = f \circ g \circ f$ for every fuzzy quasi-ideal f and every fuzzy left ideal g of S.

(3) $f \cap g = f \circ g \circ f$ for every fuzzy bi-ideal f and every left ideal g of S.

(4) $f \cap g = f \circ g \circ f$ for every fuzzy generalized bi-ideal f and every fuzzy left ideal g of S.

Proof. First assume that (1) holds. Let f and g be a fuzzy generalized bi-ideal and a fuzzy left ideal of S, respectively. Then

$$f \circ g \circ f \subseteq f \circ S \circ f \subseteq f.$$

On the other hand, since the fuzzy left ideal g is a fuzzy bi-ideal of S, it follows from Theorem 4.2.2 that g is a fuzzy right ideal of S. Thus

$$f \circ g \circ f \subseteq S \circ g \circ S \subseteq g \circ S \subseteq g.$$

Therefore, we have

$$f \circ g \circ f \subseteq f \cap g.$$

Let a be any element of S. Then it follows from Theorem 4.2.1, there exist elements $x, y \in S$ such that $a = axa$ and $ax = ya$. Hence

$$ax = axaxax = axax(ya) = (axa)(xya).$$

Thus

$$\begin{aligned}
(f \circ g \circ f)(a) &= \vee_{a=pq}((f \circ g)(p) \wedge f(q)) \\
&\geq (f \circ g)(ax) \wedge f(a) \\
&= (\vee_{ax=pq}(f(p) \wedge g(q))) \wedge f(a) \\
&\geq (f(axa) \wedge g(xya)) \wedge f(a) \\
&\geq (f(a) \wedge g(a)) \wedge f(a) \\
&= (f \cap g)(a),
\end{aligned}$$

and so $f \circ g \circ f \supseteq f \cap g$. Therefore, $f \circ g \circ f = f \cap g$, and (1) implies (4). It is clear that (4) $\Rightarrow$ (3) $\Rightarrow$ (2).

Assume that (2) holds. Let f be any quasi-ideal of S. Then S itself is a fuzzy left ideal of S and so by assumption,

$$f = f \cap S = f \circ S \circ f.$$

Thus it follows from Theorem 3.1.3 that S is regular. On the other hand, let g be any fuzzy left ideal of S. Then by the assumption,

$$g = S \cap g = S \circ g \circ S.$$

Hence g is a fuzzy interior ideal of S. Since S is regular, it follows from Lemma 2.5.5 that g is a fuzzy two-sided ideal of S. Therefore, we obtain that every fuzzy left ideal of S is a fuzzy two-sided ideal of S. Thus it follows from Lemma 2.3.1 that every left ideal of S is a two-sided ideal of S. Let a be any element of S. Then, since S is regular and the left ideal Sa is a right ideal of S, we have

$$aS \subseteq (aSa)S = a((Sa)S) \subseteq a(Sa) = (aS)a \subseteq Sa.$$

Therefore, it follows from Theorem 4.2.1 that S is a semilattice of left groups and (2) implies (1). ∎

The left-right dual of Theorem 4.2.6 reads as follows:

Theorem 4.2.7 *For a semigroup S, the following conditions are equivalent.*

(1) S is a semilattice of right groups.

(2) $f \cap g = f \circ g \circ f$ for every fuzzy quasi-ideal f and every fuzzy right ideal g of S.

(3) $f \cap g = f \circ g \circ f$ for every fuzzy bi-ideal f and every fuzzy right ideal g of S.

(4) $f \cap g = f \circ g \circ f$ for every fuzzy generalized bi-ideal f and every fuzzy right ideal g of S.

Theorem 4.2.8 *For a semigroup S, the following conditions are equivalent.*

(1) S is a semilattice of left groups.

(2) $f \cap g = f \circ S \circ g$ for every fuzzy quasi-ideal f and every fuzzy left ideal g of S.

(3) $f \cap g = f \circ S \circ g$ for every fuzzy bi-ideal f and every fuzzy left ideal g of S.

(4) $f \cap g = f \circ S \circ g$ for every fuzzy generalized bi-ideal f and every fuzzy left ideal g of S.

Proof. First assume that (1) holds. Let f and g be a fuzzy generalized bi-ideal and a fuzzy left ideal of S, respectively. Then it follows from Theorem 4.2.1 that S is regular. Thus by Theorem 3.1.11, we have

$$f \circ S \circ g \subseteq f \circ g \subseteq S \circ g \subseteq g.$$

On the other hand, it follows from Theorem 4.2.2 that f is a fuzzy right ideal of S. Hence

$$f \circ S \circ g = (f \circ S) \circ g \subseteq f \circ g \subseteq f \circ S \subseteq f.$$

Therefore, we have $f \circ S \circ g \subseteq f \cap g$.

Let a be any element of S. Then it follows from Theorem 4.2.1 that there exist elements $x, y \in S$ such that $a = axa$ and $ax = ya$. Thus we have

$$\begin{aligned}(f \circ S \circ g)(a) &= \vee_{a=pq}((f \circ S)(p) \wedge g(q)) \\ &\geq (f \circ S)(ax) \wedge g(a) \\ &= (\vee_{ax=pq}(f(p) \wedge S(q))) \wedge g(a) \\ &\geq (f(axa) \wedge S(x)) \wedge g(a) \\ &= (f(a) \wedge 1) \wedge g(a) \\ &= f(a) \wedge g(a) \\ &= (f \cap g)(a),\end{aligned}$$

and so $f \circ S \circ g \supseteq f \cap g$. Therefore, $f \circ S \circ g = f \cap g$, and (1) implies (4).

It is clear that (4) $\Rightarrow$ (3) $\Rightarrow$ (2).

Assume that (2) holds. Let f and g be a fuzzy quasi-ideal and a fuzzy left ideal of S, respectively. Then by the assumption, we have

$$f \cap g = f \circ S \circ g = f \circ (S \circ g) \subseteq f \circ g.$$

Hence it follows from Theorem 3.1.11 that S is regular. Let g be any fuzzy left ideal of S. Then, since g is a fuzzy quasi-ideal of S and S itself is a fuzzy left ideal of S, by the assumption, we have

$$g = g \cap S = g \circ S \circ S.$$

Let L be any left ideal of S. Let a be any element of L. Then C_L is a fuzzy left ideal of S be Lemma 2.3.1. Thus by Lemma 2.3.4,

$$C_{LSS}(a) = (C_L \circ C_S \circ C_S)(a) = C_L(a) = 1,$$

and so $a \in LSS$, and so $L \subseteq LSS$.

Conversely, let a be any element of LSS. Then

$$C_L(a) = (C_L \circ C_S \circ C_S)(a) = C_{LSS}(a) = 1,$$

and so $a \in L$. Thus $LSS \subseteq L$ and so $L = LSS$. Since Sa is a left ideal of S, we have

$$aS \subseteq aSaSaS = a(Sa)(SSS) \subseteq a((Sa)SS) \subseteq a(Sa) = (aS)a \subseteq Sa.$$

Hence it follows from Theorem 4.2.1 that S is a semilattice of left groups, and (4) implies (1). ∎

The left-right dual of Theorem 4.2.8 reads as follows:

Theorem 4.2.9 *For a semigroup S, the following conditions are equivalent.*

(1) S is a semilattice of right groups.

(2) $f \cap g = g \circ S \circ f$ for every fuzzy quasi-ideal f and every fuzzy right ideal g of S.

(3) $f \cap g = g \circ S \circ f$ for every fuzzy bi-ideal f and every fuzzy right ideal g of S.

(4) $f \cap g = g \circ S \circ f$ for every fuzzy generalized bi-ideal f and every fuzzy right ideal g of S. ∎

Theorem 4.2.10 *For a semigroup S, the following conditions are equivalent.*

(1) S is a semilattice of left groups.

(2) $f \cap h \cap g = f \circ h \circ g$ for every fuzzy quasi-ideal f, every fuzzy two-sided ideal h and every fuzzy left ideal g of S.

(3) $f \cap h \cap g = f \circ h \circ g$ for every fuzzy bi-ideal f, every fuzzy two-sided ideal h and every fuzzy left ideal g of S.

(4) $f \cap h \cap g = f \circ h \circ g$ for every fuzzy generalized bi-ideal f, every fuzzy two-sided ideal h and every fuzzy left ideal g of S.

Proof. First assume that (1) holds. Let f be any fuzzy generalized bi-ideal, h any fuzzy two-sided ideal, and g any fuzzy left ideal of S. Then we have

$$f \circ h \circ g \subseteq S \circ (S \circ g) \subseteq S \circ g \subseteq g,$$

and

$$f \circ h \circ g \subseteq S \circ h \circ S \subseteq h.$$

It follows from Theorem 4.2.2 that the fuzzy generalized bi-ideal f is fuzzy right ideal of S. Thus

$$f \circ h \circ g \subseteq (f \circ S) \circ S \subseteq f \circ S \subseteq f.$$

Therefore,

$$f \circ h \circ g \subseteq f \cap h \cap g.$$

Let a be any element of S. Then by Theorem 4.2.1, $a = axa$ and $ax = ya$ for some $x, y \in S$. Then

$$ax = axaxax = axax(ya) = (axa)(xya).$$

Thus

$$\begin{aligned}
(f \circ h \circ g)(a) &= \vee_{a=pq}((f \circ h)(p) \wedge g(q)) \\
&\geq (f \circ h)(ax) \wedge g(a) \\
&= (\vee_{ax=pq}(f(p) \wedge h(q))) \wedge g(a) \\
&\geq (f(axa) \wedge h(xya)) \wedge g(a) \\
&\geq (f(a) \wedge h(a)) \wedge g(a) \\
&= (f \cap h \cap g)(a),
\end{aligned}$$

and so $f \circ h \circ g \supseteq f \cap h \cap g$. Therefore, we obtain that $f \circ h \circ g = f \cap h \cap g$, and (1) implies (4).

It is clear that (4) $\Rightarrow$ (3) $\Rightarrow$ (2).

Assume that (2) holds. Let f be any fuzzy quasi-ideal and g any fuzzy left ideal of S. Then, since S itself is a fuzzy two-sided ideal of S, we have by (2) that

$$f \cap g = f \cap S \cap g = f \circ S \circ g = f \circ (S \circ g) \subseteq f \circ g.$$

Then it follows from Theorem 3.1.11 that S is regular. It can be seen in a similar way as in the proof of (4) $\Rightarrow$ (1) in the above theorem that $aS \subseteq Sa$. Therefore, we have (4) implies (1). ∎

4.3 A Semilattices of Groups

We recall that a semigroup S is called a **semilattice of groups** if it is the set-theoretical union of a family of mutually disjoint subgroups G_i $(i \in I)$ such that for all $i, j \in I$ the products G_iG_j and G_jG_i are both contained in the same subgroup G_k $(k \in I)$.

The following characterization of such a semigroup is due to M. Petrich [107, Corollary II, 4.10, p.46], and S. Lajos [73] and Clifford and Preston[17, p.129].

Theorem 4.3.1 *For a semigroup S, the following conditions are equivalent.*

(1) S is a semilattice of groups.

(2) S is regular and $aS = Sa$ for all $a \in S$.

(3) $L \cap R = LR$ for every left ideal L and every right ideal R of S.

(4) S is regular and every one-sided ideal of S is a two-sided ideal of S.

∎

Now we give the characterization of a semigroup that is a semilattice of groups in terms of fuzzy ideals.

Theorem 4.3.2 *Let S be a semigroup that is a semilattice of groups. Then every fuzzy (generalized) bi-ideal of S is a fuzzy two-sided ideal of S.*

Proof. Let f be any fuzzy generalized bi-ideal of S, and a and b any elements of S. Then it follows from Theorem 4.3.1 that

$$ab \in (aSa)S = (aS)(aS) = (aS)(Sa) = a(SS)a \subseteq aSa.$$

Thus there exists an element $x \in S$ such that $ab = axa$. Hence

$$f(ab) = f(axa) \geq f(a) \wedge f(a) = f(a).$$

Thus f is a fuzzy right ideal of S. Dually, we have that f is a fuzzy left ideal of S and so f is a fuzzy two-sided ideal of S. ∎

Theorem 4.3.3 *For a semigroup S, the following conditions are equivalent.*

(1) S is a semilattice of groups.

(2) $f \cap g = f \circ g$ for every fuzzy left ideal f and every fuzzy right ideal g of S.

(3) $f \cap g = f \circ g$ for every fuzzy left ideal f and every fuzzy quasi-ideal g of S.

(4) $f \cap g = f \circ g$ for every fuzzy left ideal f and every fuzzy bi-ideal g of S.

(5) $f \cap g = f \circ g$ for every fuzzy left ideal f and every fuzzy generalized bi-ideal g of S.

(6) $f \cap g = f \circ g$ for every fuzzy quasi-ideal f and every fuzzy right ideal g of S.

(7) $f \cap g = f \circ g$ for all fuzzy quasi-ideals f and g of S.

(8) $f \cap g = f \circ g$ for every fuzzy quasi-ideal f and every fuzzy bi-ideal g of S.

(9) $f \cap g = f \circ g$ for every fuzzy quasi-ideal f and every fuzzy generalized bi-ideal g of S.

(10) $f \cap g = f \circ g$ for every fuzzy bi-ideal f and every fuzzy right ideal g of S.

(11) $f \cap g = f \circ g$ for every fuzzy bi-ideal f and every fuzzy quasi-ideal g of S.

(12) $f \cap g = f \circ g$ for all fuzzy bi-ideals f and g of S.

(13) $f \cap g = f \circ g$ for every fuzzy bi-ideal f and every fuzzy generalized bi-ideal g of S.

(14) $f \cap g = f \circ g$ for every fuzzy generalized bi-ideal f and every fuzzy right ideal g of S.

(15) $f \cap g = f \circ g$ for every fuzzy generalized bi-ideal f and every fuzzy quasi-ideal g of S.

(16) $f \cap g = f \circ g$ for every fuzzy generalized bi-ideal f and every fuzzy bi-ideal g of S.

(17) $f \cap g = f \circ g$ for all fuzzy generalized bi-ideals f and g of S.

(18) S is regular and every fuzzy one-sided ideal of S is a fuzzy two-sided ideal of S.

(19) The set of all fuzzy quasi-ideals of S is a semilattice under the multiplication of fuzzy subsets.

(20) The set of all fuzzy bi-ideals of S is a semilattice under the multiplication of fuzzy subsets.

(21) The set of all fuzzy generalized bi-ideals of S is a semilattice under the multiplication of fuzzy subsets.

Proof. First we show that (1) $\Rightarrow$ (17). Let f and g be any fuzzy generalized bi-ideals of S. Then we have $f \circ g \subseteq S \circ g \subseteq g$ and $f \circ g \subseteq f \circ S \subseteq f$. Thus $f \circ g \subseteq f \cap g$.

To see inclusion in the other direction, let a be any element of S. Then, since $a \in aSa$ and $aS = Sa$ by Theorem 4.3.1, we have

$$\begin{aligned} a \in aSa &\subseteq (aSa)S(aSa) \\ &= aS(aS)aSa \\ &= aS(Sa)aS \\ &\subseteq (aSa)(aSa), \end{aligned}$$

and so $a = (axa)(aya)$ for some $x, y \in S$. Hence

$$\begin{aligned}(f \circ g)(a) &= \vee_{a=pq} f(p) \wedge g(q)) \\ &\geq f(axa) \wedge g(aya) \\ &\geq f(a) \wedge g(a) \\ &= (f \cap g)(a),\end{aligned}$$

and so $f \circ g \supseteq f \cap g$. Thus we have that $f \circ g = f \cap g$. Therefore, (1) implies (17).

It is clear that (17) $\Rightarrow$ (16) $\Rightarrow$ (15) $\Rightarrow$ (14) $\Rightarrow$ (10) $\Rightarrow$ (6) $\Rightarrow$ (2), (17) $\Rightarrow$ (13) $\Rightarrow$ (12) $\Rightarrow$ (11) $\Rightarrow$ (10), (13) $\Rightarrow$ (9) $\Rightarrow$ (8) $\Rightarrow$ (7) $\Rightarrow$ (6) and (9) $\Rightarrow$ (5) $\Rightarrow$ (4) $\Rightarrow$ (3) $\Rightarrow$ (2).

Assume that (2) holds. Let L and R be any left ideal and any right ideal of S, respectively. Then by Lemma 2.3.1, the characteristic functions C_L and C_R are a fuzzy left ideal and a fuzzy right ideal of S, respectively. Let a be any element of $R \cap L$. Then by Lemma 2.3.4 and by the assumption, we have

$$\begin{aligned}C_{LR}(a) &= (C_L \circ C_R)(a) \\ &= (C_L \cap C_R)(a) \\ &= C_{L \cap R}(a) \\ &= 1,\end{aligned}$$

and so $a \in LR$. Thus $L \cap R \subseteq LR$.

Conversely, let a be any element of LR. Then

$$\begin{aligned}C_{L \cap R}(a) &= (C_L \cap C_R)(a) \\ &= (C_L \circ C_R)(a) \\ &= C_{LR}(a) \\ &= 1,\end{aligned}$$

and so $a \in L \cap R$. Thus we have $LR \subseteq L \cap R$. Therefore, $L \cap R = LR$. Hence it follows from Theorem 4.3.1 that S is a semilattice of groups and so (2) implies (1).

Assume that (1) holds. Then, as shown above, (17) holds. Thus it is clear that (21) holds. It is also clear that (21) $\Rightarrow$ (20) $\Rightarrow$ (19). Assume that (19) holds. Then, since every fuzzy quasi-ideal of S is idempotent, it follows from Theorem 3.3.13 that S is regular. Let L and R be any left ideal and any right ideal of S, respectively. Then, since L and R are quasi-ideals of S, C_L and C_R are fuzzy quasi-ideals of S. Thus

$$C_{LR} = C_L \circ C_R = C_R \circ C_L = C_{RL}.$$

This implies that $LR = RL$. Then since S is regular, by Theorem 3.1.2, we have

$$R \cap L = RL = LR.$$

Hence it follows from Theorem 4.3.1 that S is a semilattice of groups, and (19) implies (1). Next assume that (1) holds. To see that (18) holds, let f be any fuzzy left ideal of S. Now S itself is a fuzzy right ideal of S and so we have

$$f = f \cap S = f \circ S.$$

Thus f is a fuzzy right ideal of S. Similarly, it can be proved that every fuzzy right ideal of S is a fuzzy left ideal of S. As shown above, S is regular. Thus we obtain that (1) implies (18). It can be similarly proved as in the proof of (19) $\Rightarrow$ (1) that (18) implies (1). ■

Theorem 4.3.4 *For a semigroup S, the following conditions are equivalent.*

(1) S is a semilattice of groups.

(2) $f \cap g = g \circ f \circ g$ for every fuzzy quasi-ideal f and every fuzzy two-sided ideal g of S.

(3) $f \cap g = g \circ f \circ g$ for every fuzzy quasi-ideal f and every fuzzy interior ideal g of S.

(4) $f \cap g = g \circ f \circ g$ for every fuzzy bi-ideal f and every fuzzy two-sided ideal g of S.

(5) $f \cap g = g \circ f \circ g$ for every fuzzy bi-ideal f and every fuzzy interior ideal g of S.

(6) $f \cap g = g \circ f \circ g$ for every fuzzy generalized bi-ideal f and every fuzzy two-sided ideal g of S.

(7) $f \cap g = g \circ f \circ g$ for every fuzzy generalized bi-ideal f and every fuzzy interior ideal g of S.

Proof. First assume that (1) holds. Let f and g be any fuzzy generalized bi-ideal and any fuzzy interior ideal of S, respectively. Then it follows from Theorem 4.3.2 that f is a fuzzy two-sided ideal of S. Thus

$$g \circ f \circ g \subseteq S \circ f \circ S \subseteq f.$$

Since S is regular by Theorem 4.3.1, it follows from Lemma 2.5.5 that the fuzzy interior ideal g is a fuzzy two-sided ideal of S. Hence

$$g \circ f \circ g \subseteq g \circ S \circ g \subseteq g \circ g \subseteq g \circ S \subseteq g.$$

Therefore, we have $g \circ f \circ g \subseteq f \cap g$. To see the inclusion in the other direction, let a be any element of S. Since S is regular by Theorem 4.3.1, there exists an element $x \in S$ such that $a = axa$. Hence

$$\begin{aligned}
(g \circ f \circ g)(a) &= \vee_{a=pq}((g \circ f)(p) \wedge g(q)) \\
&\geq (g \circ f)(a) \wedge g(xa) \\
&\geq \vee_{a=uv}(g(u) \wedge f(v))) \wedge g(a) \\
&\geq (fg(ax) \wedge f(a)) \wedge g(a)) \\
&\geq f(a) \wedge g(a) \\
&= (f \cap g)(a),
\end{aligned}$$

and so $g \circ f \circ g \supseteq f \cap g$. Therefore, we have that $g \circ f \circ g = f \cap g$, and (1) implies (7).

It is clear that (7) $\Rightarrow$ (6) $\Rightarrow$ (4) $\Rightarrow$ (2) and (7) $\Rightarrow$ (5) $\Rightarrow$ (3) $\Rightarrow$ (2).

Assume that (2) holds. Let Q and J be any quasi-ideal and any two-sided ideal of S, respectively. We note that the characteristic functions C_Q and C_J are a fuzzy quasi-ideal and a fuzzy two-sided ideal of S, respectively. Let a be any element of $Q \cap J$. Then by Lemma 2.3.4 and by the assumption, we have

$$\begin{aligned} C_{JQJ}(a) &= (C_J \circ C_Q \circ C_J)(a) \\ &= (C_J \cap C_Q)(a) \\ &= C_{J \cap Q}(a) \\ &= 1, \end{aligned}$$

and so $a \in JQJ$. Thus $J \cap Q \subseteq JQJ$.

Now, let a be any element of JQJ. Then by Lemma 2.3.4, we have

$$\begin{aligned} C_{J \cap Q}(a) &= (C_J \cap C_Q)(a) \\ &= (C_J \circ C_Q \circ C_J)(a) \\ &= C_{JQJ}(a) \\ &= 1, \end{aligned}$$

and so $a \in J \cap Q$. Thus $JQJ \subseteq J \cap Q$. Therefore, we have that $JQJ = J \cap Q$.

Let L be any left ideal of S. Then, since L is a quasi-ideal of S, and since S itself is a two-sided ideal of S, we have

$$L = L \cap S = SLS.$$

Hence

$$LS = (SLS)S = (SL)(SS) \subseteq SLS = L.$$

Thus L is a right ideal of S. Similarly, for any right ideal R, it can be seen that $R = SRS$ and $SR \subseteq R$. Therefore, we have that any one-sided ideal of S is a two-sided ideal of S. On the other hand, we note that a subset of S is a quasi-ideal of S if and only if it is the intersection of a right ideal and a left ideal of S, (see [17, p.85]). Let Q be any quasi-ideal of S. Then $Q = R \cap L$ for some right ideal R and some left ideal of S. Hence

$$\begin{aligned} QSQ &= (R \cap L)S(R \cap L) \\ &= RSR \cap RSL \cap LSR \cap LSL \\ &= R \cap L \\ &= Q. \end{aligned}$$

Thus Theorem 3.1.2 that S is regular. Hence it follows from Theorem 4.3.1 that S is a semilattice of groups. ∎

Theorem 4.3.5 *For a semigroup S, the following conditions are equivalent.*

(1) S is a semilattice of groups.

(2) For every fuzzy quasi-ideal f of S, $f(a^2) = f(a)$ and $f(ab) = f(ba)$ for all $a, b \in S$.

(3) For every fuzzy bi-ideal f of S, $f(a^2) = f(a)$ and $f(ab) = f(ba)$ for all $a, b \in S$.

(4) For every generalized bi-ideal f of S, $f(a^2) = f(a)$ and $f(ab) = f(ba)$ for all $a, b \in S$.

Proof. First assume that (1) holds. Let f be any fuzzy generalized bi-ideal of S, and a and b any elements of S. Then, since S is regular by Theorem 4.3.1, there exists an element $x \in S$ such that $a = axa(= axaxaxa)$. Since $aS = Sa$ by Theorem 4.3.1, there exist elements $y, z \in S$ such that $xa = ay$ and $ax = za$. Thus we have $a = axa = a(xaxaxa) = a(xa)x(ax)a$ $= a(ay)x(za)a = a^2(yxz)a^2$. Hence since f is a fuzzy generalized bi-ideal,

$$\begin{aligned} f(a) &= f(a^2(yxz)a^2) \\ &\geq f(a^2) \wedge f(a^2) \\ &= f(a^2) \\ &= f(a(axa)) \\ &= f(a(ax)a) \\ &\geq f(a) \wedge f(a) \\ &= f(a), \end{aligned}$$

and so $f(a) = f(a^2)$. On the other hand, by Theorem 4.3.1, we have

$$(ab)^4 \in (Sba)S(baS) = (baS)S(Sba).$$

Thus there exists an element $w \in S$ such that $(ab)^4 = bawba$. Hence

$$f(ab) = f((ab)^2) = f((ab)^4) = f((ba)w(ba)) \geq f(ba) \wedge f(ba) = f(ba).$$

Similarly, we have $f(ba) \geq f(ab)$, and so $f(ab) = f(ba)$. Thus we obtain that (1) implies (4).

It is clear that (4) $\Rightarrow$ (3) $\Rightarrow$ (2).

Assume that (2) holds. Then it follows from Theorem 3.2.8 that S is completely regular. Let a be any element of S. To see $aS = Sa$, let $ax(x \in S)$ be any element of aS. Then by Lemma 2.6.4 the characteristic function $C_{Q[xa]}$ of the principal quasi-ideal $Q[xa]$ is a fuzzy quasi-ideal of S. Thus by the assumption, we have

$$C_{Q[xa]}(ax) = C_{Q[xa]}(xa) = 1,$$

and so

$$ax \in Q[xa] = \{xa\} \cup (xaS \cap Sxa).$$

If $ax = xa$, then $ax = xa \in Sa$, and so $aS \subseteq Sa$. If $ax \in xaS$, then

$$ax \in xaS \cap Sxa \subseteq Sxa \subseteq Sa,$$

and so $aS \subseteq Sa$. In any case, we have $aS \subseteq Sa$. Similarly, $Sa \subseteq aS$. Thus $aS = Sa$. Hence it follows from Theorem 4.3.1 that S is a semilattice of groups. Thus (2) implies (1). ∎

4.4 Fuzzy Normal Semigroups

Definition 4.4.1 *Let G be a group with identity e. A fuzzy subset f of G is called a* ***fuzzy subgroup*** *of S if*
(FG:1) $f(ab) \geq f(a) \wedge f(b)$ for all $a, b \in G$;
(FG:2) $f(a^{-1}) \geq f(a)$ for all $a \in G$;

(FG:3) $f(e) = 1$.

Definition 4.4.2 *A fuzzy subgroup f of G is called* ***fuzzy normal*** *if*
(FG:4) $f(ab) = f(ba)$ for all $a, b \in G$.

Now we introduce the concept of the fuzzy normality in a semigroup which is an extended concept of a fuzzy normal subgroup in a group.

Definition 4.4.3 *A semigroup S is called* ***normal*** *if $aS = Sa$ for all $a \in S$.*

Definition 4.4.4 *A fuzzy quasi-ideal f of a semigroup S is called* ***Q-normal*** *if $f(ab) = f(ba)$ for all $a, b \in S$.*

Definition 4.4.5 *A semigroup S is called* ***fuzzy Q^*-normal*** *if every fuzzy quasi-ideal of S is Q-normal.*

Definition 4.4.6 *A fuzzy bi-ideal f of S is called* ***B-normal*** *if $f(ab) = f(ba)$ for all $a, b \in S$.*

Definition 4.4.7 *A semigroup S is called* ***fuzzy B^*-normal*** *if every fuzzy bi-ideal of S is B-normal.*

Theorem 4.4.8 *Any fuzzy Q^*-normal semigroup is normal.*

Proof. Let S be a fuzzy Q^*-normal semigroup and a any element of S. Let ax $(x \in S)$ be any element of aS. Then it follows from Lemma 2.6.4 that the characteristic function $C_{Q[xa]}$ of the principal quasi-ideal $Q[xa]$ generated by the element xa is a fuzzy quasi-ideal of S. Thus it is fuzzy Q-normal. Since $xa \in Q[xa]$, we have

$$C_{Q[xa]}(ax) = C_{Q[xa]}(xa) = 1,$$

and so

$$ax \in Q[xa] = \{xa\} \cup (xsS \cap Sxa) \subseteq Sa.$$

Thus $aS \subseteq Sa$. Similarly, we have $Sa \subseteq aS$. Thus $aS = Sa$. Hence S is normal. ■

The following theorem shows that the converse of Theorem 4.4.8 holds for a regular semigroup.

Theorem 4.4.9 *For a regular semigroup S, the following conditions are equivalent.*
(1) S is fuzzy Q^*-normal.
(2) S is normal.

Proof. It suffices to show that (2) implies (1). Assume that (2) holds. Let f be any fuzzy quasi-ideal of S. Then, since S is regular and normal, we have $ab \in (aSa)(bSb) = (aS)(ab)(Sb) \subseteq (aS)(abSab)(Sb) = (aSa)b(Sa)(bSb) \subseteq (Sb)(Sa)S = (Sb)(aS)S = S(ba)SS = baSSS \subseteq baS$. Thus there exists an element $x \in S$ such that $ab = bax$. Hence

$$\begin{aligned}(f \circ S)(ab) &= \vee_{ab=pq}(f(p) \wedge S(q)) \\ &\geq f(ba) \wedge S(x) \\ &= f(ba) \wedge 1 \\ &= f(ba).\end{aligned}$$

It can be similarly proved that

$$(S \circ f)(ab) \geq f(ba).$$

Thus since f is a fuzzy quasi-ideal of S, we have

$$\begin{aligned}f(ab) &\geq ((f \circ S) \cap (S \circ f))(ab) \\ &= (f \circ S)(ab) \wedge (S \circ f)(ab) \\ &\geq f(ba) \wedge f(ba) \\ &= f(ba).\end{aligned}$$

Similarly, it can be similarly proved that $f(ba) \geq f(ab)$. Thus $f(ab) = f(ba)$ and so S is fuzzy Q^*-normal. Therefore, (2) implies (1). ■

Theorem 4.4.10 *Any fuzzy B^*-normal semigroup is normal.*

Proof. Let f be a fuzzy (generalized) bi-ideal of a fuzzy B^*-normal semigroup S. Let a be any element of S. Let ax ($x \in S$) be any element of aS. Then it follows from Lemma 2.4.1 that the characteristic function $C_{B[xa]}$ of the principal bi-ideal $B[xa]$ generated by xa is a fuzzy bi-ideal of S. Thus it is fuzzy normal. Then, since $xa \in B[xa]$, we have

$$C_{B[xa]}(ax) = C_{B[xa]}(xa) = 1,$$

and so

$$ax \in B[xa] = \{xa\} \cup \{xaxa\} \cup xaSxa \subseteq Sa.$$

Thus we have $aS \subseteq Sa$. Similarly, the inclusion $Sa \subseteq aS$ holds. Hence $aS = Sa$. Thus S is normal. ∎

Theorem 4.4.11 *For a regular semigroup S, the following conditions are equivalent.*

(1) S is B^*-normal.

(2) S is normal.

Proof. It suffices to prove that (2) implies (1). Assume that (2) holds. Let f be any fuzzy bi-ideal of S, and a and b any elements of S. Then, since S is regular, we have $ab \in (aSa)(bSb) = (aS)(ab)(Sb) \subseteq (aS)(abSab)(Sb) = (aSa)b(Sa)(bSb) \subseteq (Sb)(aS)S = S(ba)SS = (ba)SSS \subseteq baS = (baSba)S = (baS)(Sba) = (baS)(Sba) = ba(SS)(ba) \subseteq baSba$. This implies that there exists an element $x \in S$ such that $a = baxba$. Hence since f is a fuzzy bi-ideal of S, we have

$$\begin{aligned} f(ab) &= f((ba)x(ba)) \\ &\geq f(ba) \wedge f(ba) \\ &= f(ba). \end{aligned}$$

It can be similarly seen that $f(ba) \geq f(ab)$. Thus $f(ab) = f(ba)$ and so that S is fuzzy B^*-normal. Hence (2) implies (1). ∎

Theorem 4.4.12 *For an idempotent semigroup S, the following conditions are equivalent.*

(1) S is commutative.

(2) S is fuzzy Q^*-normal.

(3) S is fuzzy B^*-normal.

Proof. It is clear that (1) $\Rightarrow$ (3) $\Rightarrow$ (2). Assume that (2) holds. Then it follows from Theorem 4.4.8 that S is normal. Let a and b be any elements of S. Then $ab \in Sb = bS$, and so there exists an element $x \in S$ such that $ab = bx$. Similarly, we have $ba = yb$ for some $b \in S$. Then, since S is idempotent, we have $ab = bx = (bb)x = b(bx) = b(ab) = (ba)b = (yb)b = y(bb) = yb = ba$. Therefore, we have that S is commutative. Thus (2) implies (1). ∎

We recall that a semigroup S is called archimedean if, for all $a, b \in S$, there exists a positive integer n such that $a^n \in SbS$.

Definition 4.4.13 *A semigroup S is called* ***weakly commutative*** *if for all $a, b \in S$, there exists a positive integer n such that $(ab)^n \in bSa$.*

The following is due to M. Petrich [107, Corollary II.5.6].

Theorem 4.4.14 *Every weakly commutative semigroup is a semilattice of archimedean semigroups.*

Theorem 4.4.15 *Any fuzzy B^*-normal semigroup is a semilattice of archimedean semigroups.*

Proof. Let S be a fuzzy B^*-normal semigroup. Let a and b be any elements of S, and f any fuzzy bi-ideal of S. Then it follows from Lemma 2.4.1 that the characteristic function $C_{B[ba]}$ of the principal bi-ideal $B[ba]$ generated by the element ba is a fuzzy bi-ideal of S. Thus it is fuzzy B-normal. Then, since $ba \in B[ba]$, we have

$$C_{B[ba]}(ab) = C_{B[ba]}(ba) = 1,$$

and so

$$ab \in B[ba] = \{ba\} \cup \{baba\} \cup baSba.$$

This implies that $(ab)^2 \in baSba \subseteq bSa$. Therefore, S is weakly commutative. Hence it follows from Theorem 4.4.14 that S is a semilattice of archimedean semigroups. ■

Similarly, we can prove the following theorem.

Theorem 4.4.16 *Every fuzzy Q^*-normal semigroup is a semilattice of archimedean semigroup.*

Theorem 4.4.17 *For a fuzzy B^*-normal semigroup S, the following conditions are equivalent.*

(1) S is regular.
(2) S is right regular.
(3) S is left regular.
(4) S is intra-regular.
(5) S is completely regular.

Proof. Since S is fuzzy B^*-normal, it follows from Theorem 4.4.10 that S is normal. Thus it can be easily see that the conditions (1), (2), (3), (4) and (5) are equivalent. ■

Similarly we can prove the following theorem.

Theorem 4.4.18 *For a fuzzy Q^*-normal semigroup S, the following conditions are equivalent.*

(1) S is regular.
(2) S is right regular.
(3) S is left regular.
(4) S is intra-regular.
(5) S is completely regular. ■

Theorem 4.4.19 *For a completely regular semigroup S, the following conditions are equivalent.*

(1) S is fuzzy Q^*-normal.

(2) S is fuzzy B^*-normal.

(3) For each elements a and b of S, there exists a positive integer n such that $(ab)^n \in baSba$.

Proof. It is clear that (2) implies (1). Assume that (1) holds. Then it follows from Theorem 4.4.8 that S is normal. Let a and b be any elements of S. Then we have $(ab)^3 = ababab = a(ba)(ba)b \subseteq (Sba)(baS) = (baS)(Sba) = (ba)SS(ba) \subseteq baSba$. Thus (1) implies (3). Assume that (3) holds. To see that (2) holds, let f be any fuzzy bi-ideal of S, and a and b any elements of S. Then, by the assumption, there exists a positive integer n such that $(ab)^n = baxba$. Since S is completely regular, for this positive integer n, there exists an element $y \in S$ such that

$$ab = (ab)^n y (ab)^n.$$

Since f is a fuzzy bi-ideal of S, we have

$$\begin{aligned} f(ab) &= f((ab)^n y (ab)^n) \\ &\geq f((ab)^n) \wedge f((ab)^n) \\ &= f((an)^n) \\ &= f((baxba) \\ &\geq f(ba) \wedge f(ba) \\ &= f(ba). \end{aligned}$$

It can be seen in a similar way that $f(ba) \geq f(ab)$. Thus $f(ab) = f(ba)$ and so S is fuzzy B^*-normal. Thus (3) implies (2). ■

Theorem 4.4.20 *For a semigroup S, the following conditions are equivalent.*

(1) S is a semilattice of groups.

(2) S is regular and fuzzy Q^*-normal.

(3) S is right regular and fuzzy Q^*-normal.

(4) S is left regular and fuzzy Q^*-normal.

(5) S is intra-regular and fuzzy Q^*-normal.

(6) S is completely regular and fuzzy Q^*-normal.

(7) S is regular and fuzzy B^*-normal.

(8) S is right regular and fuzzy B^*-normal.

(9) S is left regular and fuzzy B^*-normal.

(10) S is intra-regular and fuzzy B^*-normal.

(11) S is completely regular and fuzzy B^*-normal.

(12) S is completely regular and for every elements a and b of S there exists a positive integer n such that $(ab)^n \in baSba$.

Proof. It follows from Theorem 4.4.8 and Theorem 4.4.18 that (1) and (6) are equivalent. And it follows from Theorem 4.4.8 and Theorem 4.4.17

that (1), (7) and (11) are equivalent. It follows from Theorem 4.4.17 and Theorem 4.4.18 that (1) and (12) are equivalent. ■

4.5 Convexity and Green's Relations

The results in the remainder of the chapter are essentially from [93]. In the remaining sections of this chapter, we describe the convex structure of the set of all fuzzy ideals of a semigroup and investigate some connections between fuzzy ideals and Green's relations.

We begin by recalling some definitions. For the basics of semigroup theory and definitions of the concepts used in our examples, the reader is referred to [41] or [107].

Example 4.5.1 *A function from a semilattice L into $[0,1]$ is a fuzzy ideal of L if and only if it is order reversing.*

Example 4.5.2 *Let S be an inverse semigroup. Then by definition [41, p. 129] for each $a \in S$ there is a unique $a^* \in S$ such that $aa^* = a$ and $a^*aa^* = a^*$. Let E be the set of idempotents of S, ordered by the relation*

$$a \leq b \Leftrightarrow ab = a,$$

and let f be a fuzzy subset of S. Then f is a fuzzy right ideal if and only if the restriction of f to E is order reversing and $f(a) = f(aa^*)$ for all $a \in S$.

To prove this, first suppose that f is a fuzzy right ideal. Then, for all $a \in S$, $f(aa^*) \geq f(a) = f(aa^*a) \geq f(aa^*)$, and for any $p, q \in E$,

$$p \leq q \Rightarrow p = qp \Rightarrow f(p) = f(qp) \geq f(q).$$

Conversely, suppose that f is order reversing and that $f(a) = f(aa^*)$ for all $a \in S$. Then for any $x, y \in S$, we have $xyy^*x^* \leq xx^*$ in E and so

$$f(xy) = f(xyy^*x^*) \geq f(xx^*) = f(x).$$

Example 4.5.3 *Let S be a semilattice of groups $\{G_r \mid r \in L\}$ [41, p. 89] and for each r in the semilattice L, let 1_r be the identity of the group G_r. Then by definition, S is the disjoint union of the groups $G_r (r \in L)$ and for each r, s in the semilattice L, we have $G_rG_s \subseteq G_{rs}$. If f is a fuzzy subset of S, then the following conditions are equivalent.*

(1) f is a fuzzy right ideal.

(2) f is a fuzzy left ideal.

(3) f is constant on each G_r and $r \leq s \Rightarrow f(1_r) \geq f(1_s)$ for all $r, s \in L$.

We prove the equivalence of (1) and (3). The remaining equivalences follow from a similar argument. First suppose that f is a fuzzy right ideal. If $x \in G_r$ and x^{-1} is the inverse of x in G_r then

$$f(1_r) \leq f(1_rx) = f(x) \leq f(xx^{-1}) = f(1_r),$$

so f is constant on G_r. Suppose that $r \leq s$. Then $G_r = G_{sr}$. It is easily verified that $G_s G_r \subseteq G_{sr}$ and therefore $1_s 1_r \in G_r$. Since we have just shown that f is constant on G_r, we have $f(1_r) = f(1_s 1_r) \geq f(1_s)$.

Conversely suppose f satisfies (3) and let $x, y \in S$. Then $x \in G_r$ and $y \in G_s$ for some $r, s \in L$. Now $xy \in G_{rs}$ so $f(xy) = f(1_{rs}) \geq f(1_r) = f(x)$. Hence f is a fuzzy right ideal.

4.6 The Compact Convex Set of Fuzzy Ideals

We now show that an arbitrary fuzzy ideal can be approximated by a suitable linear combination of characteristic functions of ideals. To make this statement precise, we provide the set F of all fuzzy ideals of a semigroup S with two topologies. The topology of pointwise convergence on F is the subspace topology on F induced by the product topology on $\mathcal{FP}(S)$. The norm topology on F is the subspace topology induced on F by the Banach space $l^\infty(S)$ of all bounded real valued functions on S, equipped with the supremum norm defined by

$$\|f\| = \vee\{|f(x)| \mid x \in S\} \qquad \forall f \in l^\infty(S).$$

Proposition 4.6.1 *Let F be the set of all fuzzy right ideals of a semigroup S. Then F is a compact convex set in the topology of pointwise convergence and the extreme points of F are the characteristic functions of the right ideals of S together with the zero function. If f is a fuzzy right ideal of S, then for every $\varepsilon > 0$ there is a convex combination g of extreme points of F with $0 \leq f(x) - g(x) \leq \varepsilon$ for all $x \in S$.*

Proof. It is easy to see that F is a closed convex set in the topology of pointwise convergence and since $\mathcal{FP}(S)$ is compact, it follows that F is compact.

Let R be either a right ideal of S or the empty set and suppose that

$$C_R = t\phi + (1-t)\psi,$$

for some $\phi, \psi, \in F$ and some $t \in (0,1)$. Then

$$\begin{aligned} x \in R \quad &\Rightarrow \quad 1 = t\phi(x) + (1-t)\psi(x) \\ &\Rightarrow \quad \phi(x) = 1 = \psi(x) \end{aligned}$$

and

$$\begin{aligned} x \in S/R \quad &\Rightarrow \quad 0 = t\phi(x) + (1-t)\psi(x) \\ &\Rightarrow \quad \phi(x) = 0 = \psi(x), \end{aligned}$$

so $\phi = C_R = \psi$ showing that C_R is an extreme point of F. It will now be shown that every extreme point of F has this form. Let $f \in F$ and suppose that f takes some value $t \in (0,1)$. Then it is easily verified that the maps ϕ and ψ from S to $[0,1]$ defined by

$$\phi(x) = \begin{cases} f(x)/t & \text{if } f(x) < t, \\ 1 & \text{otherwise,} \end{cases}$$

$$\psi(x) = \begin{cases} (f(x) - t)/(1-t) & \text{if } f(x) \geq t, \\ 0 & \text{otherwise,} \end{cases}$$

are distinct fuzzy right ideals with $f = t\phi + (1-t)\psi$. Hence an extreme point e of F has range contained in $\{0,1\}$ and so e is the characteristic function of $\{x \in S \mid e(x) = 1\}$ which is clearly a right ideal or the empty set.

To prove the final assertion of the theorem, let $f \in F$ and $\varepsilon > 0$. Choose a partition $0 = t_0 < t_1 < \ldots < t_n = 1$ of the interval $[0,1]$ with $t_{i+1} - t < \varepsilon$ for each i. Then each of the sets R_i defined by

$$R_i = \{x \in S \mid f(x) \geq t_i\}$$

is either empty or a right ideal, and

$$R_n \subseteq R_{n-1} \subseteq \ldots \subseteq R_1 \subseteq R_0 = S.$$

Let g be the fuzzy right ideal defined by

$$g = \sum_{i=1}^{n} (t_i - t_{i-1}) C_{R_i}.$$

Then it follows that $0 \leq f(x) - g(x) \leq \varepsilon$ for all $x \in S$. ■

Corollary 4.6.2 *The set of all fuzzy right ideals of a semigroup is the closed convex hull of its extreme points in the supremum norm.*

It is easy to verify that a similar proposition and corollary hold for fuzzy left ideals and for fuzzy ideals.

4.7 Fuzzy Ideals and Green's Relations

We begin this section by introducing some notation and reviewing the definitions of Green's relations (or equivalences). Further details may be found in [41, 107].

Let S be a semigroup. Then for every $a \in S$, we define

$$aS^1 = \{a\} \cup \{ax \mid x \in S\};$$

$$S^1a = \{a\} \cup \{xa \mid x \in S\};$$

$$S^1aS^1 = S^1a \cup aS^1 \cup \{xay \mid x, y \in S\}.$$

Green's relations are the equivalence relations R, L, J, H, and D defined for all $a, b \in S$ by

$$aRb \Longleftrightarrow aS^1 = bS^1,$$

$$aLb \Longleftrightarrow S^1a = S^1b,$$

$$aJb \Longleftrightarrow S^1aS^1 = S^1bS^1,$$

$$H = L \cap R \text{ and } D = L \vee R.$$

We also consider the relations $R_{\leq}$, $L_{\leq}$, and $J_{\leq}$ defined for all $a, b \in S$ by

$$aR_{\leq}b \Longleftrightarrow aS^1 \subseteq bS^1.$$

$$aL_{\leq}b \Longleftrightarrow S^1a \subseteq S^1b,$$

$$aJ_{\leq}b \Longleftrightarrow S^1aS^1 \subseteq S^1bS^1$$

(see [38, page 29]).

Example 4.7.1 *Let M be a monoid with identity 1 and define*

$$\mathcal{R} = \{x \in M \mid xy = 1 \text{ for some } y \in M\},$$

$$\mathcal{L} = \{y \in M \mid xy = 1 \text{ for some } x \in M\}.$$

If M is left cancellative, then for each $a \in M$ the set $a\mathcal{R}$ is the R-class containing a. If M is right canellative, then for each $a \in M$ the set $\mathcal{L}a$ is the L-class containing a.

To prove this, suppose that M is left cancellative and let $a, x \in M$. If $x \in a\mathcal{R}$, then there are $g, h \in M$ with $x = ag$ and $gh = 1$ and consequently $a = xh$. It follows that xRa. Conversely suppose that xRa. Then there are $u, v \in M$ with $x = au$ and $a = xv$. Hence $a = auv$ and by the left cancellation law, we have $uv = 1$ so $u \in \mathcal{R}$, giving $x \in a\mathcal{R}$. A similar argument can be used with the case where M is right cancellative.

The following is easily proved.

Proposition 4.7.2 *Let f be a fuzzy subset of the semigroup S. Then f is a fuzzy right ideal if and only if for every $x, y \in S$*

$$xR_{\leq}y \Rightarrow f(x) \geq f(y).$$

If f is a fuzzy right ideal, then f is constant on the R-classes.

If S is a semigroup and $a \in S$, then aS^1 is the smallest right ideal of S containing a. Right ideals of this form are called **principal right ideals.**

Proposition 4.7.3 *If a and b are elements of a semigroup S, then the following assertions are equivalent.*

(1) $aR_{\leq}b$.
(2) $a \in bS^1$.
(3) $C_B(a) = 1$, where $B = bS^1$.
(4) $C_I(a) \geq C_I(b)$ for all principal right ideals I of S.
(5) $C_I(a) \geq C_I(b)$ for all right ideals I of S.
(6) $f(a) \geq f(b)$ for all fuzzy right ideals f of S.

Proof. Clearly, (6) $\Rightarrow$ (5) $\Rightarrow$ (4) $\Rightarrow$ (3) $\Rightarrow$ (2) $\Leftrightarrow$ (1).

First, we prove (2) $\Rightarrow$ (5). Suppose $a \in bS^1$ and let I be a right ideal of S. If $C_I(b) = 1$, then $b \in I$ so $a \in bS^1 \subseteq I$ giving $C_I(a) = 1$. Hence $C_I(a) \geq C_I(b)$.

It is now sufficient to show that (5) $\Rightarrow$ (6). Assume that (5) holds. Then clearly $f(a) \geq f(b)$ for any fuzzy right ideal f which is a convex combination of characteristic functions of right ideals. Let g be any fuzzy right ideal of S and let $\varepsilon \geq 0$. Then, by Proposition 4.6.1, there is a fuzzy right ideal f with $f(a) \geq f(b)$ and

$$0 \leq g(x) - f(x) \leq \varepsilon$$

for all $x \in S$.

It follows that $g(a) \geq g(b) - \varepsilon$ for all $\varepsilon > 0$ and so $g(a) \geq g(b)$. ■

The following corollaries are routine consequences of Proposition 4.7.3.

Corollary 4.7.4 *If a and b are elements of a semigroup S, then the following assertions are equivalent.*

(1) aRb.
(2) $C_A(b) = 1 = C_B(a)$ where $A = aS^1$ and $B = bS^1$.
(3) $C_I(a) = C_I(b)$ for all principal right ideals I of S.
(4) $C_I(a) = C_I(b)$ for all right ideals I of S.
(5) $f(a) = f(b)$ for all fuzzy right ideals f of S.

We say that a set $\mathcal{F}$ of functions on a semigroup S is **separating** if, for all distinct x and y in S, there is an $\alpha \in \mathcal{F}$ with $\alpha(x) \neq \alpha(y)$.

Corollary 4.7.5 *Let S be a semigroup. Then the following conditions are equivalent.*

(1) The relation $R_{\leq}$ is an order on S.

(2) Each R-class is a singleton.

(3) The set of all characteristic functions of principal right ideals of S is separating.

(4) The set of all characteristic functions of right ideals of S is separating.

(5) The set of all fuzzy right ideals of S is separating.

There are corresponding propositions and corollaries connecting the relation L with fuzzy left ideals and connecting the relation J with fuzzy ideals. We summarize a few of these results in the next proposition.

Proposition 4.7.6 *Let a and b be elements of a semigroup S. Then*

(1) aJb if and only if $f(a) = f(b)$ for all fuzzy ideals f of S;

(2) aDb if and only if there is an $x \in S$ with $g(a) = g(x)$ for all fuzzy right ideals g of S and $h(x) = h(b)$ for all fuzzy left ideals h of S;

(3) aHb if and only if $f(a) = f(b)$ whenever f is a fuzzy left ideal or a fuzzy right ideal of S.

Proof. To prove (2), use the fact that $D = R \circ L$ (see [41, p. 39]). Everything else is a consequence of the remarks above. ∎

Example 4.7.7 *The fuzzy right ideals of a left cancellative monoid are separating if and only if e (the identity) is the only element having a right inverse. (This follows from Corollary 4.7.5 and Proposition* 4.7.6.

Example 4.7.8 *Let S be a cancellative semigroup without identity. Then it is known that the R-classes of S are singletons and that there is a particular cancellative semigroup S without identity for which there is only one J-class (see [41, Exercise 1, p. 53]). It follows from Corollary 4.7.5 that the fuzzy right ideals of S are separating, while Proposition 4.7.6(1), implies that the fuzzy ideals of S need not be separating. If S is a free semigroup, then it is easily verified that the J-classes of S are singletons and it follows that, in this particular case, the fuzzy ideals of S are separating.*

5

Fuzzy Congruences on Semigroups

In this chapter, we introduce the notion of a fuzzy congruence relation μ on a group and a semigroup S, and a factor semigroup S/μ induced by the congruence μ. Moreover, we give homomorphism theorems for fuzzy semigroups. Finally, we consider some congruences, for example, an idempotent-separating fuzzy congruence, a group fuzzy congruence on an inverse semigroup, and a fuzzy Rees congruence.

5.1 Fuzzy Congruences on a Semigroup

Definition 5.1.1 *Let S be a semigroup. A function μ from $S \times S$ to the unit interval $[0,1]$ is called a **fuzzy relation** on S. Let μ and ν be two fuzzy relations on S. Then the **product** $\mu \circ \nu$ of μ and ν is defined by*

$$\mu \circ \nu(a, b) = \vee_{x \in S}\{\mu(a, x) \wedge \nu(x, b)\}$$

for all $a, b \in S$.

Definition 5.1.2 *A fuzzy relation μ on S is called a **fuzzy equivalence relation** on S if*

*(1) (**fuzzy reflexive**) $\mu(a, a) = 1$ for all $a \in S$,*
*(2) (**fuzzy symmetric**) $\mu(a, b) = \mu(b, a)$ for all $a, b \in S$,*
*(3) (**fuzzy transitive**) $\mu \circ \mu \subseteq \mu$.*

The concept of the fuzzy equivalence relation on a semigroup S, which is called a **similarity relation** by Zadeh [159], is an extension of the concept of an equivalence relation on S.

Definition 5.1.3 *A fuzzy relation μ on S is called* ***compatible*** *if*

$$\mu(ax, bs) \geq \mu(a,b) \qquad \text{and} \qquad \mu(xa, xb) \geq \mu(a,b)$$

for all $a, b, x \in S$. A fuzzy equivalence relation on a semigroup S which is compatible is called a ***fuzzy congruence relation*** *on S.*

Theorem 5.1.4 *Let μ and ν be fuzzy compatible relations on a semigroup S. Then the product $\mu \circ \nu$ is a fuzzy compatible relation on S.*

Proof. Let a, b, x be any elements of S. Then

$$\begin{aligned}(\mu, \nu)(xa, xb) &= \vee_{y \in S}(\mu(xa, y) \wedge \nu(y, xb)) \\ &\geq \mu(xa, xy) \wedge \nu(xy, xb) \\ &\geq \mu(a, y) \wedge \nu(y, b).\end{aligned}$$

Hence

$$\begin{aligned}(\mu \circ \nu)(xa, xb) &\geq \vee_{y \in S}(\mu(a, y) \wedge \nu(y, b)) \\ &= (\mu \circ \nu)(a, b).\end{aligned}$$

Similarly, we have

$$(\mu \circ \nu)(ax, xb) \geq (\mu \circ \nu)(a, b).$$

Thus we obtain that the product $\mu \circ \nu$ is a fuzzy compatible relation on S. ■

Theorem 5.1.5 *Let μ and ν be fuzzy congruences on a semigroup S. Then the following conditions are equivalent.*

(1) $\mu \circ \nu$ is a fuzzy congruence.
(2) $\mu \circ \nu$ is a fuzzy equivalence.
(3) $\mu \circ \nu$ is fuzzy symmetric.
(4) $\mu \circ \nu = \nu \circ \mu$.

Proof. It is clear that (1) $\Rightarrow$ (2) $\Rightarrow$ (3). Assume that (3) holds. Let a, b be any elements of S. Since μ and ν are fuzzy symmetric, we have

$$\begin{aligned}(\mu \circ \nu)(a, b) &= \vee_{x \in S}(\mu(a, x) \wedge \nu(x, b)) \\ &= \vee_{x \in S}(\nu(x, b) \wedge \mu(a, x)) \\ &= \vee_{x \in S}(\nu(b, x) \wedge \mu(x, a)) \\ &= (\nu \circ \mu)(b, a) \\ &= (\nu \circ \mu)(a, b),\end{aligned}$$

and so $\mu \circ \nu = \nu \circ \mu$. Thus (3) implies (4). Assume that (4) holds. We now prove that (1) holds. Let a be any element of S. Since μ and ν are fuzzy reflexive,

$$\begin{aligned}(\mu \circ \nu)(a, a) &= \vee_{x \in S}(\mu(a, x) \wedge \nu(x, a)) \\ &\geq \mu(a, a) \wedge \nu(a.a) \\ &= 1 \wedge 1 \\ &= 1,\end{aligned}$$

and so $(\mu \circ \nu)(a, a) = 1$. Hence $\mu \circ \nu$ is fuzzy reflexive. In order to see that $\mu \circ \nu$ is fuzzy symmetric, let a and b be any elements of S. Then, since μ and ν are fuzzy symmetric, we have

$$\begin{aligned}(\mu \circ \nu)(a, b) &= (\nu \circ \mu)(a, b) \\ &= \vee_{z \in S}(\nu(a, z) \wedge \mu(z, b)) \\ &= \vee_{z \in S}(\mu(z, b) \wedge \nu(a, z)) \\ &= \vee_{z \in S}(\mu(b, z) \wedge \nu(z, a)) \\ &= (\mu \circ \nu)(b, a).\end{aligned}$$

Thus $\mu \circ \nu = \nu \circ \mu$ is fuzzy symmetric. Since μ and ν are fuzzy transitive, we have

$$\begin{aligned}(\mu \circ \nu) \circ (\mu \circ \nu) &= \{\mu \circ (\nu \circ \mu)\} \circ \nu \\ &= \{\mu \circ (\mu \circ \nu)\} \circ \nu \\ &= (\mu \circ \mu) \circ (\nu \circ \nu) \\ &\subseteq \mu \circ \nu,\end{aligned}$$

and so $\mu \circ \nu$ is fuzzy transitive. Therefore, $\mu \circ \nu$ is a fuzzy equivalence on S. Since μ and ν are fuzzy congruences on S, it follows from Theorem 5.1.4 that $\mu \circ \nu$ is a fuzzy congruence on S. Hence (1) holds. ■

5.2 Fuzzy Congruences on a Group

Let G be a group with identity e. We recall that a fuzzy subset f of G is called a **fuzzy subgroup** of G if

(1) $f(ab) \geq f(a) \wedge f(b)$ for all $a, b \in G$.
(2) $f(a^{-1}) \geq f(a)$ for all $a \in G$.
(3) $f(e) = 1$.

We note that since

$$f(a^{-1}) \geq f(a) \geq f((a^{-1})^{-1}) \geq f(a^{-1}),$$

$$f(a^{-1}) = f(a).$$

Thus the above condition (2) is equivalent to the condition,
(2′) $f(a^{-1}) = f(a)$ for all $a \in G$.

A fuzzy subgroup f of G is called a **fuzzy normal subgroup** of G if
(4) $f(ab) = f(ba)$ for all $a, b \in G$.

Let f be a fuzzy normal subgroup of a group G. Let a, b be any elements of G. Then

$$f(a^{-1}ba) = f((a^{-1}(ba)) = f((ba)a^{-1}) = f(b(aa^{-1}) = f(be) = f(b).$$

Conversely, let f be a fuzzy subgroup of G such that $f(b) = f(a^{-1}ba)$ for all $a, b \in G$. Then we have

$$f(ab) = f(a^{-1}(ab)a) = f((a^{-1}a)(ba)) = f(e(ba)) = f(ba).$$

Thus the condition (4) is equivalent to the condition,
(4′) $f(b) = f(a^{-1}ba)$ for all $a, b \in G$.

Theorem 5.2.1 *Let μ be a fuzzy congruence on a group G. Then*

$$\mu(a^{-1}, b^{-1}) = \mu(a, b)$$

for all $a, b \in G$.

Proof. Let a and b be any elements of G. Then $\mu(a^{-1}, b^{-1}) = \mu(ea^{-1}, b^{-1}e) = \mu(b^{-1}ba^{-1}, b^{-1}aa^{-1}) \geq \mu(b, a) = \mu(a, b) = \mu(ea, be) = \mu(bb^{-1}a, ba^{-1}a) \geq \mu(b^{-1}, a^{-1}) = \mu(a^{-1}, b^{-1})$ and so $\mu(a^{-1}, b^{-1}) = \mu(a, b)$. ■

Theorem 5.2.2 *Let f be a fuzzy normal subgroup of a group G. Then the fuzzy relation μ_f defined by $\mu_f(a, b) = f(ab^{-1})$ for all $a, b \in G$ is a fuzzy congruence on G.*

Proof. Let a be any element of G. Then since

$$\mu_f(a, a) = f(aa^{-1}) = f(e) = 1,$$

μ_f is fuzzy reflexive. Let a and b be any elements of G. Then

$$\mu_f(a, b) = f(ab^{-1}) = f((a^{-1})^{-1}) = f(ba^{-1}) = \mu_f(b, a).$$

Hence μ_f is fuzzy symmetric. To see that μ_f is fuzzy transitive, let a and b be any elements of G. Thus

$$\begin{aligned}
(\mu_f \circ \mu_f)(a, b) &= \vee_{x \in G}\{\mu_f(a, x) \wedge \mu_f(x, b)\} \\
&= \vee_{x \in G}\{f(ax^{-1}) \wedge f(xb^{-1})\} \\
&\leq \vee_{x \in G} f((ax^{-1})(xb^{-1}) \\
&= f(a(xx^{-1})b^{-1}) \\
&= f(ab^{-1}) \\
&= \mu_f(a, b),
\end{aligned}$$

and so we have $\mu_f \circ \mu_f \subseteq \mu_f$. Hence μ_f is fuzzy transitive and so μ_f is a fuzzy equivalence on G. To see that μ_f is a fuzzy congruence, let a, b, x be any elements of G. Then, since f is a fuzzy normal, we have $\mu_f(xa, xb) = f((xa)(xb)^{-1}) = f((xb)^{-1}(xa)) = f(b^{-1}x^{-1})(xa)) = f(b^{-1}(x^{-1}x)a) = f(b^{-1}ea) = f(b^{-1}a) = f(ab^{-1}) = \mu_f(a, b)$. Similarly,

$$\mu_f(ax, bx) = \mu_f(a, b).$$

Therefore, μ_f is a congruence on G. ■

Theorem 5.2.3 *Let μ be a fuzzy congruence on a group G. Then the fuzzy subset f_μ of G is defined by*

$$f_\mu(a) = \mu(a, e)$$

for all $a \in G$ is a fuzzy normal subgroup of G.

Proof. Let a and b be any elements of G. Then, since μ is fuzzy transitive, we have

$$\begin{aligned} f_\mu(ab) &= \mu(ab, e) \\ &= \mu(ab, b^{-1}b) \\ &\geq \mu(a, b^{-1}) \\ &\geq (\mu \circ \mu)(a, b^{-1}) \\ &= \vee_{x \in G}\{\mu(a, x) \wedge \mu(x, b^{-1})\} \\ &\geq \mu(a, e) \wedge \mu(e, b^{-1})) \\ &= \mu(a, e) \wedge \mu(b^{-1}b, b^{-1}e)) \\ &\geq \mu(a, e) \wedge \mu(b, e) \\ &= f_\mu(a) \wedge f_\mu(b). \end{aligned}$$

Since μ is fuzzy symmetric, we have

$$f_\mu(a^{-1}) = \mu(a^{-1}, e) = \mu(a^{-1}e, a^{-1}a) \geq \mu(e, a) = \mu(a, e) = f_\mu(a).$$

Then, since

$$f_\mu(e) = \mu(e, e) = 1,$$

f_μ is a fuzzy subgroup of G. Now

$$\begin{aligned} f_\mu(ab) &= \mu(ab, e) \\ &= \mu(e(ab), e(b^{-1}b)) \\ &= \mu((b^{-1}b)(ab), b^{-1}eb) \\ &= \mu(b^{-1}(ba)b, b^{-1}eb) \\ &\geq \mu(ba, e) \\ &= f_\mu(ba). \end{aligned}$$

Similarly, we have $f_\mu(ba) \geq f_\mu(ab)$. Thus $f_\mu(ab) = f_\mu(ba)$ and so f_μ is a fuzzy normal subgroup of G. ∎

Theorem 5.2.4 *For fuzzy normal subgroups f and g of a group G, $\mu_f \circ \mu_g = \mu_{fog}$.*

Proof. Let a and b be any elements of G. Then for any element z of G, we put $az^{-1} = x$ and $zb^{-1} = y$. Hence

$$xy = (az^{-1})(zb^{-1}) = a(z^{-1}z)b^{-1} = aeb^{-1} = ab^{-1}.$$

Thus

$$\begin{aligned} (\mu_f \circ \mu_g)(a, b) &= \vee_{z \in G}\{\mu_f(a, z) \wedge \mu_g(z, b)\} \\ &= \vee_{z \in G}\{f(az^{-1}) \wedge g(zb^{-1})\} \\ &= \vee_{ab^{-1} = xy}\{f(x) \wedge g(y)\} \\ &= (f \circ g)(ab^{-1}) \\ &= \mu_{fog}(a, b), \end{aligned}$$

and so $\mu_f \circ \mu_g = \mu_{fog}$. ∎

Theorem 5.2.5 *The set $FC(G)$ of all fuzzy congruences on a group G is a semilattice under the multiplication of fuzzy relations.*

Proof. Let μ and ν be any fuzzy congruences on G and a, b any elements of G. Then

$$\begin{aligned}(\mu \circ \nu)(a,b) &= \vee_{z\in G}\{\mu(a,z) \wedge \nu(z,b)\}\\ &= \vee_{z\in G}\{\nu(z,b) \wedge \mu(a,z)\}\\ &= \vee_{z\in G}\{\nu(ze,eb) \wedge \mu(ae,ez)\}\\ &= \vee_{z\in G}\{\nu(ze,zz^{-1}b) \wedge \mu(az^{-1}z,ez)\}\\ &\geq \vee_{z\in G}\{\nu(e,z^{-1}b) \wedge \mu(az^{-1},e)\}\\ &= \vee_{z\in G}\{\nu(a^{-1}a,a^{-1}az^{-1}b) \wedge \mu(az^{-1}bb^{-1},bb^{-1})\}\\ &\geq \vee_{az^{-1}b\in G}\{\nu(a,az^{-1}b) \wedge \mu(az^{-1}b,b)\}\\ &= (\nu \circ \mu)(a,b),\end{aligned}$$

and so we have $\mu \circ \nu \supseteq \nu \circ \mu$. Similarly, $\nu \circ \mu \supseteq \mu \circ \nu$. Thus

$$\mu \circ \nu = \nu \circ \mu.$$

Hence it follows from Theorem 5.1.5 that $\mu \circ \nu$ is a fuzzy congruence on G, that is, $\mu \circ \nu \in FC(G)$.

On the other hand,

$$\begin{aligned}(\mu \circ \mu)(a,b) &= \vee_{z\in G}\{\mu(a,z) \wedge \mu(z,b)\}\\ &\geq \mu(a,a) \wedge \mu(a,b)\\ &= 1 \wedge \mu(a,b)\\ &= \mu(a,b),\end{aligned}$$

and so $\mu \circ \mu \supseteq \mu$. Since μ is fuzzy transitive, the converse inclusion holds. Thus $\mu \circ \mu = \mu$. Therefore, $FC(G)$ is a semilattice. ■

Theorem 5.2.6 *For fuzzy subgroups f and g of a group G, the following conditions are equivalent.*

(1) $f \circ g$ is a fuzzy subgroup of G.

(2) $f \circ g = g \circ f$.

Proof. First assume that (1) holds. Let a be any element of G. Then

$$\begin{aligned}(f \circ g)(a) &= \vee_{a=xy}\{f(x) \wedge g(y)\}\\ &= \vee_{a^{-1}=y^{-1}x^{-1}}\{g(y^{-1}) \wedge f(x^{-1})\}\\ &= (g \circ f)(a^{-1})\\ &= (g \circ f)(a),\end{aligned}$$

and so $f \circ g = g \circ f$. Thus (1) implies (2).

Assume that (2) holds. Then

$$\begin{aligned}(f \circ g) \circ (f \circ g) &= \{f \circ (g \circ f)\} \circ g\\ &= \{f \circ (f \circ g)\} \circ g\\ &= (f \circ f) \circ (g \circ g)\\ &\subseteq f \circ g.\end{aligned}$$

Let a be any element of S. Then

$$\begin{aligned}(f\circ g)(a^{-1}) &= \vee_{a^{-1}=xy}\{f(x)\wedge g(y)\}\\ &= \vee_{a=y^{-1}x^{-1}}\{g(y^{-1})\wedge f(x^{-1})\}\\ &= (g\circ f)(a)\\ &= (f\circ g)(a).\end{aligned}$$

On the other hand,

$$\begin{aligned}(f\circ g)(e) &= \vee_{e=xy}\{f(x)\wedge g(y)\}\\ &\geq f(e)\wedge g(e)\\ &= 1\wedge 1\\ &= 1,\end{aligned}$$

and so $(f\circ g)(e)=1$. Thus we obtain that $f\circ g$ is a fuzzy subgroup of G, and so (2) implies (1). ■

Theorem 5.2.7 *Let f be a fuzzy subgroup of a group G, and g a fuzzy normal subgroup of G. Then $f\circ g=g\circ f$, that is, $f\circ g$ is a fuzzy subgroup of G.*

Proof. Let a be any element of G. Then, since g is fuzzy normal, we have

$$\begin{aligned}(f\circ g)(a) &= \vee_{a=xy}\{f(x)\wedge g(y)\}\\ &= \vee_{a=xy}\{f(x)\wedge g(x^{-1}a)\}\\ &= \vee_{a=xy}\{g(ax^{-1})\wedge f(x)\}\\ &= (g\circ f)(a),\end{aligned}$$

and so $f\circ g=g\circ f$. Hence it follows from Theorem 5.2.6 that $f\circ g$ is a fuzzy subgroup of G. ■

Theorem 5.2.8 *Let f and g be fuzzy normal subgroups of a group G. Then $f\circ g$ is a fuzzy normal subgroup of G.*

Proof. It follows from Theorem 5.2.7 that $f\circ g$ is a fuzzy subgroup of G. Let a and b be any elements of G. Then, there exist elements x and y in G such that $ab=xy$. Then, since $b=a^{-1}xy$, we have

$$ba=(a^{-1}xy)a=(a^{-1}xey)a=(a^{-1}x(aa^{-1})a)a=(a^{-1}xa)(a^{-1}ya).$$

Hence since f and g are fuzzy normal, we have

$$\begin{aligned}(f\circ g)(ab) &= \vee_{ab=xy}\{f(x)\wedge g(y)\}\\ &= \vee_{ba=(a^{-1}xa)(a^{-1}ya)}\{f(a^{-1}xa)\wedge g(a^{-1}ya)\}\\ &= (f\circ g)(ba).\end{aligned}$$

Thus $f\circ g$ is a fuzzy normal subgroup of G. ■

Theorem 5.2.9 *The set $FN(G)$ of all fuzzy normal subgroups of a group G is a semilattice under the multiplication of fuzzy subsets.*

Proof. Let f be a fuzzy normal subgroup of G. Then for any element a of G,

$$\begin{aligned}(f \circ f)(a) &= \vee_{a=xy}\{f(x) \wedge f(y)\} \\ &\geq f(a) \wedge f(e) \\ &= f(a),\end{aligned}$$

and so $f \circ f \supseteq f$. Since $f \circ f \subseteq f$, we have $f \circ f = f$. On the other hand, it follows from Theorem 5.2.7 that $f \circ g = g \circ f$ for all $f, g \in FN(G)$. Therefore, $FN(G)$ is a semilattice. ∎

Theorem 5.2.10 *Let G be a group, $FN(G)$ the set of all fuzzy normal subgroups of G, and $FC(G)$ the set of all fuzzy congruences on G. Then there exists a one-to-one mapping ψ from $FN(G)$ onto $FC(G)$.*

Proof. We define a mapping $\psi : FN(G) \longrightarrow FC(G)$ as follows:

$$\psi(f) = \mu_f$$

for all $f \in FN(G)$. Then it follows from Theorem 5.2.2 that μ_f is a fuzzy congruence on G, that is,

$$\psi(f) = \mu_f \in FC(G).$$

If $f = g$, then for any elements a and b of G, we have

$$\mu_f(a, b) = f(ab^{-1}) = f(ab^{-1}) = \mu_g(a, b).$$

Thus

$$\psi(f) = \mu_f = \mu_g = \psi(g).$$

This implies that ψ is well-defined.

If $\psi(f) = \psi(g)$ $(f, g \in FN(G))$, then for every element a of G, $f(a) = f(ae) = f(ae^{-1}) = \mu_f(a, e) = \psi(f)(a, e) = \psi(g)(a, e) = \mu_g(a, e) = g(ae^{-1}) = g(ae) = g(a)$ and so $f = g$. Hence ψ is one-to-one.

Let ν be any fuzzy congruence on G. Then it follows from Theorem 5.2.3 that there exists a fuzzy normal subgroup f_ν of G such that

$$f_\nu(a) = \nu(a, e)$$

for all $a \in G$. Hence for this fuzzy normal subgroup f_μ of G, it follows from Theorem 5.2.2 that there exists a fuzzy congruence μ on G such that

$$\mu_{f_\nu}(a, b) = f_\nu(ab^{-1})$$

for all $a \in G$. Hence for all $a, b \in G$,

$$\mu_{f_\nu}(a,b) = f_\nu(ab^{-1)} = \nu(ab^{-1}, e) = \nu(a,b).$$

Thus

$$\psi(f_\nu) = \mu_{f_\nu} = \nu.$$

Hence ψ is onto. Therefore, we have that ψ is a one-to-one function from $FN(G)$ onto $FC(G)$. ∎

5.3 Fuzzy Factor Semigroups

Let S be a semigroup. We denote by C_R the characteristic function of a binary relation R on S. We let denote $Con(S)$ and $Con_f(S)$ denote the set of all congruences and the set of all fuzzy congruences on S, respectively. Then, as is easily seen, R is an equivalence relation on S if and only if C_R is a fuzzy equivalence relation on S.

Theorem 5.3.1 *Let R be a binary relation on a semigroup S. Then R is a congruence on S if and only if C_R is a fuzzy congruence on S.*

Proof. First assume that R is a congruence on S. Then, as is stated above, C_R is a fuzzy equivalence on S. Let a, b and x be any elements of S. Let $(a,b) \in R$. Since R is a congruence on S, we have $(ax, bx) \in R$ and $(xa, xb) \in R$. Thus

$$C_R(ax, bx) = 1 = C_R(a,b).$$

If $(a,b) \notin R$, then

$$C_R(ax, bx) \geq 0 = C_R(a,b)$$

and

$$C_R(xa, xb) \geq 0 = C_R(a,b).$$

Hence C_R is a fuzzy congruence on S.

Conversely, assume that C_R is a fuzzy congruence on S. Let $(a,b) \in R$. Then

$$C_R(ax, bx) \geq C_R(a,b) = 1$$

and

$$C_R(xa, xb) \geq C_R(a,b) = 1,$$

and so

$$C_R(ax, bx) = 1 \text{ and } C_R(xa, xb) = 1.$$

Thus we obtain that $(ax, bx) \in R$ and $(xa, xb) \in R$. Hence R is a congruence on S. ■

Let μ be a fuzzy equivalence on a semigroup S. For each $a \in S$, we define a fuzzy subset μ_a of S as follows:

$$\mu_a(x) = \mu(a, x)$$

for all $x \in S$.

Theorem 5.3.2 *Let μ be a fuzzy equivalence relation on a semigroup S and let $a, b \in S$. Then $\mu_a = \mu_b$ if and only if $\mu(a, b) = 1$.*

Proof. If $\mu_a = \mu_b$, then since μ is a fuzzy equivalence, we have

$$\mu(a, b) = \mu_a(b) = \mu_b(b) = \mu(b, b) = 1.$$

Conversely, assume that $\mu(a, b) = 1$. Then for any $x \in S$,

$$\begin{aligned} \mu_a(x) &= \mu(a, x) \\ &\geq (\mu \circ \mu)(a, x) \\ &= \vee_{y \in S}(\mu(a, y) \wedge \mu(y, x)) \\ &\geq \mu(a, b) \wedge \mu(b, x) \\ &= 1 \wedge \mu(b, x) \\ &= \mu(b, x) \\ &= \mu(x), \end{aligned}$$

and so $\mu_a \supseteq \mu_b$. By symmetry and since $\mu(b, a) = \mu(a, b) = 1$, we also have $\mu_a \subseteq \mu_b$. Thus $\mu_a = \mu_b$. ■

The fuzzy subset μ_a of a semigroup S is called a **fuzzy equivalence class** of μ containing $a \in S$.

Let μ be a fuzzy congruence on S. Then

$$S/\mu = \{\mu_a \mid a \in S\}.$$

Let $a, b \in S$ and let μ_a and μ_b be two fuzzy congruence classes of μ. Let $x \in S$. We set

$$S' = \{yb \mid y \in S, yz = x\}.$$

Then we have

$$\begin{aligned} (\mu_a \circ \mu_b)(x) &= \vee_{x=yz}(\mu_a(y) \wedge \mu_b(z)) \\ &= \vee_{x=yz}(\mu(a, y) \wedge \mu(b, z)) \\ &\leq \vee_{x=yz}(\mu(ab, yb) \wedge \mu(yb, yz)) \\ &= \vee_{t \in S'}(\mu(ab, t) \wedge \mu(t, x)) \\ &\leq \vee_{t \in S'}(\mu(ab, t) \wedge \mu(t, x)) \\ &= (\mu \circ \mu)(ab, x) \\ &\leq \mu(ab, x) \\ &= \mu_{ab}(x), \end{aligned}$$

and so

$$\mu_a \circ \mu_b \subseteq \mu_{ab}.$$

Define the binary relation $*$ on S/μ as follows:

$$\mu_a * \mu_b = \mu_{ab}$$

for all $\mu_a, \mu_b \in S/\mu$.

Theorem 5.3.3 *The binary relation $*$ on S/μ is well-defined.*

Proof. Assume that $\mu_a = \mu_b$ and $\mu_c = \mu_d$. Then by Theorem 5.3.2, we have

$$\mu(a,b) = \mu(c,d) = 1.$$

Thus

$$\begin{aligned}\mu(ac,bd) &\geq (\mu \circ \mu)(ac,bd)\\ &= \vee_{x \in S}(\mu(ac,x) \wedge \mu(x,bd))\\ &\geq \mu(ac,bc) \wedge \mu(bc,bd)\\ &\geq \mu(a,b) \wedge \mu(c,d)\\ &= 1 \wedge 1\\ &= 1,\end{aligned}$$

and so $\mu(ac,bd) = 1$. Thus by Theorem 5.3.2, we have

$$\mu_a * \mu_c = \mu_{ac} = \mu_{bd} = \mu_b * \mu_d.$$

Hence the binary relation $*$ is well-defined. ■

Theorem 5.3.4 *Let μ be a fuzzy congruence on a semigroup S. Then*

$$\mu^{-1}(1) = \{(a,b) \in S \times S \mid \mu(a,b) = 1\}$$

is a congruence on S.

Proof. $\mu^{-1}(1)$ is clearly reflexive and symmetric. To prove its transitivity, let $(a,b), (b,c) \in \mu^{-1}(1)$. Then, since $\mu(a,b) = \mu(b,c) = 1$, we have

$$\begin{aligned}\mu(a,c) &\geq (\mu \circ \mu)(a,c)\\ &= \vee_{x \in S}(\mu(a,x) \wedge \mu(x,c))\\ &\geq \mu(a,b) \wedge \mu(b,c))\\ &= 1 \wedge 1\\ &= 1,\end{aligned}$$

and so $(a,c) \in \mu^{-1}(1)$. Thus $\mu^{-1}(1)$ is an equivalence relation on S. Let $(a,b) \in \mu^{-1}(1)$, and $x \in S$. Then, since μ is a fuzzy congruence on S,

$$\mu(ax,bx) \geq \mu(a,b) = 1,$$

which implies that $\mu(ax,bx) = 1$, that is, $(ax,bx) \in \mu^{-1}(1)$. Similarly we can see that $(xa,xb) \in \mu^{-1}(1)$. Thus $\mu^{-1}(1)$ is a congruence on S. ■

5.4 Homomorphism Theorems

Let S and T be semigroups and f be a homomorphism of S into T. Then, as is well-known (see for example, J.M. Howie [41, p.22, Theorem 5.3]), the relation

$$Ker(f) = \{(a,b) \in S \times S \mid f(a) = f(b)\}$$

is a congruence on S. Then it follows from Theorem 5.3.1 that the characteristic function $C_{K_{er}(f)}$ of $Ker(f)$ is a fuzzy congruence on S. Now we set

$k(f) = C_{K_{er}(f)}$.

We call this the **fuzzy kernel** of f. It follows that

$$k(f)(a,b) = \begin{cases} 1, & \text{if } f(a) = f(b), \\ 0, & \text{if } f(a) \neq f(b). \end{cases}$$

Theorem 5.4.1 *If μ a fuzzy congruence on a semigroup S, then S/μ is a semigroup with respect to the binary operation $*$. The function*

$$\mu^\natural : S \longrightarrow S/\mu$$

defined by

$$\mu^\natural(a) = \mu_a$$

for all $a \in S$, is a homomorphism. If $f : S \longrightarrow T$ is a homomorphism, where S and T are semigroups, then the fuzzy relation $k(f)$ is a fuzzy congruence on S, and there is a monomorphism

$$g : S/k(f) \longrightarrow T$$

such that

$$g(k(f))^\natural = f.$$

Proof. Let $a, b \in S$. By the definition of $\mu^\natural$ and $*$, we have

$$\mu^\natural(ab) = \mu_{ab} = \mu_a * \mu_b = \mu^\natural(a) * \mu^\natural(b).$$

Now define $g : S/k(f) \longrightarrow T$ by

$$g((k(f))a) = f(a)$$

for all $a \in S$. Then g is well-defined.

In fact, assume that for any $a, b \in S$,

$$(k(f))_a = (k(f))_b.$$

Then it follows from Theorem 5.3.2 that

$$C_{K_{er}(f)}(a,b) = k(a,b) = 1,$$

and so $(a,b) \in Ker(f)$. Thus we have

$$g((k(f))_a) = f(a) = f(b) = g((k(f))_b).$$

Hence g is well-defined.

To see that g is one-to-one, assume that $f(a) = f(b)$. Then $(a,b) \in Ker(f)$. Thus we have

$$k(f)(a,b) = C_{Ker(f)}(a,b) = 1.$$

Hence it follows from Theorem 5.3.2 that

$$(k(f))_a = (k(f))_b.$$

Thus g is one-to-one.

To see that g is a homomorphism, let a, b be any elements of S. Then

$$\begin{aligned} g((k(f))_a) * (k(f))_b) &= g((k(f))_{ab}) \\ &= f(ab) \\ &= f(a)f(b) \\ &= g((k(f))_a)g((k(f))_b). \end{aligned}$$

Now let a be any element of S. Then we have

$$(g(k(f))^{\natural})(a) = g((k(f))_a) = f(a),$$

and so $g(k(f))^{\natural} = f$. ∎

Theorem 5.4.2 *Let μ and ν be fuzzy congruences on a semigroup S such that $\mu \subseteq \nu$. Then there is a unique homomorphism $g : S/\mu \longrightarrow S/\nu$ such that $\mu^{\natural}g = \nu^{\natural}$, and $(S/\mu)/k(g)$is isomorphic to S/ν.*

Proof. Define $g : S/\mu \longrightarrow S/\nu$ by $g(\mu_a) = \nu_a$ for all $a \in S$. We shall show that g is well-defined. Assume that $\mu_a = \mu_b$.Then it follows from Theorem 5.3.2 and the assumption $\mu \subseteq \nu$ that

$$1 = \mu(a,b) \geq \nu(a,b).$$

Hence $\nu(a,b) = 1$, that is, $\nu_a = \nu_b$. Therefore, g is well-defined. Next to see that g is homomorphism, let $a, b \in S$. Then

$$g(\mu_a * \mu_b) = g(\mu_{ab}) = \nu_{ab} = \nu_a * \nu_b = g(\mu_a) * g(\mu_b).$$

The remainder of the proof can be easily seen. ∎

5.5 Idempotent-separating Fuzzy Congruences

We recall that a semigroup S is regular if for each element a of S there exists an element x in S such that $a = axa$.

Theorem 5.5.1 *For a regular semigroup S, $\mu \in Con_f$ and $a \in S$, the following conditions are equivalent.*
(1) $\mu_a \in E(S/\mu)$.
(2) $\mu_a = \mu_e$ for some $e \in E(S)$ such that $Se \subseteq Sa$ and $eS \subseteq aS$.
(3) $\mu_a = \mu_e$ for some $e \in E(S)$.

Proof. First assume that (1) holds. Then

$$\mu_a = \mu_a \mu_a = \mu_{a^2}.$$

Let x be inverse of a^2 in S, that is, $a^2 = a^2 x a^2$ and $x = x a^2 x$. Set $e = axa$. Then we have

$$e^2 = (axa)(axa) = a(xa^2x)a = axa = e,$$

and so $e \in E(S)$. Since $\mu_a \in E(S/\mu)$, we have

$$\mu_e = \mu_{a^2} * \mu_x * \mu_{a^2} = \mu_{a^2xa^2} = \mu_{a^2} = (\mu_a)^2 = \mu_a.$$

On the other hand, we have

$$eS = (axa)S = a(xaS) \subseteq aS,$$

and

$$Se = S(axa) = (Sax)a \subseteq Sa.$$

Thus (1) implies (2). It is clear that (2) implies (3). Assume that (3) holds, that is, $\mu_a = \mu_e$ for some $e \in E(S)$. Then

$$\mu_a * \mu_a = \mu_e * \mu_e = \mu_{ee} = \mu_e = \mu_a,$$

and so μ_a is idempotent. Thus (3) implies (1). ∎

We say that a fuzzy congruence μ on a semigroup S is **idempotent-separating** if for all $e, f \in E(S)$ the equality $\mu_e = \mu_f$ implies $e = f$.

A semigroup S is called an **inverse semigroup** if every element a of S possesses a unique inverse, i.e., there exists a unique element a^{-1} of S such that

$$aa^{-1}a = a \quad \text{and} \quad a^{-1} = a^{-1}aa^{-1}.$$

The following is due to J.M. Howie [41, p.130,Theorem 1.2].

Lemma 5.5.2 *A semigroup S is an inverse semigroup if and only if S is regular and idempotents commute.*

The following is due to also J.M. Howie [41, p.141,Theorem 3.2].

Lemma 5.5.3 *If S is an inverse semigroup with semilattice of idempotents $E(S)$, then the relation*

$$\eta = \{(a,b) \in S \times S \mid a^{-1}esa = b^{-1}eb \text{ for all } e \in E(S)\}$$

is the greatest idempotent-separating congruence on S.

Theorem 5.5.4 *Let μ be a fuzzy congruence on an inverse semigroup S. Then S/μ is an inverse semigroup and $\mu(a^{-1}, b^{-1}) = \mu(a,b)$ for all $a, b \in S$.*

Proof. Let a be any element of S. Then, since S is regular, there exists an element x in S such that $a = axa$. Hence

$$\mu_a = \mu_{axa} = \mu_a * \mu_x * \mu_a.$$

Therefore, S/μ is regular. Let μ_a and μ_b $(a, b \in S)$ be any idempotents of S/μ. Then it follows from Theorem 5.5.1 that there exist idempotents e and f of S such that

$$\mu_a = \mu_e \qquad \text{and} \qquad \mu_b = \mu_f.$$

Since S is an inverse semigroup, by Lemma 5.5.2, we have $ef = fe$. Thus

$$\mu_a * \mu_b = \mu_e * \mu_f = \mu_{ef} = \mu_{fe} = \mu_f * \mu_e = \mu_b * \mu_a.$$

Hence S/μ is an inverse semigroup. Thus we have $(\mu_a)^{-1} = \mu_{a^{-1}}$.

Let $a, b \in S$. Hence $\mu(a^{-1}, b^{-1}) = \mu_{a^{-1}}(b^{-1}) = (\mu_{a^{-1}}(b))^{-1} = (\mu(a^{-1}, b))^{-1} = (\mu(b, a^{-1}))^{-1} = (\mu_b(a^{-1}))^{-1} = ((\mu_b(a))^{-1})^{-1} = \mu_b(a) = \mu(b,a) = \mu(a,b)$. ■

The following theorem can be easily seen.

Theorem 5.5.5 *Let S be an inverse semigroup. Then the fuzzy relation C_η is an idempotent-separating fuzzy congruence on S.*

Theorem 5.5.6 *Let μ be a fuzzy congruence on an inverse semigroup S. Then μ is idempotent-separating if and only if $\mu^{-1}(1) \subseteq \eta$.*

Proof. Assume that μ is an idempotent-separating congruence on S. Let $(a, b) \in \mu^{-1}(1)$. Then $\mu(a,b) = 1$. Let $e \in E(S)$. Thus since μ is a fuzzy congruence on S, we have

$$\begin{aligned}
\mu(a^{-1}ea, b^{-1}eb) &\geq (\mu \circ \mu)(a^{-1}ea, b^{-1}eb) \\
&= \vee_{x \in S}(\mu(a^{-1}ea, x) \wedge \mu(x, b^{-1}eb)) \\
&\geq \mu(a^{-1}ea, b^{-1}ea) \wedge \mu(b^{-1}ea, b^{-1}eb)) \\
&\geq \mu(a^{-1}, b^{-1}) \wedge \mu(a,b)) \\
&= 1 \wedge 1 \\
&= 1,
\end{aligned}$$

and so

$$\mu(a^{-1}ea, b^{-1}eb) = 1.$$

Hence it follows from Theorem 5.3.2 that

$$\mu_{a^{-1}ea} = \mu_{b^{-1}eb}.$$

Since μ is idempotent-separating and since $a^{-1}ea$ and $b^{-1}eb$ are both idempotents, we have

$$a^{-1}ea = b^{-1}eb,$$

and so $(a, b) \in \eta$. Therefore, $\mu^{-1}(1) \subseteq \eta$.

Conversely, assume that $\mu^{-1}(1) \subseteq \eta$. Let $\mu_e = \mu_f$ for some $e, f \in E(S)$. Then $\mu(e, f) = 1$, and $(e, f) \in \mu^{-1}(1) \subseteq \eta$. Since η is idempotent-separating, we have $e = f$. Thus μ is an idempotent-separating fuzzy congruence on S. ■

5.6 Group Fuzzy Congruences

A fuzzy congruence μ on a semigroup S is called a **group congruence** if S/μ is a group under the binary operation $*$. If S is an inverse semigroup S/μ is an inverse semigroup by Theorem 5.5.4. Since, as it may be easily proved, a group is an inverse semigroup having only one idempotent, μ is a group fuzzy congruence if and only if $\mu_e = \mu_f$ for all $e, f \in E(S)$.

The following is due to J.M. Howie [41, p.139,Theorem 3.1].

Lemma 5.6.1 *If S is an inverse semigroup with semilattice of idempotents $E(S)$, then the relation*

$$\sigma = \{(a, b) \in S \times S \mid ea = eb \text{ for some } e \in E(S)\}$$

is the least group congruence on S.

The following can be easily seen.

Theorem 5.6.2 *Let S be an inverse semigroup. Then C_σ is a group fuzzy congruence on S.*

Theorem 5.6.3 *Let S be an inverse semigroup and μ is a fuzzy congruence on S. Then μ is a group fuzzy congruence if and only if $\sigma \subseteq \mu^{-1}(1)$.*

Proof. Let μ be a group fuzzy congruence on S. Let $(a, b) \in \sigma$. Then $ea = eb$ for some $e \in E(S)$. Since s/μ is a group, we have

$$\mu_a = \mu_e * \mu_a = \mu_{ea} = \mu_{eb} = \mu_e * \mu_b = \mu_b.$$

Hence it follows from Theorem 5.3.2 that $\mu(a,b) = 1$ and so $(a,b) \in \mu^{-1}(1)$. Thus we have that $\sigma \subseteq \mu^{-1}(1)$.

Conversely, assume that $\sigma \subseteq \mu^{-1}(1)$. Let $e, f \in E(S)$. Then, since S is an inverse semigroup, $E(S)$ is commutative by Lemma 5.5.2. Thus we have $efe \in E(S)$ and also

$$(efe)e = (efe)f.$$

This implies that $(e,f) \in \sigma$. Hence from $\sigma \subseteq \mu^{-1}(1)$, we have $(e,f) \in \mu^{-1}(1)$ and so $\mu_e = \mu_f$. This implies that S/μ is a group, and so μ is a group fuzzy congruence on S. ■

5.7 The Lattice of Fuzzy Congruence Relations on a Semigroup

In this section, we examine a fuzzy equivalence relation generated by a given fuzzy relation on a set and the fuzzy congruence relation generated by a given fuzzy relation on a semigroup. We also study the lattice of fuzzy equivalence relations and fuzzy congruence relations on a semigroup. The results are from [117].

Let S denote a semigroup. Let $S^1 = S$ if S has an identity and $S^1 = S \cup \{1\}$ otherwise, where $s1 = s = 1s\ \forall s \in S \cup \{1\}$.

Definition 5.7.1 *For a set X, we denote the family of all fuzzy equivalence relations on X by $FE(X)$. The elements Δ and ∇ in $FE(X)$ are defined by $\Delta(x,y) = 0$ if $x \neq y$ and $\Delta(x,y) = 1$ if $x = y$, and $\nabla(x,y) = 1$ for all $x, y \in X$.*

Proposition 5.7.2 *[103]. If θ is a fuzzy equivalence relation on X, then $\theta \circ \theta = \theta$.* ■

In this section, I denotes an index set.

Proposition 5.7.3 *[103]. Let $\{\theta_i\}_{i \in I}$ be a nonempty family of fuzzy equivalence relations on a set X. Then $\cap_{i \in I} \theta_i$ is a fuzzy equivalence relation on X.* ■

Definition 5.7.4 *Let θ be a fuzzy relation on a set X and $\{\theta_i \mid \theta \subseteq \theta_i, i \in I\}$ be the family of all the fuzzy equivalence relations on X containing θ. The fuzzy equivalence relation generated by θ, denoted by θ^e, is defined as*

$$\theta^e = \cap_{i \in I, \theta \subseteq \theta_i} \theta_i.$$

It it clear that θ^e is the smallest fuzzy equivalence relation containing θ.

Definition 5.7.5 *[25]. Let θ be a fuzzy relation on X. Then the transitive closure of θ, denoted by θ^∞, is defined as*

$$\theta^\infty = \cup_{n=1}^{\infty}\theta^n, \text{ where } \theta^n = \theta \circ \theta \circ ... \circ \theta \ (n \text{ factors}).$$

Proposition 5.7.6 *[25]. Let θ be a fuzzy relation on X. Then θ^∞ is the smallest transitive fuzzy relation on X containing θ.* ■

Proposition 5.7.7 *If θ is symmetric, then so is θ^∞.*

Proof. For $n \in \mathbb{N}$ and $x, y \in X$,

$$\begin{aligned} \theta^n(x,y) &= \vee_{z_1,...,z_{n-1}\in S}\{\theta(x,z_1) \wedge \theta(z_1,z_2) \wedge ... \wedge \theta(z_{n-1},y)\} \\ &= \vee_{z_{n-1},...,z_1\in S}\{\theta(y,z_{n-1}) \wedge \theta(z_{n-1},z_{n-2}) \wedge ... \wedge \theta(z_1,x)\} \\ &= \theta^n(y,x). \end{aligned}$$

Hence θ^n is symmetric for any $n \in \mathbb{N}$ and so θ^∞ is symmetric. ■

Proposition 5.7.8 *Let θ, ϕ, and γ be fuzzy relations on a set X. Then the following assertions hold.*

(1) $\theta \circ (\phi \cup \gamma) = (\theta \circ \phi) \cup (\theta \circ \gamma)$.

(2) If $\theta \subseteq \phi$, then $\theta \circ \gamma \subseteq \phi \circ \gamma$.

(3) If $\theta \subseteq \phi$, then $\theta^\infty \subseteq \phi^\infty$.

(4) If $\theta \circ \phi = \phi \circ \theta$, $\theta, \phi \in FE(X)$, then $(\theta \circ \phi)^\infty = \theta \circ \phi$.

Proof. Let $x, y \in X$.

(1)

$$\begin{aligned} \theta \circ (\phi \cup \gamma)(x,y) &= \vee_{z\in S}\theta(x,z) \wedge (\phi(z,y) \vee \gamma(z,y)) \\ &= (\vee_{z\in S}\theta(x,z) \wedge \phi(z,y)) \vee (\vee_{z\in S}\theta(x,z) \vee \gamma(z,y)) \\ &= ((\theta \circ \phi) \cup (\theta \circ \gamma))(x,y). \end{aligned}$$

(2)

$$\begin{aligned} (\theta \circ \gamma)(x,y) &= \vee_{z\in S}\theta(x,z) \vee \gamma(z,y) \\ &\le \vee_{z\in S}\phi(x,z) \wedge \gamma(z,y) \\ &= (\phi \circ \gamma)(x,y). \end{aligned}$$

(3) For $n \in \mathbb{N}$,

$$\begin{aligned} \theta^n(x,y) &= \vee_{z_1,...,z_{n-1}\in S}\{\theta(x,z_1) \wedge \theta(z_1,z_2) \wedge ... \wedge \theta(z_{n-1},y)\} \\ &\le \vee_{z_1,...,z_{n-1}\in S}\{\phi(x,z_1) \wedge \phi(z_1,z_2) \wedge ... \wedge \phi(z_{n-1},y)\} \\ &= \phi^n(x,y). \end{aligned}$$

Hence $\theta^n \subseteq \phi^n$ for all $n \in \mathbb{N}$ and so $\theta^\infty \subseteq \phi^\infty$.

(4) For $n \in \mathbb{N}$,

$$\begin{aligned} (\theta \circ \phi)^n &= (\theta \circ \phi) \circ (\theta \circ \phi) \circ ... \circ (\theta \circ \phi) \\ &= (\theta \circ \theta \circ ... \circ \theta) \circ (\phi \circ \phi \circ ... \circ \phi) \\ &= \theta \circ \phi. \end{aligned}$$

Thus $(\theta \circ \phi)^n = \theta \circ \phi$ for all $n \in \mathbb{N}$ and so

$$(\theta \circ \phi)^\infty = \theta \circ \phi.$$

■

Theorem 5.7.9 *If θ is a fuzzy relation on a set X, then $\theta^e = [\theta \cup \theta^{-1} \cup \Delta]^\infty$.*

Proof. Let $\phi = [\theta \cup \theta^{-1} \cup \Delta]^\infty$. By Proposition 5.7.6, ϕ is transitive and $\theta \subseteq \phi$. Since $\Delta \subseteq \theta \cup \theta^{-1} \cup \Delta \subseteq \phi$, for any $x \in X$, $\phi(x,x) \geq \Delta(x,x) = 1$. Hence $\phi(x,x) = 1$ and ϕ is reflexive. Clearly, $\theta \cup \theta^{-1} \cup \Delta$ is symmetric and so by Proposition 5.7.7, ϕ is symmetric. Thus $\phi \in FE(X)$ and $\theta \subseteq \phi$. If $\gamma \in FE(X)$ with $\theta \subseteq \gamma$, then $\Delta \subseteq \gamma$ and $\theta^{-1} \subseteq \gamma^{-1} = \gamma$. Consequently, $\theta \cup \theta^{-1} \cup \Delta \subseteq \gamma$. Hence $\phi \subseteq \gamma$. Therefore, $\theta^e = \phi = [\theta \cup \theta \cup \Delta]^\infty$. ■

Definition 5.7.10 *Let θ be a fuzzy relation of a semigroup S. Then θ is called* ***fuzzy left (right) compatible*** *if $\theta(x,\ y) \leq \theta(zx,\ zy)$ ($\theta(x,\ y) \leq \theta(xz,\ yz)$) for all $x, y, z \in S$. θ is called* ***fuzzy compatible*** *if $\theta(x,\ y) \wedge \theta(z,\ t) \leq \theta(xz,\ yt)$ for all $x, y, z, t \in S$.*

Definition 5.7.11 *Let $\theta \in FE(S)$. Then θ is called a* ***fuzzy left congruence (right congruence, congruence)*** *if it is left compatible (right compatible, compatible).*

We denote the set of all fuzzy congruences on a semigroup S by $FC(S)$.

Proposition 5.7.12 *Let $\theta \in FE(S)$. Then $\theta \in FC(S)$ if and only if θ is fuzzy left and right compatible.*

Proof. If $\theta \in FC(S)$, then for all $x, y, z \in S$, $\theta(x,y) = \theta(x,y) \wedge \theta(z,z) \leq \theta(xz, yz)$. Also $\theta(x,y) = \theta(z,z) \wedge \theta(x,y) \leq \theta(zx, zy)$. Hence θ is left and right compatible.

Conversely, if θ is left and right compatible, then

$$\begin{aligned} \theta(x,y) \wedge \theta(z,t) &= \theta(x,y) \wedge \theta(z,z) \wedge \theta(y,y) \wedge \theta(z,t) \\ &\leq \theta(xz,yz) \wedge \theta(yz,yt) \\ &\leq \theta(xz,yt)\ \forall x,y,z,t \in S. \end{aligned}$$

Therefore, $\theta \in FC(S)$. ■

Let θ be a fuzzy relation on a semigroup S and $\{\theta_i \mid \theta \leq \theta_i\ i \in I\}$ be the family of all fuzzy congruences on S containing θ. Then the fuzzy relation defined by $\hat{\theta} = \cap_{i \in I, \theta \subseteq \theta_i} \theta_i$ is clearly the smallest fuzzy congruence on S containing θ. $\hat{\theta}$ is called the **fuzzy congruence relation** on S **generated by** θ.

Definition 5.7.13 *If θ is a fuzzy relation on a semigroup S, then define the fuzzy relation θ^* on S by $\forall c, d \in S$,*

$$\theta^*(c,d) = \vee_{x,y \in S^1,\ xay=c,\ xby=d}\, \theta(a,b).$$

Proposition 5.7.14 *Let θ and ϕ be two fuzzy relations on a semigroup S. Then*

(1) $\theta \subseteq \theta^*$;
(2) $(\theta^*)^{-1} = (\theta^{-1})^*$;
(3) $\theta \subseteq \phi \Rightarrow \theta^* \subseteq \phi^*$;
(4) $(\theta^*)^* = \theta^*$;
(5) $(\theta \cup \phi)^* = \theta^* \cup \phi^*$;
(6) $\theta = \theta^*$ *if and only if* θ *is left and right compatible.*

Proof. The proofs of (1) and (2), and (3) are clear. To prove (6), notice first that from (1) and (3), we have

$$\theta^* \subseteq (\theta^*)^*. \tag{5.1}$$

On the other hand, if $c, d \in S$, then

$$\begin{array}{rcl}(\theta^*)^*(c,d) & = & \vee_{x,y \in s^1, xay=c, xby=d} \theta^*(a,b) \\ & = & \vee_{x,y \in s^1, xay=c, xby=d} \vee_{z,t \in s^1, zpt=a, zqt=b} \theta(p,q) \\ & \leq & \vee_{xz,ty \in s^1, xzpty=c, xzqty=d} \theta(p,q) \\ & = & \theta^*(c,d).\end{array} \tag{5.2}$$

From Equations (5.1) and (5.2), we conclude that $\theta^* = (\theta^*)^*$.

To prove (5), notice first that by (3), $\theta^* \subseteq (\theta \cup \phi)^*$ and $\phi^* \subseteq (\theta \cup \phi)^*$. Hence

$$\theta^* \cup \phi^* \subseteq (\theta \cup \phi)^*. \tag{5.3}$$

Now for $c, d \in S$,

$$\begin{array}{rcl}(\theta \cup \phi)^*(c,d) & = & \vee_{x,y \in s^1, xay=c, xby=d} (\theta \cup \phi)(a,b) \\ & = & \vee_{x,y \in s^1, xay=c, xby=d} \theta(a,b) \vee \phi(a,b) \\ & \leq & (\vee_{x,y \in s^1, xay=c, xby=d} \theta(a,b)) \vee (\vee_{x,y \in s^1, xay=c, xbd=d} \phi(a,b)) \\ & = & \theta^*(c,d) \vee \phi^*(c,d).\end{array} \tag{5.4}$$

From Equations (5.3) and (5.4), we obtain $\theta^* \cup \phi^* = (\theta \cup \phi)^*$.

To prove (6), suppose that $\theta = \theta^*$. Then for $c, d, e \in S$,

$$\theta(ec, ed) = \theta^*(ec, ed) = \vee_{x,y \in S^1, xay=ec, xby=ed} \theta(a,b) \geq \theta(c,d).$$

Similarly, $\theta(ce, de) \geq \theta(c,d)$. Thus θ is left and right compatible.

Conversely, assume θ is left and right compatible. Then for $c, d \in S$,

$$\theta^*(c,d) = \vee_{x,y \in S^1, xay=c, xby=d} \theta(a,b).$$

However since θ is left and right compatible, we have

$$\theta(a,b) \leq \theta(xay, xby) = \theta(c,d).$$

Thus $\theta^*(c,d) \leq \theta(c,d)$. Since $\theta(c,d) \leq \theta^*(c,d)$ by (1), we have $\theta = \theta^*$. ■

Proposition 5.7.15 *If θ is a fuzzy relation on a semigroup S that is left and right compatible, then so is θ^{∞}.*

Proof. For $a, b, c \in S$ and $n \in \mathbb{N}$,

$$\begin{aligned}\theta^n(a,b) &= \vee_{z_1,\ldots,z_{n-1}\in S}\{\theta(a,z_1)\wedge\theta(z_1,z_2)\wedge\ldots\wedge\theta(z_{n-1},b)\}\\ &\leq \vee_{z_1c,\ldots,z_{n-1}c}\{\theta(ac,z_1c)\wedge\theta(z_1c,z_2c)\wedge\ldots\wedge\theta(z_{n-1}c,bc)\}\\ &= \theta^n(ac,bc).\end{aligned}$$

Similarly, $\theta^n(a,b) \leq \theta^n(ca,cb)$. Hence θ^n is left and right compatible for any $n \in \mathbb{N}$. Therefore, θ^{∞} is left and right compatible. ■

Theorem 5.7.16 *If θ is a fuzzy relation on a semigroup S, then $\hat{\theta} = (\theta^*)^e$.*

Proof. From Theorem 5.7.9, we know that $(\theta^*)^e$ is a fuzzy equivalence relation containing θ^*. Hence it contains θ. To show that $(\theta^*)^e$ is a fuzzy congruence on S, notice that by Proposition 5.7.14(2), (5), $\theta^*\cup(\theta^*)^{-1}\cup\Delta = (\theta\cup\theta^{-1}\cup\Delta)^*$. Thus by Proposition 5.7.14(6), $\theta^*\cup(\theta^*)^{-1}\cup\Delta$ is left and right compatible. By applying Proposition 5.7.15, we get $(\theta^*)^e = (\theta^*\cup(\theta^*)^{-1}\cup\Delta)^{\infty}$ is left and right compatible and so by Proposition 5.7.12, $(\theta^*)^e$ is a congruence on S. Now if $\phi \in FC(S)$ and $\theta \subseteq \phi$, then by Proposition 5.7.14 (3, 6), $\theta^* \subseteq \phi^* = \phi$. Therefore, $(\theta^*)^e \subseteq \phi$ and so $\hat{\theta} = (\theta^*)^e$.

If S is a semigroup, then Murali [103] proved that $\langle FE(S), \subseteq\rangle$ is a complete lattice with Δ and ∇ as the least and the greatest elements of $FE(S)$. Hence $\langle FC(S), \subseteq\rangle$ is a sublattice of $\langle FE(S), \subseteq\rangle$. If θ and ϕ are in $FE(S)$, then so are $\theta\cap\phi$ and $(\theta\cup\phi)^e$. Thus we can define addition and multiplication on $FE(S)$ as

$$\theta + \phi = (\theta\cup\phi)^e \text{ and } \theta\cdot\phi = \theta\cap\phi.$$

Similarly, if θ and ϕ are in $FC(S)$, then so are $\theta\cap\phi$ and $(\theta\cup\phi)^{\wedge}$. Hence addition and multiplication on $FC(S)$ are defined as

$$\theta + \phi = (\theta\cup\phi)^{\wedge} \text{ and } \theta\cdot\phi = \theta\cap\phi.$$

Notice that if θ and ϕ are in $FC(S)$, then by Proposition 5.7.14, $(\theta\cup\phi)^* = \theta^*\cup\phi^* = \theta\cup\phi$. Hence by Theorem 5.7.16, $(\theta\cup\phi)^{\wedge} = (\theta\cup\phi)^e$; that is, the addition of θ and ϕ in the lattice $\langle FC(S), +, \cdot\rangle$ coincides with the addition of θ and ϕ in the lattice $\langle FE(S), +, \cdot\rangle$. ■

The following theorem gives another description for the addition $\theta + \phi$ of two fuzzy equivalence (congruence) relations.

Theorem 5.7.17 *If θ and ϕ are in $FE(S)$ (or in $FC(S)$) for a semigroup S, then*

$$\theta + \phi = (\theta\circ\phi)^{\infty}.$$

Proof. By Theorem 5.7.9, we have

$$\theta + \phi = (\theta \cup \phi)^e = [(\theta \cup \phi) \cup (\theta \cup \phi)^{-1} \cup \Delta]^\infty.$$

However since θ and ϕ are in $FE(S)$, we have

$$(\theta \cup \phi) \cup (\theta \cup \phi)^{-1} \cup \Delta = \theta \cup \phi. \tag{5.5}$$

Since $\theta \subseteq \theta \cup \phi$ and $\phi \subseteq \theta \cup \phi$, $\theta \circ \phi \subseteq (\theta \cup \phi) \circ (\theta \cup \phi) = \theta \cup \phi$ by Proposition 5.7.8(2). Thus applying Proposition 5.7.8(3) yields

$$(\theta \circ \phi)^\infty \subset (\theta \vee \phi)^\infty. \tag{5.6}$$

Since θ and ϕ are in $FE(S)$, $\theta \subseteq \theta \circ \phi$ and $\phi \subseteq \theta \circ \phi$ and so $\theta \cup \phi \subseteq \theta \circ \phi$. It follows from Proposition 5.7.8(3) that

$$(\theta \cup \phi)^\infty \subseteq (\theta \circ \phi)^\infty. \tag{5.7}$$

Therefore, from Equations (5.5), (5.6) and (5.7), we conclude that

$$\theta + \phi = (\theta \cup \phi)^\infty = (\theta \circ \phi)^\infty.$$

■

Corollary 5.7.18 *If θ and ϕ are in $FE(S)$ [or in $FC(S)$] for a semigroup S such that $\theta \circ \phi = \phi \circ \theta$, then $\theta + \phi = \theta \circ \phi$.*

Proof. Follows at once from Theorem 5.7.175.7.16 and Proposition 5.7.8(4). ■

Proposition 5.7.19 *If S is a group, then $\theta \circ \phi = \phi \circ \theta$ for all $\theta, \phi \in FC(S)$.*

Proof. For $x, y \in S$,

$$(\theta \circ \phi)(x, y) = \vee_{z \in S} \theta(x, z) \wedge \phi(z, y),$$

but

$$\theta(x, z) = \theta(y, y) \wedge \theta(z^{-1}, z^{-1}) \wedge \theta(x, z) \leq \theta(y^{-1}xz, y).$$

Similarly, $\phi(z, y) \leq \phi(x, xy^{-1}x)$. Thus

$$(\theta \circ \phi)(x, y) \leq \vee_{yz^{-1}x} \phi(x, y^{-1}zx) \wedge \theta(y^{-1}zx, y) = (\phi \circ \theta)(x, y).$$

Using the same line of proof we can show that $(\phi \circ \theta)(z, y) \leq (\theta \circ \phi)(x, y)$. Therefore, $\theta \circ \phi = \phi \circ \theta$. ■

Definition 5.7.20 *A lattice $\langle L, +, \cdot \rangle$ is called **modular** if for all $x, y, z \in L$ with $x \leq z$,*

$$(x + y) \cdot z \leq x + (y \cdot z).$$

Theorem 5.7.21 *Let S be a semigroup and H be any sublattice of $\langle FC(S), +, \cdot\rangle$ such that $\theta \circ \phi = \phi \circ \theta$ for all $\theta, \phi \in H$. Then H is a modular lattice.*

Proof. Let $\theta, \phi, \gamma \in H$ be such that $\theta \subseteq \gamma$. We must show that $(\theta+\phi)\cdot\gamma \subseteq \theta+(\phi\cdot\gamma)$. Since $\theta\circ\phi = \phi\circ\theta$ for all $\theta, \phi \in H$, by Corollary 5.7.18, it suffices to show that $(\theta \circ \phi) \cap \gamma \subseteq \theta \circ (\phi \cap \gamma)$. To see this, let $x, y \in S$. Then

$$
\begin{aligned}
((\theta \circ \phi) \cap \gamma)(x, y) &= \vee_{z\in S}\{\theta(x, z) \wedge \phi(z, y) \wedge \gamma(x, y)\} \\
&= \vee_{z\in S}\{\theta(x, z) \wedge \theta(x, z) \wedge \phi(z, y) \wedge \gamma(x, y)\} \\
&\leq \vee_{z\in S}\{\theta(x, z) \wedge \gamma(x, z) \wedge \phi(z, y) \wedge \gamma(x, y)\} \\
&\quad (\theta \leq \gamma) \\
&\leq \vee_{z\in S}\{\theta(x, z) \wedge \phi(z, y) \wedge \gamma(z, y)\} \\
&= (\theta \circ (\phi \cap \gamma))(x, y).
\end{aligned}
$$

Therefore, H is modular. ■

Corollary 5.7.22 *If S is a group then $\langle FC(S), +, \cdot\rangle$ is modular.*

Proof. The proof follows from Proposition 5.7.19 and Theorem 5.7.21. ■

5.8 Fuzzy Congruence Pairs of Inverse Semigroups

If a and b are elements of a semigroup S, then b is called an **inverse** of a if $a = aba$ and $b = bab$. A semigroup in which every element has a unique inverse is called an **inverse semigroup**. Every regular element of a semigroup has an inverse. In this section, we assume that S is an inverse semigroup. We let E_S denote the set of all idempotents of S. The inverse of an element x of S is denoted by x^{-1}.

We carry on the investigation of fuzzy congruences on semigroups by introducing the fuzzy analogs of the kernel and trace of a congruence. We introduce fuzzy congruence pairs and prove in Proposition 5.8.10 that each pair defines a fuzzy congruence. In Theorem 5.8.11, we prove that there is a one-one correspondence between fuzzy congruences on an inverse semigroup and fuzzy congruence pairs. The results are from [7].

Definition 5.8.1 *[120] Let $x \in S$ and θ be a fuzzy congruence on S. The fuzzy congruence class determined by x and θ, denoted by $x\theta$, is the fuzzy subset of S defined by $x\theta(y) = \theta(x, y)$ $\forall y \in S$. Let S/θ denote the set of all fuzzy congruence classes, that is, $S/\theta = \{x\theta \mid x \in S\}$.*

Proposition 5.8.2 *[120] S/θ is an inverse semigroup with binary operation defined as*

$$(x\theta)(y\theta) = (xy)\theta \ \forall x, y \in S. \ ■$$

Proposition 5.8.3 *If θ is a fuzzy congruence on S, then*

$$\theta(x^{-1}, y^{-1}) = \theta(x, y)\ \forall x, y \in S.$$

Proof. By Proposition 5.8.2, S/θ is an inverse semigroup with $x^{-1}\theta = (x\theta)^{-1}\ \forall x \in S$. Now, for $x, y \in S$, $\theta(x^{-1}, y^{-1}) = x^{-1}\theta(y^{-1}) = (x\theta(y^{-1}))^{-1} = (y^{-1}\theta(x))^{-1} = y\theta(x) = \theta(y, x) = \theta(x, y)$. ∎

A fuzzy subset k of S is a fuzzy subsemigroup of S if for all $x, y \in S$, $k(xy) \geq k(x) \wedge k(y)$ and $k(x^{-1}) = k(x)$.

Definition 5.8.4 *Let θ be a congruence on S. Let K_θ be the fuzzy subset of S defined by*

$$K_\theta(x) = \vee_{e \in E_s}\theta(x, e) \forall x \in S.$$

*Then K_θ is called the **fuzzy kernel** of θ. The **fuzzy trace** of θ, denoted by T_θ, is defined by $T_\theta(e, f) = \theta(e, f)\ \forall e, f \in E_s$.*

Definition 5.8.5 *Let k be a fuzzy subset of S. Then k is called **fuzzy full** if $k(e) = 1\ \forall e \in E_S$. k is called **fuzzy self conjugate** if $k(s^{-1}xs) \geq k(x)$ $\forall x, s \in S$. A fuzzy subsemigroup of S is called **fuzzy normal** if it is fuzzy full and fuzzy self conjugate.*

Definition 5.8.6 *A fuzzy congruence τ on E_S is called a **fuzzy normal congruence** on E_S if $\tau(s^{-1}es, s^{-1}fs) \geq \tau(e, f)\ \forall e, f \in E_S$, $s \in S$.*

Definition 5.8.7 *The pair (k, τ) is called a **fuzzy congruence pair** for S if k is a fuzzy normal subsemigroup of S, τ is a fuzzy normal congruence on E_S and $\forall x \in S$, $e \in E_S$ the following conditions hold:*
(i) $k(x) \geq k(xe) \wedge \tau(e, x^{-1}x)$,
(ii) $\tau(xx^{-1}, x^{-1}x) \geq k(x)$.

Lemma 5.8.8 *If θ is a fuzzy congruence on S, then (K_θ, T_θ) is a fuzzy congruence pair for S.*

Proof. (1) For $x, y \in S$, $K_\theta(xy) = \vee_{e \in E_S}\theta(xy, e)$. Since θ is a fuzzy congruence, we have

$$\theta(xy, e_1e_2) \geq \theta(x, e_1) \wedge \theta(y, e_2),\ e_1, e_2 \in E_S.$$

Since $e_1e_2 \in E_S$, we have $K_\theta(xy) \geq \theta(x, e_1) \wedge \theta(x, e_2)$. Taking the sup over $e_1, e_2 \in E_S$ independently, we get $K_\theta(xy) \geq K_\theta(x) \wedge K_\theta(y)$. Moreover,

$$\begin{aligned} K_\theta(x) &= \vee_{e \in E_S}\theta(x, y) \\ &= \vee_{e \in E_S}\theta(x^{-1}, e^{-1}) \text{ (by Proposition 5.8.3)} \\ &= \vee_{e \in E_S}(x^{-1}, e) \\ &= K_\theta(x^{-1}). \end{aligned}$$

Thus K_θ is a fuzzy subsemigroup of S. It is clear that $K_\theta(e) = 1\ \forall e \in E_S$. Also

$$\begin{aligned} K_\theta(s^{-1}xs) &= \vee_{e\in E_S}\theta(s^{-1}xs, e) \\ &\geq \vee_{e\in E_S}\theta(s^{-1}xs, s^{-1}es) \text{ (because } s^{-1}E_S s \subset E_S) \\ &\geq \vee_{e\in E_S}\theta(x, e) \text{ (because } \theta \text{ is a congruence).} \end{aligned}$$

This proves that K_θ is a fuzzy normal subsemigroup of S.

(2) $T_\theta(s^{-1}es, s^{-1}fs) = \theta(s^{-1}es, s^{-1}fs) \geq \theta(e, f)$ since θ is a congruence, and $\theta(e, f) = T_\theta(e, f)\ \forall e, f \in E_S$, $s \in S$. Thus T_θ is a fuzzy normal congruence on E_S.

(3) Since

$$K_\theta(x) \geq \theta(x, e) \geq \theta(xx^{-1}x, xe_0) \wedge \theta(xe_0, e) \geq \theta(x^{-1}x.e_0) \wedge \theta(xe_0, e),$$

it follows that $K_\theta(x) \geq \theta(x^{-1}x, e_0) \wedge \vee_{e\in E_S}\theta(xe_0, e)$. Thus

$$K_\theta(x) \geq T_\theta(x^{-1}x, e_0) \wedge K_\theta(xe_0)$$

for all $x \in S$, $e_0 \in E_S$.

(4) We have

$$\begin{aligned} T_\theta(xx^{-1}, x^{-1}x) &= \theta(xx^{-1}, x^{-1}x) \\ &\geq \theta(xx^{-1}, e) \wedge \theta(e, x^{-1}x) \\ &= \theta(xx^{-1}, ee) \wedge \theta(ee, x^{-1}x) \\ &\geq \theta(x, e) \wedge \theta(x^{-1}, e) \wedge \theta(e, x^{-1}) \wedge \theta(e, x) \\ &= \theta(x, e) \wedge \theta(x^{-1}e) \\ &= \theta(x, e) \qquad (\theta(x^{-1}, e^{-1}) = \theta(x, e)) \end{aligned}$$

Thus $T_\theta(xx^{-1}, x^{-1}x) \geq K_\theta(x)\ \forall x \in S$. ∎

The following proposition give some properties of fuzzy congruence pairs.

Proposition 5.8.9 *Let (k, τ) be a fuzzy congruence pair for S. Then $\forall a, b \in S$, $e \in E_S$,*

(1) $k(ab) \geq k(aeb) \wedge \tau(e, a^{-1}a)$,

(2) $k(aeb) \geq k(ab)$,

(3) $\tau(a^{-1}ea, b^{-1}eb) \geq k(ab^{-1}) \wedge \tau(a^{-1}a, b^{-1}b)$.

Proof. (1) Note first that $b^{-1}eb \in E_S$. Now

$$k(ab) \geq k(abb^{-1}eb) \wedge \tau(b^{-1}eb, b^{-1}a^{-1}ab) \text{ (Definition 5.8.7).}$$

Since τ is fuzzy normal, we have $\tau(b^{-1}eb, b^{-1}a^{-1}ab) \geq \tau(e, a^{-1}a)$. Since E_S is commutative, $abb^{-1}eb = aebb^{-1} = aeb$. Thus $k(ab) \geq k(aeb) \wedge \tau(e, a^{-1}a)$.

(2) Since $b^{-1}eb \in E_S$ and k is fuzzy full, $k(b^{-1}eb) = 1$. Now

$$k(aeb) = k(aebb^{-1}b) = k(abb^{-1}eb) \geq k(ab) \wedge k(b^{-1}eb) = k(ab).$$

(3) We have

$$\begin{aligned}\tau(ab^{-1}ba^{-1}, ba^{-1}ab^{-1}) &= \tau(a^{-1}ea(a^{-1}a)a^{-1}ea, a^{-1}eab^{-1}ba^{-1}ea)\\ &\geq \tau(a^{-1}a, b^{-1}b)\end{aligned} \tag{5.8}$$

and

$$\tau(ab^{-1}ba^{-1}, ba^{-1}ab^{-1}) \geq k(ab^{-1}) \text{ by Definition 5.8.7.}$$

Hence

$$\tau(a^{-1}eab^{-1}ba^{-1}ea, a^{-1}eba^{-1}ab^{-1}ea) \geq k(ab^{-1}) \tag{5.9}$$

since τ is a fuzzy normal congruence.

Also

$$\begin{aligned}\tau(a^{-1}ea, a^{-1}(eba^{-1})(ab^{-1}e)a) &\geq \tau(a^{-1}ea, a^{-1}eab^{-1}ba^{-1}ea)\wedge\\ &\quad\ \tau(a^{-1}eab^{-1}ba^{-1}ea, a^{-1}eba^{-1}ab^{-1}ea)\\ &\geq \tau(a^{-1}a, b^{-1}b) \wedge k(ab^{-1}).\end{aligned} \tag{5.10}$$

By Definition 5.8.7 and since τ is normal, we have

$$\begin{aligned}&\tau(a^{-1}(eba^{-1})(ab^{-1}e)a, a^{-1}(ab^{-1}e)(eba^{-1})a)\\ \geq\ &\tau((ab^{-1}e)^{-1}(ab^{-1}e), (ab^{-1}e)(ab^{-1}e)^{-1})\\ =\ &\tau((ab^{-1}e)(ab^{-1}e)^{-1}, (ab^{-1}e)^{-1}(ab^{-1}e))\\ \geq\ &k(ab^{-1}e).\end{aligned} \tag{5.11}$$

Now, from (5.10) and (5.11), we have

$$\begin{aligned}&\tau(a^{-1}ea, a^{-1}ab^{-1}eba^{-1}a)\\ =\ &\tau(a^{-1}ea, a^{-1}(ab^{-1}e)(eba^{-1})a)\\ \geq\ &\tau(a^{-1}ea, a^{-1}(eba^{-1})(ab^{-1}e)a)\wedge\\ &\tau(a^{-1}(eba^{-1})(ab^{-1}e)a, a^{-1}(ab^{-1}e)(eba^{-1})a)\\ \geq\ &\tau(a^{-1}a, b^{-1}b) \wedge k(ab^{-1}) \wedge k(ab^{-1}e)\\ \geq\ &\tau(a^{-1}a, b^{-1}b) \wedge k(ab^{-1}) \wedge k(ab^{-1}) \wedge k(e)\\ =\ &\tau(a^{-1}a, b^{-1}b) \wedge k(ab^{-1})\end{aligned} \tag{5.12}$$

and

$$\begin{aligned}&\tau(a^{-1}ab^{-1}, b^{-1}eb)\\ \geq\ &\tau(a^{-1}ea, a^{-1}ab^{-1}eba^{-1}a) \wedge \tau(a^{-1}ab^{-1}eba^{-1}a, b^{-1}eb)\\ \geq\ &\tau(a^{-1}a, b^{-1}b) \wedge \tau(b^{-1}eb, b^{-1}eb)\\ =\ &\tau(a^{-1}a, b^{-1}b).\end{aligned} \tag{5.13}$$

Hence

$$\begin{aligned}&\tau(a^{-1}ea, b^{-1}eb)\\ \geq\ &\tau(a^{-1}ea, a^{-1}ab^{-1}eba^{-1}a) \wedge \tau(a^{-1}ab^{-1}eba^{-1}a, b^{-1}eb)\\ \geq\ &\tau(a^{-1}a, b^{-1}b) \wedge k(ab^{-1}) \wedge \tau(a^{-1}a, b^{-1}b)\\ =\ &\tau(a^{-1}a, b^{-1}b) \wedge k(ab^{-1})\end{aligned} \tag{5.14}$$

by using (5.12) and (5.13) above. ■

Proposition 5.8.10 *If (k, τ) is a fuzzy congruence pair for S, then the fuzzy relation $\mu_{(k,\tau)}$ defined by $\mu_{(k,\tau)}(a,b) = \tau(a^{-1}a, b^{-1}b) \wedge k(ab^{-1})$ $\forall a, b \in S$ is a fuzzy congruence on S.*

Proof. Let $\theta = \mu_{(K,\tau)}$. Let $a, b, c \in S$.

(1) Since k is fuzzy full and $aa^{-1} \in E_S$, we have $k(aa^{-1}) = 1$. Hence

$$\theta(a,a) = \tau(a^{-1}a, a^{-1}a) \wedge k(aa^{-1}) = 1.$$

(2) Since k is a fuzzy subsemigroup of A, $k(ba^{-1}) = k((ba^{-1})^{-1}) = k(ab^{-1})$. Thus

$$\theta(b,a) = \tau(b^{-1}b, a^{-1}a) \wedge k(ba^{-1}) = \tau(a^{-1}a, b^{-1}b) \wedge k(ab^{-1}) = \theta(a,b).$$

(3) Note that $\tau(a^{-1}a, c^{-1}c) \geq \tau(a^{-1}a, b^{-1}b) \wedge \tau(b^{-1}b, c^{-1}c)$. By Proposition 5.8.9(1), we have $k(ac^{-1}) \geq k(a(b^{-1}b)c^{-1}) \wedge \tau(b^{-1}b, a^{-1}a)$. Now

$$\begin{aligned}\theta(a,c) &= \tau(a^{-1}a, c^{-1}c) \wedge k(ac^{-1}) \\ &\geq \tau(a^{-1}a, b^{-1}b) \wedge \tau(b^{-1}b, c^{-1}c) \wedge k(ab^{-1}bc^{-1}) \wedge \tau(b^{-1}b, a^{-1}a) \\ &\geq \tau(a^{-1}a, b^{-1}b) \wedge \tau(b^{-1}b, c^{-1}c) \wedge k(ab^{-1}) \wedge k(bc^{-1}) \\ &= \theta(a,b) \wedge \theta(b,c).\end{aligned}$$

From (1), (2) and (3), we conclude that θ is a fuzzy equivalence relation on S. We must show the θ is fuzzy left and right compatible. To show it is fuzzy right compatible, we let $a, b, c \in S$. Since τ is a fuzzy normal congruence on E_S and $a^{-1}a, b^{-1}b \in E_S$, we have $\tau(c^{-1}a^{-1}ac, c^{-1}b^{-1}bc) \geq \tau(a^{-1}a, b^{-1}b)$, and by Proposition 5.8.9(2), $k(acc^{-1}b^{-1}) \geq k(ab^{-1})$. Thus

$$\begin{aligned}\theta(ac, bc) &= \tau(c^{-1}a^{-1}ac, c^{-1}b^{-1}bc) \wedge k(acc^{-1}b^{-1}) \\ &\geq \tau(a^{-1}a, b^{-1}b) \wedge k(ab^{-1}) \\ &= \theta(a,b).\end{aligned}$$

Finally, since k is a fuzzy normal, we have $k(cab^{-1}c^{-1}) \geq k(ba^{-1})$, and by Proposition 5.8.9(3), $\tau(a^{-1}c^{-1}ca, b^{-1}c^{-1}cb) \geq k(ab^{-1}) \wedge \tau(a^{-1}a, b^{-1}b)$. Thus

$$\begin{aligned}\theta(ca, cb) &= \tau(a^{-1}c^{-1}ca, b^{-1}c^{-1}cb) \wedge k(cab^{-1}c^{-1}) \\ &\geq k(ab^{-1}) \wedge \tau(a^{-1}a, b^{-1}b) \wedge k(ab^{-1}) \\ &= \tau(a^{-1}a, b^{-1}b) \wedge (ab^{-1}) \\ &= \theta(a,b).\end{aligned}$$

Therefore, $\theta = \mu_{(k,\tau)}$ is a fuzzy congruence on S. ■

We now prove the main theorem which gives a characterization of fuzzy congruences in terms of fuzzy kernels and fuzzy traces.

Theorem 5.8.11 *Let S be an inverse semigroup. If (k, τ) is a fuzzy congruence pair for S, then $\theta = \mu_{(k,\tau)}$ is the unique fuzzy congruence on S for which $K_\theta = k$ and $T_\theta = \tau$. Conversely, if μ is a fuzzy congruence on S, then (K_μ, T_μ) is a fuzzy congruence pair for S and $\mu_{(K_\mu, T_\mu)} = \mu$.*

Proof. Assume first that (k, τ) is a fuzzy congruence pair for S. By Proposition 5.8.10, we know that $\theta = \mu_{(k,\tau)}$ is a fuzzy congruence on S. To prove $K_\theta = k$, let $x \in S$. Then $K_\theta(x) = \vee_{e \in E_S} \theta(x, e)$. However,

$$\begin{aligned}\theta(x,e) &= \tau(x^{-1}x, e^{-1}e) \wedge k(xe^{-1}) \\ &= \tau(x^{-1}x, e) \wedge k(xe) \\ &\leq k(x) \text{ by Definition 5.8.7.}\end{aligned}$$

Thus $K_\theta(x) \leq k(x)$. On the other hand,

$$\begin{aligned}k(x) &= \tau(x^{-1}x, x^{-1}x) \wedge k(x) \\ &= \tau(x^{-1}x, x^{-1}xx^{-1}x) \wedge k(xx^{-1}x) \\ &= \tau(x^{-1}x, x^{-1}x(x^{-1}x)^{-1}) \wedge k(xx^{-1}x) \\ &= \theta(x, x^{-1}x) \\ &= k_\theta(x).\end{aligned}$$

Therefore, $K_\theta = k$.

To prove $T_\theta = \tau$, let $e, f \in E_S$. Then

$$T_\theta(e,f) = \theta(e,f) = \tau(e^{-1}e, f^{-1}f) \wedge k(ef^{-1}) = \tau(e,f) \wedge 1 = \tau(e,f).$$

To prove uniqueness, let φ be a fuzzy congruence on S such that $K_\varphi = k$ and $T_\varphi = \tau$. We want to show that $\varphi = \theta$.

$$\begin{aligned}\varphi(a,b) &= \varphi(a,b) \wedge \varphi(a,b) \\ &= \varphi(a,b) \wedge \varphi(a^{-1}, b^{-1}) \\ &\leq \varphi(a^{-1}a, b^{-1}b) \\ &= \tau(a^{-1}a, b^{-1}b).\end{aligned}$$

Also,

$$k(ab^{-1}) = K_\varphi(ab^{-1}) = \vee_{e \in E_S} \varphi(ab^{-1}, e) \geq \varphi(ab^{-1}, bb^{-1}) \geq \varphi(a,b).$$

Therefore,

$$\theta(a,b) = \tau(a^{-1}a, b^{-1}b) \wedge k(ab^{-1}) \geq \varphi(a,b) \wedge \varphi(a,b) = \varphi(a,b).$$

To prove the other inequality,

$$\tau(a^{-1}a, b^{-1}b) = \varphi(a^{-1}a, b^{-1}b) \leq \varphi(aa^{-1}a, ab^{-1}b) = \varphi(a, ab^{-1}b), \quad (5.15)$$

$$\varphi(ab^{-1}, e) \leq \varphi(ab^{-1}b, eb), \quad (5.16)$$

$$\varphi(ab, eb) \geq \varphi(a, ab^{-1}b) \wedge \varphi(ab^{-1}b, eb) \geq \varphi(a^{-1}a, b^{-1}b) \wedge \varphi(ab^{-1}, e), \quad (5.17)$$

$$\varphi(a^{-1}a, b^{-1}b) \leq \varphi(ba^{-1}a, bb^{-1}b) = \varphi(ba^{-1}a, b), \tag{5.18}$$

$$\varphi(ab^{-1}, e) = \varphi(ba^{-1}, e^{-1}) = \varphi(ba^{-1}, e) = \varphi(ba^{-1}a, ea). \tag{5.19}$$

Thus

$$\begin{aligned} \varphi(b, ea) &\geq \varphi(b, ba^{-1}a) \wedge \varphi(ba^{-1}a, ea) \\ &\geq \varphi(a^{-1}a, b^{-1}b) \wedge \varphi(ab^{-1}, e) \text{ (by (5.18) and(5.19))}, \end{aligned} \tag{5.20}$$

$$\begin{aligned} \varphi(a, ea) &\geq \varphi(a, eb) \wedge \varphi(eb, ea) \\ &= \varphi(a, eb) \wedge \varphi(eb, eea) \\ &\geq \varphi(a, eb) \wedge \varphi(b, ea) \\ &\geq \varphi(a^{-1}a, b^{-1}b) \wedge \varphi(ab^{-1}, e) \text{ (by (5.17) and (5.20))}, \end{aligned} \tag{5.21}$$

$$\begin{aligned} \varphi(a, b) &\geq \varphi(a, ea) \wedge \varphi(ea, b) \\ &\geq \varphi(a^{-1}a, b^{-1}b) \wedge \varphi(ab^{-1}, e) \text{ (by (5.20) and (5.21))}. \end{aligned} \tag{5.22}$$

But e is arbitrary in E_S. Hence

$$\begin{aligned} \varphi(a, b) &\geq \varphi(a^{-1}a, b^{-1}b) \wedge \vee_{e \in E_S} \varphi(ab^{-1}, e) \\ &= \tau(a^{-1}a, b^{-1}b) \wedge K_\theta(ab^{-1}) \\ &= \tau(a^{-1}a, b^{-1}b) \wedge k(ab^{-1}) \\ &= \theta(a, b). \end{aligned}$$

Therefore, $\varphi = \theta$.

Conversely, let μ be a fuzzy congruence on S. Then by Lemma 5.8.8, (K_μ, T_μ) is a fuzzy congruence pair for S. Thus we need only to show that $\mu_{(K_\mu, T_\mu)} = \mu$. However, this follows from the uniqueness argument proved earlier in the proof of this theorem. ■

5.9 Fuzzy Rees Congruences on Semigroups

In this section, we consider [149]. Let I be an ideal of a semigroup S. Let $\rho_I = \{(x, x) \mid x \in S\} \cup (I \times I)$. Then the relation ρ_I is a congruence. Its congruence classes consist of the ideal I and singletons. It is called a **Rees congruence** and its quotient S/ρ_I (or S/I) is called the **Rees quotient** of S by I. In this section, we study fuzzy Rees congruences on a semigroup. Clearly, the set of all fuzzy ideals of S, denoted by $FI(S)$, is a distributive lattice with respect to the meet $\cap$ and union $\cup$. The lattice $FI(S)$ has a greatest element, namely the fuzzy subset 1_S of S defined by $1_S(x) = 1$ $\forall x \in S$. $FI(S)$ has the least element Φ, where $\Phi(x) = 0 \ \forall x \in S$ if S has no zero element, otherwise, the least element is 0_S, where $0_S(x) = 0 \ \forall x \in S$, $x \neq 0$.

Let f be a fuzzy ideal of S. If S has a zero element 0, then it is not difficult to show that $f(0) \geq f(x)\ \forall x \in S$. In this section, we assume $f(0) = 1$.

We denote the set of all fuzzy congruences on a semigroup S by $FC(S)$. Then from Section 5.7 it follows that $FC(S)$ is a complete lattice with respect to $\cdot$ and $+$ defined as follows: $\forall \theta, \phi \in FC(S)$,

$$\theta + \phi = (\theta \cup \phi)^e,\ \theta \cdot \phi = \theta \cap \phi.$$

Furthermore,

$$\theta + \phi = (\theta \circ \phi)^\infty.$$

Let elements ∇_S and Δ_S be the fuzzy congruences on S defined by $\nabla_S(x, y) = 1$ for all $x, y \in S$, and $\Delta_S(x, y) = 0$ if $x \neq y$ and $\Delta_S(x, y) = 1$ if $x = y$. Then it is clear that ∇_S and Δ_S are the greatest element and least element of the lattice $(FC(S), +, \cdot)$ respectively.

Definition 5.9.1 *A semigroup S is called fuzzy congruences free, if S has no fuzzy congruences other than ∇_S and Δ_S.*

Definition 5.9.2 *A semigroup S is called fuzzy 0-simple, if $S^2 \neq \{0\}$ and 0_S and 1_S are the only fuzzy ideals.*

Let θ be a fuzzy congruence on S. We denote by S/θ the set $\{a\theta \mid a \in S\}$ of all fuzzy congruence classes of S. We define a binary operation $*$ on S/θ as follows:

$$(\forall a, b \in S)\ a\theta * b\theta = (ab)\theta.$$

Then S/θ is a semigroup [7, 79].

For other definition and terminology not given in this section, we employ those in [10, 41, 110, 117].

Let f be a fuzzy ideal of a semigroup S. Then, we assign a fuzzy relation θ_f on S defined as follows: $\forall x, y \in S$,

$$\theta_f(x, y) = \begin{cases} f(x) \wedge f(y), & \text{if } x \neq y, \\ 1, & \text{if } x = y. \end{cases}$$

Clearly, θ_f is well-defined. Moreover, we have the following result.

Proposition 5.9.3 *Let f be a fuzzy ideal of a semigroup S. Then θ_f is a fuzzy congruence on S.*

Proof. By the definition of θ_f, it is clear that θ_f is reflexive and symmetric. For all $x, y \in S$,

$$\theta_f \circ \theta_f(x, y) = \vee_{z \in S}(\theta_f(x, z) \wedge \theta_f(z, y)).$$

We consider the following cases.

(1) If $x = y$, then

$$\theta_f \circ \theta_f(x,x) = \vee_{z\in S}\theta_f(x,z) \geq \theta_f(x,x) = 1.$$

Thus $\theta_f \circ \theta_f(x,x) = \theta_f(x,x)$.

(2) If $x \neq y$, then

$$\begin{aligned}\theta_f \circ \theta_f(x,y) &= \vee_{z\in S\setminus\{x,y\}}(\theta_f(x,z) \wedge \theta_f(y,z)) \\ &\quad \vee(\theta_f(x,x) \wedge \theta_f(x,y)) \vee (\theta_f(x,y) \wedge \theta_f(y,y)) \\ &= \theta_f(x,y) \vee \vee_{z\in S\setminus\{x,y\}}(f(x) \wedge f(z) \wedge f(z) \wedge f(y)) \\ &\leq \theta_f(x,y) \vee \vee_{z\in S\setminus\{x,y\}}(f(x) \wedge f(y)) \\ &= \theta_f(x,y) \vee \theta_f(x,y) \\ &= \theta_f(x,y).\end{aligned}$$

By (1) and (2), θ_f is transitive. Furthermore, θ_f is left compatible. In fact, for $x, y, c \in S$, if $cx = cx$, then

$$\theta_f(cx, cy) = 1 \geq \theta_f(x,y).$$

If $cx \neq cy$, then $x \neq y$. Thus

$$\theta_f(cx, cy) = f(cx) \wedge f(cy) \geq f(x) \wedge f(y) = \theta_f(x,y).$$

In the same way, we can also prove that θ_f is right compatible. ∎

We call a fuzzy congruence θ_f determined by a fuzzy ideal $f(\neq \Phi)$ of S a fuzzy **Rees congruence.**

Let f be a fuzzy ideal of a semigroup S and let

$$f_* = \{x \in S \mid f(x) = 1\}.$$

Then it is not difficult to verify that f_* is an ideal of S. Similar to the case of crisp Rees congruences on S, we have the following result.

Theorem 5.9.4 *Let f be a fuzzy ideal of a semigroup S. Let $\mathcal{A}$ be the set of ideals of S containing f_* and let $\mathcal{B}$ be the set of ideals of the quotient semigroup $(S/\theta_f, *)$. Then the mapping*

$$\alpha : J \mapsto J\theta_f \ (J \in \mathcal{A})$$

is an inclusion-preserving bijection from $\mathcal{A}$ onto $\mathcal{B}$.

Proof. Let J be an ideal of S. Then it is clear that $J\theta_f$ is an ideal of the semigroup $(S/\theta_f, *)$. Let $J_1, J_2 \in \mathcal{A}$ and $J_1 \neq J_2$. Then there exists $a \in S$ such that $a \in J_1\setminus J_2$ or $a \in J_2\setminus J_1$. If $a \in J_1\setminus J_2$, then we have $J_1\theta_f \neq J_2\theta_f$. In fact, if $J_1\theta_f = J_2\theta_f$, then there exists $b \in J_2$ such that $a\theta_f = b\theta_f$. Since $a \notin J_1$, we have $a \neq b$. Thus

$$(a\theta_f)(b) = \theta_f(a,b) = f(a) \wedge f(b) = (b\theta_f)(b) = 1.$$

Hence $f(a) = f(b) = 1$. Thus $a \in f_* \subseteq J_2$., which is impossible. In the same way, if $a \in J_2 \backslash J_1$, then we also have $J_1\theta_f \neq J_2\theta_f$. This proves that the mapping α is injective.

Let $K\theta_f$ be an ideal of S/θ_f and

$$K_1 = \{x \in S \mid x\theta_f \in K\theta_f\}.$$

Then for all $x \in K_1$,

$$\begin{aligned}(Sx)\theta_f &= S\theta_f \circ x\theta_f \\ &\subseteq S\theta_f \circ K\theta_f \\ &\subseteq K\theta_f.\end{aligned}$$

Thus $Sx \subseteq K_1$. Similarly, we have $xS \subseteq K_1$. It follows that K_1 is an ideal of S. On the other hand, $f_* \subseteq K_1$, for consider any $a \in f_*$ and let $x \in K_1$. If $a = ax$, then clearly $a \in K_1$. Suppose $a \neq ax$ for all $x \in S$. We consider the following cases.

(1) If $z \neq a$, ax, then

$$\begin{aligned}(a\theta_f)(z) &= \theta_f(a, z) \\ &= f(a) \wedge f(z) \text{ (since } f(a) = 1) \\ &= f(ax) \wedge f(z) \\ &= ((ax)\theta_f)(z).\end{aligned}$$

(2) If $z = a$, then

$$(a\theta_f)(a) = 1 = f(a) \wedge f(z) = ((ax)\theta_f)(a).$$

(3) If $z = ax$, by (2), we have $(a\theta_f)(ax) = ((ax)\theta_f)(ax)$. It follows that $(a\theta_f) = (ax)\theta_f \subseteq K\theta_f$. By definition of K_1, $a \in K_1$. Clearly, $K_1\theta_f = K\theta_f$. Thus α is onto. That α is inclusion-preserving follows easily. ■

Proposition 5.9.5 *Let S be a semigroup with* 0. *Then the relation*

$$\beta : f \mapsto \theta_f \ (f \in FI(S))$$

is an order-preserving and injective function from $FI(S)$ into $FC(S)$.

Proof. Suppose that $f_1, f_2 \in FI(S)$ and $f_1 \neq f_2$. Then there exists $x \in S$ such that $f_1(x) \neq f_2(x)$. Obviously, $x \neq 0$. Thus

$$\begin{aligned}\theta_{f_1}(x, 0) &= f_1(x) \wedge f_1(0) = f_1(x), \\ \theta_{f_2}(x, 0) &= f_2(x) \wedge f_2(0) = f_2(x).\end{aligned}$$

It follows that $\theta_{\mu_{f1}} \neq \theta_{f_2}$. Therefore, β is injective. It is follows easily that β is order-preserving. ■

Definition 5.9.6 *A semigroup S is called a* ***fuzzy Rees congruence semigroup*** *(FRC-semigroup) if every fuzzy congruence on S is a fuzzy Rees congruence.*

Proposition 5.9.7 *Let S be a FRC-semigroup. Then the following assertions hold.*

(1) S has a zero element 0.

(2) If θ is a fuzzy congruence on S, then $\theta_f = \theta$, where $f(x) = f(x, 0)$ for all $x \in S$.

Proof. To prove (1), observe that Δ_S is a fuzzy congruence on S. Since S is a FRC-semigroup, there exists a fuzzy ideal $f(\neq \Phi)$ of S such that $\Delta_S = \theta_f$. Since $f \neq \Phi$, we have $f(x) \neq 0$ for some $x \in S$. Then for any $y \in S$ such that $y \neq x$, we have

$$\Delta_S(y, x) = \theta_f(x, y) = f(x) \wedge f(y) = 0.$$

Thus $f(y) = 0$. Since f is a fuzzy ideal of S, for all $z \in S$, we have $f(zx) \geq f(x)$, $f(xz) \geq f(x)$. Hence $zx = xz = x$. This proves that x is a zero element of S.

To prove (2), let θ be a fuzzy congruence on S. Then, by hypothesis, there exists a fuzzy ideal $f(\neq \Phi)$ such that $\theta = \theta_f$. Let

$$(\forall x \in S),\ g(x) = \theta(x, 0) \text{ (by (1))}.$$

Then for all $y \in S$,

$$\begin{aligned} &g(yx) = \theta(yx, 0) \geq \theta(x, 0) = g(x), \\ &g(yx) = \theta(yx, 0) \geq \theta(y, 0) = g(y), \\ &g(0) = \theta(0, 0) = 1. \end{aligned}$$

Thus g is a fuzzy ideal of S. Moreover, we have $g = f$. In fact, since for any $y \in S$ such that $y \neq 0$,

$$g(y) = \theta(y, 0) = \theta_f(y, 0) = f(y) \wedge f(0) = f(y).$$

Hence we have $f = g$. ■

Theorem 5.9.8 *Let S be a FRC-semigroup. Let $FI(S)$ be the lattice of all fuzzy ideals of S and $FC(S)$ the lattice of all fuzzy congruences on S. Then the mapping*

$$\beta : f \longmapsto \theta_f \ (f \in FI(S))$$

is an isomorphism from $FI(S)$ onto $FC(S)$.

Proof. By Proposition 5.9.5 and 5.9.7, β is order-preserving and bijective. To prove that β is lattice-order preserving, we show that

$$\begin{aligned} (\forall f_1, f_2 \in FI(S))\ \theta_{f_1} \cap \theta_{f_2} &= \theta_{f_1 \cap f_2}, \\ \theta_{f_1} \cup \theta_{f_2} &= \theta_{f_1 \cup f_2}. \end{aligned}$$

For all $x, y \in S$ such that $x \neq y$, we have

$$\begin{aligned}(\theta_{f_1} \cap \theta_{f_2})(x,y) &= \theta_{f_1}(x,y) \wedge \theta_{f_2}(x,y) \\ &= f_1(x) \wedge f_1(y) \wedge f_2(x) \wedge f_2(y) \\ &= (f_1 \cap f_2)(x) \wedge (f_1 \cap f_2)(y) \\ &= \theta_{f_1 \cap f_2}(x,y)\end{aligned}$$

and obviously, $\theta_{f_1} \cap \theta_{f_2}(x,x) = \theta_{f_1 \cap f_2}(x,x) = 1$. This implies that $\theta_{f_1} \cap \theta_{f_2} = \theta_{f_1 \cap f_2}$.

Since $f_1, f_2 \subseteq f_1 \cup f_2$, we have $\theta_{f_1} \subseteq \theta_{f_1 \cup f_2}$ and $\theta_{f_2} \subseteq \theta_{f_1 \cup f_2}$. Thus $\theta_{f_1} \cup \theta_{f_2} \subseteq \theta_{f_1 \cup f_2}$. On the other hand, for all $x, y \in S$ such that $x \neq y$,

$$\begin{aligned}\theta_{f_1 \cup f_2}(x,y) &= (f_1 \cup f_2)(x) \wedge (f_1 \cup f_2)(y) \\ &= (f_1(x) \vee f_2(x)) \wedge (f_1(y) \vee f_2(y)) \\ &= (f_1(x) \wedge f_1(y)) \vee (f_1(x) \wedge f_2(y)) \\ &\quad \vee (f_1(y) \wedge f_2(x)) \vee (f_2(x) \wedge f_2(y)).\end{aligned}$$

Furthermore,

$$\begin{aligned}f_1(x) \wedge f_1(y) &= \theta_{f_1}(x,y) \\ &\leq (\theta_{f_1} \circ \theta_{f_2})(x,y) \\ &\leq (\theta_{f_1} \circ \theta_{f_2})^{\infty}(x,y) \\ &= (\theta_{f_1} \cup \theta_{f_2})(x,y),\end{aligned} \tag{5.23}$$

$$\begin{aligned}f_2(x) \wedge f_2(y) &= \theta_{f_2}(x,y) \\ &\leq (\theta_{f_1} \circ \theta_{f_2})(x,y) \\ &\leq (\theta_{f_1} \circ \theta_{f_2})^{\infty}(x,y) \\ &= (\theta_{f_1} \cup \theta_{f_2})(x,y),\end{aligned} \tag{5.24}$$

and

$$f_1(x) \wedge f_2(y) \leq f_1(xy) \wedge f_2(xy) \wedge f_1(x) \wedge f_2(y), \tag{5.25}$$

$$f_2(x) \wedge f_1(y) \leq f_1(xy) \wedge f_2(xy) \wedge f_2(x) \wedge f_1(y). \tag{5.26}$$

We consider (5.25) in the following cases.

(1) If $xy = x$, then

$$\begin{aligned}f_1(x) \wedge f_2(y) &\leq f_1(x) \wedge f_2(x) \wedge f_2(y) \\ &\leq f_2(x) \wedge f_2(y) \\ &\leq (\theta_{f_1} \cup \theta_{f_2})(x,y) \text{ (by (5.24))}.\end{aligned}$$

(2) If $xy = y$, then

$$\begin{aligned}f_1(x) \wedge f_2(y) &\leq f_1(y) \wedge f_2(y) \wedge f_1(x) \\ &\leq f_1(x) \wedge f_1(y) \\ &\leq (\theta_{f_1} \cap \theta_{f_2})(x,y) \text{ (by (5.23))}.\end{aligned}$$

(3) If $xy \neq x, y$, then

$$\begin{aligned} f_1(x) \wedge f_2(y) &\leq \theta_{f_1}(x, xy) \wedge \theta_{f_2}(xy, y) \\ &\leq (\theta_{f_1} \circ \theta_{f_2})(x, y) \\ &\leq (\theta_{f_1} \circ \theta_{f_2})^{\infty}(x, y) \\ &= (\theta_{f_1} \cup \theta_{f_2})(x, y) \text{ (by (5.24)).} \end{aligned}$$

In the same way, we consider (5.26).

(4) If $xy = x$, then

$$\begin{aligned} f_2(x) \wedge f_1(y) &\leq f_1(x) \wedge f_1(y) \\ &\leq (\theta_{f_1} \cup \theta_{f_2})(x, y) \text{ (by (5.23).} \end{aligned}$$

(5) If $xy = y$, then

$$\begin{aligned} f_2(x) \wedge f_1(y) &\leq f_2(x) \wedge f_2(y) \\ &\leq (\theta_{f_1} \cup \theta_{f_2})(x, y) \text{ (by (5.24)).} \end{aligned}$$

(6) If $xy \neq x, y$, then

$$\begin{aligned} f_2(x) \wedge f_1(y) &\leq \theta_{f_1}(y, xy) \wedge \theta_{f_2}(xy, x) \\ &\leq (\theta_{f_1} \cup \theta_{f_2})(y, x) \\ &\leq (\theta_{f_1} \circ \theta_{f_2})^{\infty}(y, x) \\ &= (\theta_{f_1} \cup \theta_{f_2})(x, y). \end{aligned}$$

Therefore, we have

$$\theta_{f_1 \cup f_2}(x, y) \leq (\theta_{f_1} \cup \theta_{f_2})(x, y).$$

Thus $\theta_{f_1 \cup f_2} \leq (\theta_{f_1} \cup \theta_{f_2})$. Consequently,

$$\theta_{f_1 \cup f_2} = (\theta_{f_1} \cup \theta_{f_2}).$$

■

Since $FI(S)$ is a distributive lattice, by Theorem 5.9.8, we have the following result.

Corollary 5.9.9 *Let S be an FRC-semigroup. Then the lattice $FC(S)$ is distributive.* ■

Theorem 5.9.10 *Let S be a FRC-semigroup and $S^2 \neq \{0\}$. Then S is fuzzy congruence free if and only if S is fuzzy 0-simple.*

Proof. Suppose that S is fuzzy congruences free and $f(\neq \Phi)$ is a fuzzy ideal of S. Then θ_f is a fuzzy congruence on S. By Definition 5.9.1, either $\theta_f = \nabla_S$ or $\theta_f = \Delta_S$.

If $\theta_f = \nabla_S$, then for all $x \in S (x \neq 0)$,

$$\theta_f(0, x) = \nabla_S(0, x) = 1 = f(0) \wedge f(x) = f(x).$$

Thus $f = 1_S$.

If $\theta_f = \Delta_S$, then, by Proposition 5.9.7(1), for all $x \in S$ $(x \neq 0)$,

$$\theta_f(0, x) = f(x) = \Delta_S(0, x) = 0.$$

Hence $\theta_f = 0_S$.

Suppose that S is fuzzy 0-simple and let θ be a fuzzy congruence on S. By Definition 5.9.2, there exists a fuzzy ideal $f(\neq \Phi)$ of S such that $\theta = \theta_f$. Since S is fuzzy 0-simple, either $f = 0_S$ or $f = 1_S$. If $f = 1_S$, then for all $x, y \in S$ $(x \neq y)$,

$$\theta(x, y) = \theta_f(x, y) = f(x) \wedge f(y) = 1.$$

This implies that $\theta = \nabla_S$. If $f = 0_S$, it follows that $\theta_f = \Delta_S$. ■

Proposition 5.9.11 *The homomorphic image of a FRC-semigroup is a FRC-semigroup.*

Proof. Let S be a FRC-semigroup and let $\alpha : S \mapsto T$ be a homomorphism of S onto a semigroup T and let θ be a fuzzy congruence on T. Define

$$\phi(x, y) = \theta(\alpha(x), \alpha(y)) \; (\forall x, y \in S).$$

Then ϕ is a fuzzy congruence on S. In fact, the reflexivity and symmetry of ϕ are obvious. Now

$$\begin{aligned} \phi \circ \phi(x, y) &= \vee_{z \in S}(\phi(x, z) \wedge \phi(z, y)) \\ &= \vee_{z \in S}(\theta(\alpha(x), \alpha(z)) \wedge \theta(\alpha(z), \alpha(y))) \\ &= (\theta \circ \theta)(\alpha(x), \alpha(y)) \\ &\leq \theta(\alpha(x), (\alpha(y))) \\ &= \phi(x, y). \end{aligned}$$

Thus ϕ is transitive. On the other hand,$\forall x, y, z, t \in S$,

$$\begin{aligned} \phi(xz, yt) &= \theta(\alpha(xz), \alpha(yt)) \\ &= \theta(\alpha(x)\alpha(z), \alpha(y)\alpha(t)) \\ &\geq \theta((x), \alpha(y)) \wedge \theta(\alpha(z), \alpha(t)) \\ &= \phi(x, y) \wedge \phi(y, t). \end{aligned}$$

Therefore, ϕ is compatible. Since ϕ is a fuzzy congruence on S, by hypothesis, then there exist a fuzzy ideal $f(\neq \Phi)$ of S such that $\phi = \phi_f$. Let $\alpha(f)$ be the fuzzy subset of S defined as follows:

$$\alpha(f)(x) = \vee_{z \in \alpha^{-1}(x)} f(z), \; \forall x \in S.$$

Then it is clear that $\alpha(f)$ is a fuzzy ideal of T.

We now claim that $\theta = \theta_{\alpha(f)}$. For all $x, y \in T$, if $x = y$, then $\theta(x,y) = \theta_{\alpha(f)}(x,y) = 1$ and if $x \neq y$, then there exist $a, b \in S$ such that $x = \alpha(a)$, $y = \alpha(b)$. Moreover,

$$\begin{aligned}
\theta(x,y) &= \theta(\alpha(a), \alpha(b)) \text{ (obviously, } a \neq b) \\
&= \phi(a,b) \\
&= \phi_f(a,b) \\
&= f(a) \wedge f(b) \\
&\leq (\vee_{z\in\alpha^{-1}(x)} f(z)) \wedge (\vee_{z\in\alpha^{-1}(y)} f(z)) \\
&= \alpha(f)(x) \wedge \alpha(f)(y) \\
&= \theta_{\alpha(f)}(x,y).
\end{aligned}$$

Thus we have $\theta \subseteq \theta_{\alpha(f)}$. On the other hand,

$$\begin{aligned}
\theta_{\alpha(f)}(x,y) &= \alpha(f)(x) \wedge \alpha(f)(y) \\
&= \vee_{z\in\alpha^{-1}(x)} f(z) \wedge \vee_{w\in\alpha^{-1}(y)} f(w) \\
&= \vee_{z\in\alpha^{-1}(x), w\in\alpha^{-1}(y)} (\mu(z) \wedge \mu(w)) \qquad \text{(obviously, } z \neq w) \\
&= \vee_{z\in\alpha^{-1}(x), w\in\alpha^{-1}(y)} \phi_f(z,w) \\
&= \vee_{z\in\alpha^{-1}(x), w\in\alpha^{-1}(y)} \phi(z,w) \\
&= \bigvee_{z\in\alpha^{-1}(x), w\in\alpha^{-1}(y)} \theta(\alpha(z), \alpha(w)) \\
&\leq \theta(x,y).
\end{aligned}$$

Hence $\theta_{\alpha(f)} \subseteq \theta$. Consequently, $\theta_{\alpha(f)} = \theta$. ■

5.10 Additional Fuzzy Congruences on Semigroups

In [161], the author introduces the concepts of fuzzy complete inner-unitary subsemigroups and fuzzy group congruences on a semigroup. It is shown that there is a one to one correspondence. between fuzzy complete inner-unitary subsemigroups and fuzzy group congruences on a regular semigroup.

In [130], the author shows that the set of all fuzzy congruences on a regular semigroup contained in C_H forms a modular lattice, where H is the equivalence relation$\{(a,b) \in S \times S \mid S^1a = S^1b \text{ and } aS^1 = bS^1\}$. The author also introduces idempotent separating fuzzy congruences on a semigroup and proves that the idempotent separating fuzzy congruences on a regular semigroup form a modular lattice. It is also shown that the lattice of fuzzy congruences on a regular semigroup is a disjoint union of some modular sublattices of the lattice.

In [160], the author introduces the concepts of 0-fuzzy congruence and t-fuzzy congruence on 0-simple semigroups and examines the correspondence between them. It is shown that there is an order-preserving bijection from the set of 0-fuzzy congruences onto the set of fuzzy admissible triples. The author also describes proper fuzzy congruences on completely 0-simple semigroups.

In [154], the author introduces the notions of the fuzzy ideal extension property, strongly fuzzy ideal extension property, and fuzzy principal ideal extension property of a semigroup. (A semigroup is said to have the fuzzy ideal extension property if for each fuzzy subsemigroup f of S and each fuzzy ideal g of f, there exists a fuzzy ideal h of S such that $h \cap f = g$.) The relationships of these properties with semigroups and the fuzzy congruence extension property are examined. It is shown that a semigroup has the ideal extension property if and only if it has the strongly fuzzy ideal extension property. The author applies these results to characterize t-cyclic semigroups.

We close this chapter with some comments concerning [2]. In this paper, the authors initiate the study of fuzzy subsystems of an S-system, where S a monoid with zero and S-systems are representations of S. They characterize monoids by properties of their fuzzy subsystems. The authors define pure fuzzy ideals and pure fuzzy subsystems and establish some of their basic properties. They introduce purely fuzzy maximal and purely fuzzy prime ideals of a monoid S. They show that the set of all purely fuzzy prime ideals of S admits the structure of a topological space, called the pure fuzzy spectrum of S. The authors prove, as their main result, one which provides characterizations of monoids in terms of their fuzzy ideals and fuzzy subsystems. For example, it is shown that a monoid S is right weakly regular if and only if each fuzzy subsystem of an S-system is pure.

6

Fuzzy Congruences on T^*-pure Semigroups

In this chapter, we introduce the concept of a T^*-pure semigroup and show that a semigroup S is regular and T^*-pure if and only if S is a semilattice of groups. We also show that a semigroup S is T^*-pure if and only if S^3 is a semilattice of groups. Moreover, we study group congruences and semilattice congruences on such a semigroup and give some properties of fuzzy congruences on T^*-pure semigroups.

6.1 T^*-pure Semigroups

A subsemigroup A of a semigroup S is called a **bi-ideal** of S if $ASA \subseteq A$. A bi-ideal A of S is called **two-sided pure** if

$$A \cap xSy = xAy$$

for all $x, y \in S$. A semigroup S is called **T^*-pure** if every bi-ideal of it is two-sided pure.

Theorem 6.1.1 *For any elements a and b of a T^*-pure semigroup S, the following properties hold:*

(1) $aSb = a^2Sb^2$.

(2) $abSab = baSba$.

Proof. (1) Since S is T^*-pure, the bi-ideal aSb is two-sided pure. Then

$$aSb = aSb \cap aSb = a(aSb)b = a^2Sb^2.$$

(2) Since the bi-ideal $baSba$ is a two-sided pure, by (1), we have $abSab = (ab)^2S(ab)^2 = a(ba)(bSa)(ba)b \subseteq a(ba)S(ba)b = baSba \cap aSb \subseteq baSba$. Similarly, it can be seen that $baSba \subseteq abSab$. Thus $abSab = baSba$. ■

Theorem 6.1.2 *For any element a of a T^*-pure semigroup S, $a^n (n \geq 3)$ is a completely regular element of S.*

Proof. Since $n \geq 3$, we have by Theorem 6.1.1 that

$$a^n = aa^{n-2}a \in aSa = (a^n)^2S(a^n)^2.$$

This means that a^n is a completely regular element of S. ■

We denote by $E(S)$ the set of all idempotents of a semigorup S and by $Z(S)$ the center of S. Thus we have the following result.

Theorem 6.1.3 *For a T^*-pure semigroup S, $E(S) \subseteq Z(S)$.*

Proof. Let e be any element of $E(S)$ and a any element of S. Since S is T^*-pure, the bi-ideal eSe is two-sided pure. Thus

$$ae = aeee \in a(eSe)e = eSe \cap aSe \subseteq eSe.$$

This implies that there exists an elements $x, y \in S$ such that $ae = exe$ and $ea = eye$. Hence $ae = exe = (ee)(xe) = e(exe) = e(ae) = (ea)e = (eye)e = (ey)(ee) = eye = ea$. Thus $e \in Z(S)$ and so $E(S) \subseteq Z(S)$. ■

Theorem 6.1.4 *A T^*-pure semigroup S is weakly commutative. Also S is a semilattice of archimedean semigroups.*

Proof. Let a and b be any elements of S. Then, since the bi-ideal bSb is two-sided pure, we have

$$(ab)^3 \in a(bSb)ab = aSb \cap aSab \subseteq bSb.$$

Hence S is weakly commutative. Then it follows from M. Petrich [107, p.49] that S is a semilattice of archimedean semigroup. ■

The following theorem shows that the converses of Theorem 6.1.3 and Theorem 6.1.4 hold for a regular semigroup.

Theorem 6.1.5 *For a regular semigroup S, the following conditions are equivalent.*

(1) S is T^*-pure.
(2) S is weakly commutative.
(3) $E(S) \subseteq Z(S)$.
(4) $aS = Sa$ for all $a \in S$.

Proof. It follows from Theorem 6.1.4 that (1) implies (2). Assume that (2) holds. Let e and a be any elements of $E(S)$ and S, respectively. Since S is regular, there exists an element $x \in S$ such that $ea = (ea)x(ea)$.

Since S is weakly commutative, there exists a positive integer n such that $((xe)a)^n \in aS(xe)$. Hence since $x(ea) = (xe)a \in E(S)$, we have $ea = (ea)x(ea) = (ea)((xe)a) = (ea)((xe)a)^n \in ea(aSxe) = (eaaaSx)e \subseteq Se$. This implies that there exists an element $y \in S$ such that $ea = ye$.

Similarly, there exists an element $z \in S$ such that $ae = ez$. Hence we have

$$ea = ye = y(ee) = (ea)e = e(ae) = e(ez) = (ee)z = ez = ae.$$

Thus $e \in Z(S)$ and so $E(S) \subseteq Z(S)$. Therefore, we have that (2) implies (3).

Next assume that (3) holds. Let A be any bi-ideal of S, and x, y any elements of S. Let $a = xsy$ $(a \in A, s \in S)$ be any element of $A \cap xSy$. Then, since S is regular, there exist elements x', y' and a' in S such that

$$x = xx'x, \ y = yy'y, \text{ and } a = aa'a.$$

We note that $aa', a'a \in E(S)$. Since $E(S) \subseteq Z(S)$, we have $a = xsy = (xx'x)s(yy'y) = xx'(xsy)y'y = xx'ay'y = xx'(aa'aa'a)y'y = x(x'(aa'))a((a'a)\,y')y = x((aa')x')a(y'(a'a))y = x(a(a'x'ay'a')a)y \in x(ASA)y \subseteq xAy$ and so we have $A \cap xSy \subseteq xAy$.

Conversely, let xay $(a \in A)$ be any element of xAy. Then, since S is regular, there exists an element $a' \in S$ such that $a = aa'a$ and $a', a'a \in E(S)$. Hence $xay = x(aa'aa'a)y = (x(aa'))a((a'a)y) = ((aa')x)a\,(y(a'a)) = a(a'xaya')a \in ASA \subseteq A$, and so we have $xAy \subseteq A$. Since the relation $xAy \subseteq xSy$ always holds, we have $xAy \subseteq A \cap xSy$. Therefore, we obtain that $xAy = A \cap xSy$ and so S is T^*-pure. Thus (3) implies (1).

Assume that (3) holds. Let a be any element of S, and as $(s \in S)$ be any element of aS. Then, since S is regular, there exists an element $a' \in S$ such that $a = aa'a$ and $aa', a'a \in E(S)$. Thus we have

$$as = (aa'a)s = a((a'a)s) = a(s(a'a)) = (asa')a \in Sa,$$

and so $aS \subseteq S$. Similarly, we have $Sa \subseteq aS$. Thus $aS = Sa$, and (3) implies (4).

Assume that (4) holds. Let $e \in E(S)$ and $a \in S$. Then, since $ea \in eS = Se$, there exists an element $x \in S$ such that $ea = xe$. Similarly, there exists an element $y \in S$ such that $ae = ey$. Hence $ea = xe = x(ee) = (xe)e = (ea)e = e(ae) = e(ey) = (ee)y = ey = ae$, and so $e \in Z(S)$. Thus $E(S) \subseteq Z(S)$, and (4) implies (3). ∎

Theorem 6.1.6 *For a semigroup S, the following conditions are equivalent.*

(1) S is a semilattice of groups.
(2) S is regular and $E(S) \subseteq Z(S)$.
(3) S is regular and weakly commutative.
(4) S is regular and T^*-pure.

Proof. The proof follows from Theorems 6.1.5 and 4.3.1. ■

Theorem 6.1.7 *A semigroup S is T^*-pure if and only if S^3 is a semilattice of groups.*

Proof. Assume that S is T^*-pure. Let a be any element of S^3. Then

$$a = bcd \in bSd = b^3Sd^3.$$

Thus there exists an element x in S such that $a = b^3xd^3$. Since b^3 and d^3 are completely regular elements of S by Theorem 6.1.2, there exists elements y and z in S such that

$$b^3 = b^3yb^3 \quad \text{and} \quad d^3 = d^3zd^3.$$

Hence since the bi-ideal aS^3a of S is two-sided pure, we have $a = b^3xd^3 = (b^3yb^3)x(d^3zd^3) = (b^3y)(b^3xd^3)(zd^3) = (b^3y)a(zd^3) \in (b^3y)(aSd)d^2 = (b^3y)(a^5Sd^5)d^2 \subseteq (b^3y)(aS^3a)d^7 = aS^3a \cap b^3ySd^7 \subseteq aS^3a$. Therefore, a is regular in S^3.

On the other hand, since the bi-ideal aS^3a is two-sided pure, $aS^3 \subseteq (aS^3aS^3a)S^3 = (aS^3)(aS^3a)S^3 = aS^3a \cap aS^3SS^3 \subseteq aS^3a \subseteq S^3a$.

It can be proved in a similar way that the converse inclusion holds. Thus we obtain that $aS^3 = S^3a$.

Hence it follows from Theorem 4.3.1 that S^3 is a semilattice of groups.

Conversely, assume that S^3 is a semilattice of groups. Let A be any bi-ideal of S. To see that A is two-sided pure, let x and y be any elements of S. Let $a = xsy$ $(s \in S)$ be any element of $A \cap xSy$. Then, since $a = xsy \in S^3$, and S^3 is regular, we have

$$\begin{aligned} a \in aS^3aS^3aS^3a &= (xsy)S^3aS^3aS^3(xsy) \\ &= x\{(syS^3)a\}S^3\{a(S^3xs)\}y \\ &\subseteq x(S^3a)S^3(aS^3)y \\ &= x(aS^3)S^3(S^3a)y \\ &\subseteq x(ASA)y \subseteq xAy, \end{aligned}$$

and so $A \cap xSy \subseteq xAy$.

Let xay $(a \in A)$ be any element of xAy. Then by Theorem 6.1.1, we have $xay \in xSy = x^3Sy^3$ and so there exists an elements $z \in S$ such that $xay = x^3zy^3$. Since x^3 and y^3 are regular, we have $xay = x^3zy^3 = (x^3bx^3)z(y^3cy^3) = (x^3b)(x^3zy^3)(cy^3) = (x^3b)(xay)(cy^3) \in (x^3b)(xay)S^3 (xay)\,(cy^3) = \{(x^3bx)a\}(yS^3x)\{a(ycy^3)\} \subseteq (S^3a)S(aS^3) = (aS^3)S(S^3a) = a(S^3SS^3)a \subseteq ASA \subseteq A$ and so $xAy \subseteq A$.

Since the inclusion $xAy \subseteq xSy$ always holds, we have $xAy \subseteq A \cap xSy$. Therefore, $A \cap xSy = xAy$ and so A is a two-sided pure bi-ideal of S. Thus S is T^*-pure. ■

As is stated in the Theorem 6.1.4, a T^*-pure semigroup S is a semilattice of archimedean semigroups. So we study an archimedean T^*-pure semigroup. We recall that a semigroup S is called archimedean if for each elements a and b of S there exists a positive integer n such that $a^n \in SbS$.

As is well-know (A.H. Clifford and G.B. Preston [17, p.135]), an archimedean commutative semigroup can contain at most one idempotent.

Theorem 6.1.8 *For a T^*-pure semigroup S, the following conditions are equivalent.*

(1) S is archimedean.
(2) Every bi-ideal of S is archimedean.
(3) S has exactly one idempotent.

Proof. Assume that (1) holds. In order to prove that (3) holds, let e and f be any idempotents of S. Then, since S is archimedean, there exists a positive integer n such that $e^n \in SfS$. Since S is T^*-pure, the bi-ideal fSf is two-sided pure. Thus we have

$$e = e^n \in SfS = S(fff)S \subseteq S(fSf)S = fSf \cap SSS \subseteq fSf.$$

This implies that there exists an element $x \in S$ such that $e = fxf$.

It can be seen in a similar way that there exists an element $y \in S$ such that $f = eye$. Hence we have

$$e = fxf(ff)xf = f(fxf) = fe = (eye)e = ey(ee) = eye = f.$$

Since $E(S)$ is nonempty by Theorem 6.1.2, we obtain that S has exactly one idempotent, and (1) implies (3).

Next assume that (3) holds. To see that (2) holds, let A be any bi-ideal of S, and a and b be any elements of A. Then, since S is T^*-pure, it follows from Theorem 6.1.2 that a^3 and b^3 are both regular. Thus there exist elements x and y in S such that $a^3 = a^3xa^3$ and $b^3 = b^3yb^3$. We note that a^3x and b^3y are both idempotent. Hence $a^3x = b^3y$. Thus by Theorem 6.1.1, we have $a^3 \in aSa = a^3Sa^3 = (a^3xa^3)Sa^3 = (a^3x)(a^3Sa^3) = (b^3y)(a^3Sa^3) = bb\{b(ya^3Sa^2)a\} \subseteq Ab(ASA) \subseteq AbA$. This means that A is archimedean. Therefore, (3) implies (2). It is clear that (2) implies (1). ∎

Theorem 6.1.9 *Any cancellative archimedean semigroup without zero does not properly contain any two-sided pure bi-ideal.*

Proof. Let A be any two-sided pure bi-ideal of S. Let a and s be any elements of A and S, respectively. Since S is archimedean, there exist a positive integer n and elements $x, y \in S$ such that $a^n = xsy$. Since A is two-sided pure bi-ideal of S, we have

$$a^n = xsy \in A \cap xSy = xAy.$$

This implies that there exists an element $b \in A$ such that $xsy = xby$. Since S is cancellative and since x and y are both nonzero elements, we have $s = b \in A$, and so $S \subseteq A$. Thus $S = A$. ∎

Theorem 6.1.10 *For a T^*-pure archimedean semigroup S the following conditions are equivalent.*

(1) S is a group.

(2) S is a regular semigroup.

Proof. It is clear that (1) implies (2). Assume that (2) holds. Let A be any bi-ideal of S. Let a and b be any elements of A and S, respectively. Then, since S is archimedean, there exists a positive integer n such that $b^n \in SaS$.

On the other hand, since S is T^*-pure, the bi-ideal aSa is two-sided pure. Thus by the regularity of S and by Theorem 6.1.1, we have $b \in bSb = b^n Sb^n \subseteq (SaS)S(SaS) = S\{a(SSS)a\}S \subseteq S(aSa)S = aSa \cap SSS \subseteq aSa \subseteq ASA \subseteq A$, and so we have $S \subseteq A$. Therefore, $A = S$. Hence it follows from [17, p.84] that S is a group. Thus (2) implies (1). ■

Theorem 6.1.11 *Let S be a T^*-pure archimedean semigroup with idempotent e. Then $e \in A$ for any bi-ideal A of S.*

Proof. Let A be any bi-ideal of S and a any element of A. Then, since the bi-ideal aSa is two-sided pure, and S is archimedean, there exists a positive integer n such that $e = e^n \in SaS$, and so $e = ee = e^n e^n \in (SaS)(SaS) = S(a(SS)a)S \subseteq S(aSa)S = aSa \cap SSS \subseteq aSa \subseteq ASA \subseteq A$. ■

For each idempotent e of a semigroup S, we set

$$G_e = \{a \in S \mid ae = ea = a, xa = ax = e \text{ for some } x, y \in S\}.$$

Then, as is well-known (A.H. Clifford and G.B. Preston [17, p.23]), G_e is the maximal subgroup of S.

Theorem 6.1.12 *For a T^*-pure archimedean semigroup S with idempotent e, $G_e = S^3$.*

Proof. Let a be any element of the group G_e. Then, since e is the identity of G_e, we have $a = eae \in S^3$ and so $G_e \subseteq S^3$.

Conversely, let $a = bcd$ $(b, c, d \in S)$ be any element of S^3. Then it follows from Theorem 6.1.2 that b^3 and d^3 are both regular elements of S. Thus there exists elements $x, y \in S$ such that $b^3 = b^3xb^3$ and $d^3 = d^3yd^3$. We note that b^3x and yd^3 are both idempotent. Hence it follows Theorem 6.1.8 that $b^3x = yd^3 = e$. Thus

$$\begin{aligned} a = bcd \in bSd &= b^3Sd^3 \\ &= (b^3xb^3)S(d^3yd^3) \\ &= (b^3x)(b^3Sd^3)(yd^3) \\ &\subseteq eSe. \end{aligned}$$

This implies that there exists an element $z \in S$ such that $a = eze$. Thus we have

$$ae = (eze)e = ez(ee) = eze = a.$$

Note that by Theorem 6.1.3

$$ae = ea.$$

On the other hand, a^3 is regular by Theorem 6.1.2. Hence there exists an element $u \in S$ such that $a^3 = a^3ua^3$. Thus it follows from Theorem 6.1.8 that

$$a^3u = ua^3 = e.$$

This implies that

$$a(a^2u) = (ua^2)a = e.$$

Hence $a \in G_e$, and so $S^3 \subseteq G_e$. Therefore, $G_e = S^3$. ■

6.2 Semilattice Fuzzy Congruences

A congruence β on a semigroup S is called a **semilattice congruence** if S/β is a semilattice. A fuzzy congruence β on S is called a **semilattice fuzzy congruence** if S/β is a semilattice and is called a **group congruence** if S/β is a group.

Theorem 6.2.1 *Let S be a T^*-pure semigroup and δ the binary relation on S defined by*

$$\delta = \{(a, b) \in S \times S \mid a^3 \in bSb \text{ and } b^3 \in aSa\}.$$

Then δ is the least semilattice congruence on S.

Proof. It is clear that δ is reflexive and symmetric. To see that δ is transitive, let $(a, b) \in \delta$ and $(b, c) \in \delta$. Then $a^3 \in bSb$, $b^3 \in aSa$, $b^3 \in cSc$, and $c^3 \in bSb$. Thus by Theorem 6.1.1, we have

$$a^3 \in bSb = b^3Sb^3 \subseteq (cSc)S(cSc) = c(ScScS)c \subseteq cSc$$

and

$$c^3 \in bSb = b^3Sb^3 \subseteq (aSa)S(aSa) = a(SaSaS)a \subseteq aSa.$$

This implies that $(a, c) \in \delta$. To see δ is compatible, let $(a, b) \in \delta$ and x any element of S. Then $a^3 \in bSb$ and $b^3 \in aSa$. Since $bxSbx$ is a two-sided pure bi-ideal of S, by Theorem 6.1.1, we have

$$\begin{aligned}(ax)^3 = axaxax \in ax(aSa)x &= ax(a^3Sa^3)x \\ &\subseteq ax(bSb)S(bSb)x \\ &= a(xb)(SbSbSbS)(bx) \\ &\subseteq a(xbSbx) \\ &= a(xb)^2S(bx)^2 \\ &= ax(bx)(bS)(bx)bx \\ &\subseteq ax(bxSbx)bx \\ &= bxSbx \cap axSbx \\ &\subseteq bxSbx.\end{aligned}$$

Similarly, we can prove that $(bx)^3 \in axSax$. Therefore, $(ax, bx) \in \delta$.

Similarly we can prove that $(xa, xb) \in \delta$. Thus δ is a congruence on S. To see that δ is a semilattice congruence on S, let a and b be any elements of S. Then by Theorem 6.1.1, we have

$$a^3 \in aSa = a^2Sa^2$$

and

$$(a^2)^3 = aa^4a \in aSa.$$

Hence $(a, a^2) \in \delta$. On the other hand, since the set $baSba$ is a two-sided bi-ideal of S, we have by Theorem 6.1.1 that

$$\begin{aligned}(ab)^3 \in abSab &= (ab)^2S(ab)^2 \\ &= a(ba)(bSa)(ba)b \\ &\subseteq a(baSba)b \\ &= baSba \cap aSb \\ &\subseteq baSba.\end{aligned}$$

Similarly, we can see that $(ba)^3 \in abSab$. Therefore, $(ab, ba) \in \delta$. Thus δ is a semilattice congruence on S. Let β be any semilattice congruence on S. To see that $\delta \subseteq \beta$, let $(a, b) \in \delta$. Then there exist elements $x, y \in S$ such that $a^3 = bxb$ and $b^3 = aya$. Thus $a\beta = (a\beta)^3 = a^3\beta = (bxb)\beta = (b\beta)(xb\beta) = (b\beta)^2(xb\beta) = (b\beta)\{(b\beta)(bxb\beta)\} = (b\beta)(bxb\beta) = (b\beta)(a^3\beta) = (b\beta)(a\beta)^3 = (b\beta)(a\beta) = (b\beta)^3(a\beta) = (b^3\beta)(a\beta) = (aya\beta)(a\beta) = (ay\beta)(a\beta)^2 = (ay\beta)(a\beta) = aya\beta = b^3\beta = (b\beta)^3 = b\beta$ and so $(a, b) \in \beta$. Therefore, $\delta \subseteq \beta$ and so δ is the least semilattice congruence on S. ■

The following theorem can be easily proved.

Theorem 6.2.2 *Let S be a T^*-pure semigroup and δ the least semilattice congruence on S defined in Theorem 6.2.1. Then the characteristic function C_δ of δ is a semilattice fuzzy congruence on S.*

Theorem 6.2.3 *Let μ be any fuzzy congruence on a T^*-pure semigroup S such that $C_\delta \subseteq \mu$. Then μ is a semilattice fuzzy congruence on S.*

Proof. Let a and b be any elements of S. Then, since $(a, a^2) \in \delta$, we have

$$\mu(a, a^2) \geq C_\delta(a, a^2) = 1,$$

and so

$$\mu(a, a^2) = 1.$$

Thus

$$\mu_a = \mu_{a^2} = (\mu_a)^2.$$

Since $(ab, ba) \in \delta$,

$$\mu(ab, ba) \geq C_\delta(ab, ba) = 1,$$

and so

$$\mu(ab, ba) = 1.$$

Hence

$$\mu_a * \mu_b = \mu_{ab} = \mu_{ba} = \mu_b * \mu_a.$$

Thus S/μ is a semilattice, that is, μ is a semilattice fuzzy congruence on S. ■

Theorem 6.2.4 *For a fuzzy congruence μ on a T^*-pure semigroup S, the following conditions are equivalent.*

(1) μ is a semilattice fuzzy congruence on S.

(2) $\delta \subseteq \mu^{-1}(1)$.

Proof. Assume that (1) holds. Let $(a, b) \in \delta$. Then there exist elements $x, y \in S$ such that

$$a^3 = bxb \qquad \text{and} \qquad b^3 = aya.$$

Since μ is a semilattice fuzzy congruence on S, we have $\mu_a = (\mu_a)^3 = \mu_{a^3} = \mu_{bxb} = (\mu_b) * (\mu_{xb}) = (\mu_b)^2 * (\mu_{xb}) = (\mu_b) * \{(\mu_b) * (\mu_{xb})\} = (\mu_b) * (\mu_{bxb}) = (\mu_b) * (\mu_{a^3}) = (\mu_b) * (\mu_a)^3 = (\mu_b)^3 * (\mu_a) = (\mu_{b^3}) * (\mu_a) = (\mu_{aya}) * (\mu_a) = \{(\mu_{ay}) * (\mu_a)\} * (\mu_a) = (\mu_{ay}) * (\mu_a)^2 = (\mu_{ay}) * (\mu_a) = \mu_{aya} = \mu_{b^3} = \mu_b$. Hence it follows from Theorem 5.3.2 that $\mu(a, b) = 1$ and so $(a, b) \in \mu^{-1}(1)$. Thus we have $\delta \subseteq \mu^{-1}(1)$, and (1) implies (2).

Assume that (2) holds. Let a and b be any elements of S. Since $(a, a^2) \in \delta \subseteq \mu^{-1}(1)$ and $(ab, ba) \in \delta \subseteq \mu^{-1}(1)$, we have

$$\mu_a = \mu_{a^2} = (\mu_a)^2$$

and

$$(\mu_a) * (\mu_b) = \mu_{ab} = \mu_{ba} = (\mu_b) * (\mu_a).$$

Thus S/μ is a semilattice and so μ is a semilattice fuzzy congruence on S. ■

6.3 Group Fuzzy Congruences

First we give the least group congruence on a T^*-pure semigroup.

Theorem 6.3.1 *Let S be a T^*-pure semigroup and let σ be a relation on S defined by the rule that*

$$\sigma = \{(a, b) \in S \times S \mid ea = eb \text{ for some } e \in E(S)\}.$$

Then σ is the least group congruence on S.

Proof. It follows from Theorem 6.1.2 that $E(S)$ is nonempty. Thus it is clear that σ is reflexive. It is also clear that it is symmetric. To see that σ is transitive, let $(a, b) \in \sigma$ and $(b, c) \in (a, b, c \in S)$. Then there exist idempotents e and f of S such that

$$ea = eb \quad \text{and} \quad fb = fc.$$

Since fe is idempotent by Theorem 6.1.3, we have $(fe)a = f(ea) = f(eb) = (fe)b = (ef)b = e(fb) = e(fc) = (ef)c = (fe)c$. Hence $(a, c) \in \sigma$, and σ is transitive. To see that σ is compatible, let $(a, b) \in \sigma(a, b \in S)$ and let $x \in S$. Then there exists an idempotent $e \in S$ such that $ea = eb$. Thus by Theorem 6.1.3, we have $e(xa) = (ex)a = (xe)a = x(ea) = x(eb) = (xe)b = (ex)b = e(xb)$. Therefore, $(xa, xb) \in \sigma$.

Clearly, $(ax, bx) \in \sigma$. Thus σ is congruence on S. Let $e, f \in E(S)$. Then

$$(ef)e = e(fe) = e(ef) = (ee)f = ef = e(ff) = (ef)f.$$

Since $ef \in E(S)$, $(e, f) \in \sigma$, i.e., $e\sigma = f\sigma$. Let $a \in S$ and $e \in E(S)$. Then

$$ea = (ee)a = e(ea),$$

and so

$$a\sigma = (ea)\sigma = (e\sigma)(a\sigma).$$

Let $a \in S$. Then it follows from Theorem 6.1.2 that a^3 is regular and so there exists an element $x \in S$ such that $a^3 = a^3xa^3$. We note that $xa^3 \in E(S)$. Hence

$$((xa^2)\sigma)(a\sigma) = ((xa^2)a)\sigma = xa^3\sigma = e\sigma.$$

Thus S/σ is a group. Let μ be any group congruence on S. To prove that $\sigma \subseteq \mu$, let $(a, b) \in \sigma(a, b \in S)$. Then there exists $e \in E(S)$ such that $ea = eb$. Therefore,

$$(e\mu)(a\mu) = (ea)\mu = (eb)\mu = (e\mu)(b\mu).$$

Since $e\mu \in E(S/\mu)$, $e\mu$ is the identity of S/μ. Hence we have $a\mu = b\mu$, and so $(a, b) \in \mu$. Thus $\sigma \subseteq \mu$ and so σ is the least group congruence on S. ∎

Theorem 6.3.2 *Let S be a T^*-pure semigroup with semilattice $E(S)$ of idempotents, and σ the least group congruence on S defined in Theorem 6.3.1. The the following conditions are equivalent.*

(1) $(a,b) \in \sigma$.
(2) $ae = be$ *for some* $e \in E(S)$.
(3) $xa = xb$ *for some* $x \in S$.
(4) $ax = bx$ *for some* $x \in S$.
(5) $ea = fb$ *for some* $e, f \in E(S)$.
(6) $ae = bf$ *for some* $e, f \in E(S)$.

Proof. It follows from Theorem 6.1.3 that (1) and (2) are equivalent and (5) and (6) are equivalent. It is clear that (1) implies (5). Conversely, assume that (5) holds, that is, $ea = fb$ for some $e, f \in E(S)$. Since $E(S)$ is commutative, $(ef)a = ((ee)f)a = (e(ef))a = (e(fe))a = e((fe)a) = e(f(ea)) = e(f(fb)) = e((ff)) = e(fb) = (ef)b$. Since $ef \in E(S)$, $(a,b) \in \sigma$ and so (5) implies (1).

Clearly, (1) implies (3). Assume that (3) holds. Then $xa = xb$ for some $x \in S$. Since x^3 is regular by Theorem 6.1.2, there exists an element $y \in S$ such that $x^3 = x^3yx^3$. Note that $yx^3 \in E(S)$. Hence

$$(yx^3)a = (yx^2)(xa) = (yx^2)(xb) = (yx^3)b,$$

which implies that $(a,b) \in \sigma$. Thus (3) implies (1). It can be seen in a similar way that (2) and (4) are equivalent. ∎

Theorem 6.3.3 *Let μ be a fuzzy congruence on a T^*-pure semigroup S, and $a \in S$. Then $\mu_a \in E(S/\mu)$ if and only if $\mu_a = \mu_e$ for some $e \in E(S)$.*

Proof. If μ_a is idempotent in S/μ, then $\mu = \mu_{a^4}$. Since the bi-ideal aSa is two-sided pure, we have

$$a^4 \in aSa = a^4Sa^4.$$

This means that a^4 is a regular element. Let x be an inverse of a^4 in S. Then

$$a^4 = a^4xa^4 \qquad \text{and} \qquad x = xa^4x.$$

If $e = a^2xa^2$, then

$$e^2 = (a^2xa^2)(a^2xa^2) = a^2(xa^4x)a^2 = a^2xa^2 = e,$$

and so e is idempotent. Thus we have

$$\begin{aligned} \mu_e &= \mu_{a^2xa^2} \\ &= (\mu_{a^2}) * (\mu_x) * (\mu_{a^2}) \\ &= (\mu_a)^2 * (\mu_x) * (\mu_a)^2 \\ &= (\mu_a)^4 * (\mu_x) * (\mu_a)^4 \\ &= \mu_{a^4xa^4} \\ &= \mu_{a^4} \\ &= (\mu_a)^4 \\ &= \mu_a. \end{aligned}$$

■

The following theorem can be easily seen.

Theorem 6.3.4 *Let S be a T^*-pure semigroup. Then C_σ is a group fuzzy congruence on S.*

Theorem 6.3.5 *Let μ be a fuzzy congruence on a T^*-pure semigroup S such that $C_\sigma \subseteq \mu$. Then μ is a group fuzzy congruence on S.*

Proof. Let $e, f \in E(S)$. Then, as is stated in the proof of Theorem 6.3.1, we have $(e, f) \in \sigma$. Thus

$$\mu(e, f) \geq C_\sigma(e, f) = 1,$$

and so $\mu(e, f) = 1$. Hence it follows from Theorem 5.3.2 that $\mu_e = \mu_f$. Let e be any idempotent of S, and a any element of S. Then, since

$$e(ea) = (ee)a = ea,$$

we have $(ea, a) \in \sigma$. Thus

$$\mu(ea, a) \geq C_\sigma(ea, a) = 1,$$

and so $\mu(ea, a) = 1$. Hence it follows from Theorem 5.3.2 that

$$\mu_e * \mu_a = \mu_{ea} = \mu_a.$$

Therefore, μ_e is the identity of S/μ. ■

Theorem 6.3.6 *For a fuzzy congruence μ on a T^*-pure semigroup S, the following conditions are equivalent.*

(1) μ is a group fuzzy congruence.

(2) $\sigma \subseteq \mu^{-1}(1)$.

Proof. First assume that (1) holds. Let a and b be any elements of S such that $(a, b) \in \sigma$. Then there exists an idempotent e of S such that $ea = eb$. Since μ is a group fuzzy congruence on S, μ_e is the identity of S/μ. Thus

$$\mu_a = \mu_e * \mu_a = \mu_{ea} = \mu_{eb} = \mu_e * \mu_b = \mu_b.$$

Hence it follows from Theorem 5.3.2 that $\mu(a,b) = 1$. Thus $(a,b) \in \mu^{-1}(1)$, and so $\sigma \subseteq \mu^{-1}(1)$. Therefore, (1) implies (2).

Assume that (2) holds. If $(a,b) \in \sigma$, then $C_\sigma(a,b) = 1$. On the other hand, by the assumption, $(a,b) \in \mu^{-1}(1)$. Thus $\mu(a,b) = 1$, and so $C_\sigma(a,b) = \mu(a,b)$. If $(a,b) \notin \sigma$, then

$$C_\sigma(a,b) = 0 \leq \mu(a,b).$$

Therefore, $C_\sigma \subseteq \mu$. Hence it follows from Theorem 6.3.5 that μ is a group fuzzy congruence on S. ■

7

Prime Fuzzy Ideals

In this chapter, we characterize prime fuzzy ideals of a semigroup S. Sections 7.1-7.11 are essentially from [151]. We show that a nonconstant fuzzy ideal f of a semigroup S is prime if and only if f is two-valued and there exists an element x_0 in S such that $f(x_0) = 1$ and $f_1 = \{x \in S \mid f(x) = 1\}$ is a prime ideal of S. We also consider the concepts of weakly prime fuzzy ideals, completely prime fuzzy ideals and weakly completely prime fuzzy ideals of a semigroup S. We establish relations among four classes of prime ideals.

7.1 Preliminaries

Let S be a semigroup. Recall that a nonempty subset I of S is called an ideal of S if SI, $IS \subseteq I$. Let I be an ideal of S. Then I is called **prime** if for any two ideals A, B of S, $AB \subseteq I$ implies that $A \subseteq I$ or $B \subseteq I$. I is called **completely prime** if for any two elements $a, b \in S$, $ab \in I$ implies $a \in I$ or $b \in I$.

Let $a \in S$ and $t \in [0, 1]$. Define the fuzzy subset a_t of S as follows:

$$a_t(x) = \begin{cases} t & \text{if } x = a, \\ 0 & \text{otherwise,} \end{cases}$$

$\forall x \in S$. Then a_t is called a **fuzzy point** or **fuzzy singleton.**

For any fuzzy subset f of S, it is clear that $f = \cup_{a_t \subseteq f} a_t$.

Let A be a subset of S and $t \in [0,1]$. Define the fuzzy subset tC_A of S as follows:

$$tC_A(x) = \begin{cases} t & \text{if } x \in A, \\ 0 & \text{otherwise,} \end{cases}$$

$\forall x \in S$.

Recall that a fuzzy subset f of S is called a fuzzy ideal of S if $C_S \circ f \subseteq f$ and $f \circ C_S \subseteq f$, or equivalently, if

$$f(xy) \geq f(x) \vee f(y)$$

$\forall x, y \in S$.

Definition 7.1.1 *A fuzzy ideal f of S is called* **prime** *if for any two fuzzy ideals g and h of S, $g \circ h \subseteq f$ implies $g \subseteq f$ or $h \subseteq f$.*

It can be shown that any improper fuzzy subset f (f is a constant function) is a prime fuzzy ideal of S.

By the definition of the product of two fuzzy subsets of S, we have the following result.

Lemma 7.1.2 *Let f, g and h be fuzzy subsets of S. Then $f \circ (g \cup h) = f \circ g \cup f \circ h$.* ■

Let f be a fuzzy subset of S. Let (f) denote the intersection of all fuzzy ideals of S which contain f. Then (f) is a fuzzy ideal of S, called the **fuzzy ideal generated** by f.

Lemma 7.1.3 *Let a_t be a fuzzy point of S. Then the fuzzy ideal generated by a_t denoted by (a_t), is*

$$(a_t)(x) = \begin{cases} t & \text{if } x \in (a), \\ 0 & \text{otherwise,} \end{cases}.$$

for any $x \in S$, where (a) is an ideal of S generated by a. ■

Lemma 7.1.4 *Let a_t be a fuzzy point of S. Then the following statements hold:*

(1) For all $x \in S$

$$(S \circ a_t \circ S)(x) = \begin{cases} t & \text{if } x \in SaS, \\ 0 & \text{otherwise.} \end{cases}.$$

(2) $a_t \circ b_r = (ab)_{t \wedge r}$ for all fuzzy points b_r of S.
(3) $(a_t) = a_t \cup a_t \circ S \cup S \circ a_t \cup S \circ a_t \circ S$.
(4) $(a_t)^3 = S \circ a_t \circ S$.

Proof. The proofs of (1), (2), and (3) can be found in [144]. We only prove statement (4). By (2), we have

$$(a_t)^3 = (a_t \cup a_t \circ S \cup S \circ a_t \cup S \circ a_t \circ S)^2 \circ (a_t).$$

Since

$$\begin{aligned}(a_t)^2 &= (a_t \cup a_t \circ S \cup S \circ a_t \cup S \circ a_t \circ S)\circ \\ &\quad (a_t \cup a_t \circ S \cup S \circ a_t \cup S \circ a_t \circ S) \\ &\subseteq S \circ (a_t \cup a_t \circ S \cup S \circ a_t \cup S \circ a_t \circ S) \\ &\subseteq S \circ a_t \cup S \circ a_t \circ S \text{ (by Lemma 7.1.2)},\end{aligned}$$

we have

$$\begin{aligned}(a_t)^3 &\subseteq (S \circ a_t \cup S \circ a_t \circ S) \circ (a_t \cup a_t \circ S \cup S \circ a_t \cup S \circ a_t \circ S) \\ &\subseteq (S \circ a_t \cup S \circ a_t \circ S) \\ &\subseteq S \circ a_t \circ S.\end{aligned}$$

■

As a generalization of Lemma 2.7 in [70], we have the following result.

Lemma 7.1.5 *Let A be a subset of S. Then for any $t \in (0,1]$ the following statements hold:*

(1) $tC_A \circ tC_B = tC_{AB}$.

(2) $tC_A \cap tC_B = tC_{A\cap B}$.

(3) $tC_A = \cup_{a\in A} a_t$.

(4) $S \circ tC_A = tC_{SA}$.

(5) If A is an ideal (right ideal, left ideal) of S, then tC_A is a fuzzy ideal (fuzzy right ideal, fuzzy left ideal) of S. ■

7.2 Prime Fuzzy Ideals

In this section, we assume that f is a nonconstant fuzzy ideal of a semigroup S.

Theorem 7.2.1 *Let I be an ideal of S. Then I is a prime ideal of S if and only if C_I is a prime fuzzy ideal of S.*

Proof. Let I be an ideal of S. Then by Lemma 7.1.5(5), C_I is a fuzzy ideal of S. Let f and g be fuzzy ideals of S such that $f \circ g \subseteq C_I$. If $f \not\subseteq C_I$, then there exists a fuzzy point $x_t \subseteq f (t > 0)$ such that $x_t \not\subseteq C_I$. For any $y_r \subseteq g (r \neq 0)$, since

$$(x_t) \circ (y_r) \subseteq f \circ g \subseteq C_I,$$

$\forall z \in S$,

$$(x_t) \circ (y_r)(z) = \begin{cases} t \wedge r > 0 & \text{if } z \in (x)(y), \\ 0 & \text{otherwise.} \end{cases}$$

Thus $(x)(y) \subseteq I$. By hypothesis, it follows that $(x) \subseteq I$ or $(y) \subseteq I$. Since $x_t \nsubseteq C_I$, we have $x \notin I$ and so $(x) \nsubseteq I$. Hence $(y) \subseteq I$, which implies $y_r \subseteq C_I$. Therefore, $g = \cup_{y_r \subseteq g} y_r \subseteq C_I$.

Conversely, let A and B be ideals of S. Suppose $AB \subseteq I$. Then by Lemma 7.1.5, C_A, C_B are fuzzy ideals of S and $C_A \circ C_B = C_{AB} \subseteq C_I$. By hypothesis, we have $C_A \subseteq C_I$ or $C_B \subseteq C_I$. Thus $A \subseteq I$ or $B \subseteq I$. ■

Lemma 7.2.2 *If f is a prime fuzzy ideal of S, then $|\operatorname{Im}(f)| = 2$.*

Proof. Since f is a nonconstant prime fuzzy ideal, we have $|\operatorname{Im}(f)| \geq 2$. Suppose $|\operatorname{Im}(f)| \geq 3$. Then there exist x, y and $z \in S$ such that $f(x)$, $f(y)$, and $f(z)$ are pairwise unequal. Without loss of generality, we assume that $f(x) < f(y) < f(z)$. Let $r, t \in (0,1)$ be such that $f(x) < r < f(y) < t < f(z)$. Then $\forall u \in S$,

$$(x_r) \circ (y_t)(u) = \begin{cases} r \wedge t = r & \text{if } u \in (x)(y), \\ 0 & \text{otherwise.} \end{cases}$$

If $u \in (x)(y)$, then there exist $s_1, s_2, s_3, s_4 \in S^1$ such that $u = s_1 x s_2 s_3 y s_4$. Thus

$$f(u) \geq f(s_1 x s_2 s_3 y s_4) \geq f(x) \vee f(y) = f(y) > r.$$

Therefore, $(x_r) \circ (y_t) \subseteq f$ holds, which implies $(x_r) \subseteq f$ or $(y_t) \subseteq f$, say $(x_r) \subseteq f$. Then $f(x) \geq (x_r)(x) = r$, which is impossible. ■

Theorem 7.2.3 *If f is a prime fuzzy ideal of S, then there exists an $x_0 \in S$ such that $f(x_0) = 1$.*

Proof. By Lemma 7.2.2, $|\operatorname{Im}(f)| = 2$. If for all $x \in S$, $f(x) < 1$, then $\operatorname{Im}(f) = \{t, s\}$, $t < s < 1$. Thus there exist $x, y \in S$ and $m \in (0,1]$ such that $f(x) = t < s = f(y) < m \leq 1$. Let $t_1, t_2 \in (0,1)$ be such that $t < t_1 < s < t_2 < m$. Then in a manner similar to the proof of Lemma 7.2.2, we have $(x_{t_1}) \circ (y_{t_2}) \subseteq f$. Since f is a prime fuzzy ideal of S, we have $(x_{t_1}) \subseteq f$ or $(y_{t_2}) \subseteq f$, that is, $f(x) \geq t$, or $f(y) \geq t_2$, which is impossible. ■

Theorem 7.2.4 *If f is a prime fuzzy ideal of S, then each level subset f_t, $t \in (0,1]$, is a prime ideal of S.*

Proof. Clearly, for any $t \in (0,1]$, f_t is an ideal of S. Let I and J be ideals of S such that $IJ \subseteq f_t$ and let $g = tC_I$ and $h = tC_J$. Then by Lemma 7.1.5, g and h are fuzzy ideals of S. Furthermore, $g \circ h \subseteq f$. In fact, if $g \circ h(x) = 0$ $\forall x \in S$, then the desired result holds trivially. If $g \circ h(x) \neq 0$ for some $x \in S$, then there exist $y_0, z_0 \in S$ such that $x = y_0 z_0$ and $g(y_0) \wedge h(z_0) \neq 0$. Thus $y_0 \in I$ and $z_0 \in J$, and so $x \in IJ \subseteq f_t$. That is, $f(x) \geq t$. Therefore,

$$g \circ h(x) = \vee_{x=yz}(g(y) \wedge h(z)) \leq t \leq f(x),$$

i.e., $g \circ h \subseteq f$.

Since f is a prime fuzzy ideal of S, we have $g \subseteq f$ or $h \subseteq f$, say $g \subseteq f$. Then for any $x \in I$, $g(x) = t \leq f(x)$. Thus $I \subseteq f_t$. ∎

By Theorems 7.2.3 and 7.2.4, we have the following result.

Corollary 7.2.5 *If f is a prime fuzzy ideal of S, then C_I is a prime ideal of S.*

Remark 7.2.6 *The converse of Theorem 7.2.4 is not true. For example, let I be a prime ideal of S and let*

$$f(x) = \begin{cases} t & \text{if } x \in I \\ 0 & x \notin I, \end{cases}$$

where $0 < t < 1$. Then f is a fuzzy ideal of S. But f is not a prime fuzzy ideal of S by Theorem 7.2.3.

Now we give a characterization of prime fuzzy ideals of S.

Theorem 7.2.7 *Let f be a fuzzy subset of S. Then f is a prime fuzzy ideal of S if and only if f satisfies the following conditions:*

(1) $|\mathrm{Im}(f)| = 2$.

(2) $f_1 \neq \emptyset$ *and* f_1 *is a prime ideal of* S.

Proof. Let f be a prime fuzzy ideal of S. Then by Lemma 7.2.2, Theorem 7.2.3 and Corollary 7.2.5, (1) and (2) hold.

Conversely, let $\mathrm{Im}(f) = \{t, 1\}$ $(t < 1)$. Then f is a fuzzy ideal of S. We now show that f is prime. Let g and h be fuzzy ideals of S such that $g \circ h \subseteq f$. If $g \nsubseteq f$ and $h \nsubseteq f$, then there exist $x, y \in S$ such that $g(x) > f(x)$ and $h(y) > f(y)$. In this case, $x, y \notin f_1$, which implies $xSy \nsubseteq f_1$ else $(SxS)(SyS) \subseteq f_1$ and by condition (2), we have $(SxS) \subseteq f_1$ or $(SyS) \subseteq f_1$, say $(SxS) \subseteq f_1$. Thus $(x)^3 \subseteq f_1$ and it follows that $x \in (x) \subseteq f_1$, which is impossible.

Since $xSy \nsubseteq f_1$, there exists $s \in S$ such that $xsy \notin f_1$. Let $a = xsy$. Then $f(a) = f(x) = f(y) = t$. Thus $g \circ h(a) \geq g(x) \wedge h(sy) \geq g(x) \wedge h(y) > t = f(a)$. This contradicts the fact that $g \circ h \subseteq f$. ∎

Theorem 7.2.8 *Let f be a prime fuzzy ideal of S. Then there exists a proper prime fuzzy ideal g of S such that $f \subset g$.*

Proof. Let f be a prime fuzzy ideal of S. Then by Theorem 7.2.7, there exists an element $x_0 \in S$ such that $f(x_0) = 1$ and $\mathrm{Im} f = \{t, 1\}$, where

$t < 1$. Let g be the fuzzy subset of S defined by $\forall x \in S$, $g(x) = 1$ if $x \in f_1$ and $g(x) = r$ if $x \notin f_1$, where $t < r < 1$. ■

7.3 Weakly Prime Fuzzy Ideals

Definition 7.3.1 *A fuzzy ideal f of S is called a* ***weakly prime fuzzy ideal*** *of S if for all ideals A and B of S and for all $t \in (0,1]$, $tC_A \circ tC_B \subseteq f$ implies $tC_A \subseteq f$ or $tC_B \subseteq f$.*

Theorem 7.3.2 *Let f be a fuzzy ideal of S. Then f is a weakly prime fuzzy ideal of S if and only if $f_t (t > 0)$ is a prime ideal of S for $f_t \neq \emptyset$.*

Proof. Let A and B be ideals of S. For any $t \in (0,1]$, if $AB \subseteq f_t$, then $tC_{AB} \subseteq f$, that is, $tC_A \circ tC_B = tC_{AB} \subseteq f$ (Lemma 7.1.5). Since f is a weakly prime fuzzy ideal of S, we have $tC_A \subseteq f$ or $tC_B \subseteq f$. Say $tC_A \subseteq f$. Then $A \subseteq f_t$.

Conversely, let A and B be ideals of S such that $tC_A \circ tC_B \subseteq f (t > 0)$, that is $tC_{AB} \subseteq f$. Then $AB \subseteq f_t$. Since f_t is a prime ideal of S, we have $A \subseteq f_t$ or $B \subseteq f_t$, say $A \subseteq f_t$. Then $tC_A \subseteq f$. ■

Theorem 7.3.3 *Let f be a fuzzy ideal of S. Then the following statements are equivalent.*

(1) f is a weakly prime fuzzy ideal of S.

(2) For any $x, y \in S$ and $r \in (0,1]$, if $x_r \circ S \circ y_r \subseteq f$, then $x_r \subseteq f$ or $y_r \subseteq f$.

(3) For any $x, y \in S$ and $r \in (0,1]$, if $(x_r) \circ (y_r) \subseteq f$, then $x_r \subseteq f$ or $y_r \subseteq f$.

(4) If A and B are right ideals of S such that $tC_A \circ tC_B \subseteq f$, then $tC_A \subseteq f$ or $tC_B \subseteq f$.

(5) If A and B are left ideals of S such that $tC_A \circ tC_B \subseteq f$, then $tC_A \subseteq f$ or $tC_B \subseteq f$.

(6) If A is a right ideal of S and B is left ideal of S such that $tC_A \circ tC_B \subseteq f$, then $tC_A \subseteq f$ or $tC_B \subseteq f$.

Proof. (1) $\Rightarrow$ (2): Let f be a weakly prime fuzzy ideal of S. For all $x, y \in S$ and $r \in (0,1]$, if $x_r \circ S \circ y_r \subseteq f$, then, by Lemma 7.1.4,

$$\begin{aligned} rC_{SxS} \circ rC_{SyS} &= (S \circ x_r \circ S) \circ (S \circ y_r \circ S) \\ &\subseteq S \circ (x_r \circ S \circ y_r) \circ S \\ &\subseteq S \circ f \circ S \\ &\subseteq f. \end{aligned}$$

Thus $S \circ x_r \circ S \subseteq f$ or $S \circ y_r \circ S \subseteq f$, say $S \circ x_r \circ S \subseteq f$. Then $(x_r)^3 \subseteq S \circ x_r \circ S \subseteq f$. Since f is weakly prime, $(x_r) = rC_{(x)} \subseteq f$. Hence $x_r \in (x_r) \subseteq f$.

(2) $\Rightarrow$ (3): Suppose for any $x, y \in S$ and $r \in (0,1]$, $(x_r) \circ (y_r) \subseteq f$. Then by Lemma 7.1.4, $x_r \circ S \circ y_r \subseteq (x_r) \circ (y_r) \subseteq f$. Thus we have $x_r \subseteq f$ or $y_r \subseteq f$.

(3) $\Rightarrow$ (4). Let A and B be right ideals of S. Then tC_A and tC_B are fuzzy right ideals of S. If $tC_A \circ tC_B \subseteq f$ $(\forall t > 0)$ and $tC_A \not\subseteq f$, then there exists an $x \in A$ such that $x_t \not\subseteq f$. For any $y \in B$, by Lemma 7.1.5 and hypothesis, we have

$$\begin{aligned}(x_t) \circ (y_t) &= tf_{(x)} \circ tf_{(y)} \\ &= tf_{(x)(y)} \\ &\subseteq tC_A \\ &= tC_A \cup tC_A \\ &= (tC_A \circ tC_B) \cup (S \circ tC_A \circ tC_B) \\ &\subseteq f.\end{aligned}$$

Since $x_t \not\subseteq f$, $(x_t) \not\subseteq f$, $(y_t) \subseteq f$. Thus

$$tC_B = \cup_{y \in B}(y_t) \subseteq f.$$

(4) $\Rightarrow$ (1), (5) $\Rightarrow$ (1) and (6) $\Rightarrow$ (1) are obvious.

(3) $\Rightarrow$ (5) is similar to that of (3) $\Rightarrow$ (4).

(3) $\Rightarrow$ (6): Let A be a right ideal of S and B be a left ideal of S. If $tC_A \circ tC_B \subseteq f$ $(\forall t > 0)$ and $tC_A \not\subseteq f$, then there exists an $x \in A$ such that $x_t \not\subseteq f$. For any $y \in B$, since $(x_t) \circ (y_t) = tC_{(x)(y)} \subseteq tC_{(A \cup SA)(B \cup BS)} = tC_A$ $= (tC_A \circ tC_B) \cup S \circ (tC_A \circ tC_B) \cup (tC_A \circ tC_B) \circ S \cup S \circ (tC_A \circ tC_B) \circ S \subseteq f$ and $(x_t) \not\subseteq f$, we have $(y_t) \subseteq f$. It follows that $tC_B = \cup_{y \in B}(y_t) \subseteq f$.

Similarly, it follows that if $tC_B \not\subseteq f$, then $tC_A \subseteq f$. ∎

7.4 Completely Prime and Weakly Completely Prime Fuzzy Ideals

Definition 7.4.1 *A fuzzy ideal f of S is called a* ***completely prime fuzzy ideal*** *if for any two fuzzy points x_t, y_r of S $(t, r \in (0,1])$, $x_t \circ y_r \subseteq f$ implies $x_t \subseteq f$ or $y_r \subseteq f$.*

Definition 7.4.2 *A fuzzy ideal f of S is called a* ***weakly completely prime fuzzy ideal*** *if for any fuzzy points x_t, y_t of $S(t \in (0,1])$, $x_t \circ y_t \subseteq f$ implies $x_t \subseteq f$ or $y_t \subseteq f$.*

Theorem 7.4.3 *A fuzzy ideal f of S is completely prime if and only if for any fuzzy subsets f and g of S, $g \circ h \subseteq f$ implies $g \subseteq f$ or $h \subseteq f$.*

Proof. Assume that g and h are fuzzy subsets of S satisfying $g \circ h \subseteq f$ and $g \not\subseteq f$. Then there exists an $x_t \subseteq g$ such that $x_t \not\subseteq f$. Since $g \circ h = \cup_{x_t \subseteq g,\, y_r \subseteq h} x_t \circ y_r$, $x_t \circ y_r \subseteq g \circ h \subseteq f$ for all $y_r \subseteq h$. Thus $y_r \subseteq f$. Since $h = \cup_{y_r \subseteq h} y_r$, $h \subseteq f$.

The "only if" part is obvious. ∎

Theorem 7.4.4 *A fuzzy ideal f of S is weakly completely prime if and only if* $\forall x, y \in S$, $f(xy) = f(x) \vee f(y)$.

Proof. The proof is similar to that in the rings (see [156]). ∎

Theorem 7.4.5 *Let f be a fuzzy ideal of S. Then f is weakly completely prime if and only if f_t $(0 < t \leq 1)$ is a completely prime ideal of S for $f_t \neq \emptyset$.*

Proof. Let $xy \in f_t$ be such that $xy \in f_t$. Then $f(xy) = f(x) \vee f(y) \geq t$. Hence $f(x) \geq t$ or $f(y) \geq t$. Thus $x \in f_t$ or $y \in f_t$.

Conversely, let $f(xy) = t$ for $xy \in S$. Then $xy \in f_t$. Since f_t is a completely prime ideal of S, we have $x \in f_t$ or $y \in f_t$. Hence $f(x) \geq t$ or $f(y) \geq t$. Thus $f(xy) \leq f(x) \vee f(y)$. By the hypothesis and Theorem 7.4.4, we have $f(xy) = f(x) \vee f(y)$. Hence it follows that f is weakly completely prime. ∎

7.5 Relationships

By Theorems 7.4.3 and 7.4.4 and Definition 7.4.2, we have the following result.

Theorem 7.5.1 *Let f be a fuzzy ideal of S. Then the following statements hold:*

(1) If f is a completely prime fuzzy ideal of S, then f is a prime fuzzy ideal and a weakly completely prime fuzzy ideal of S.

(2) If f is a prime fuzzy ideal of S, then f is a weakly prime fuzzy ideal of S.

(3) If f is a weakly completely prime fuzzy ideal of S, then f is a weakly prime fuzzy ideal of S. ∎

By Theorems 7.2.1 and 7.5.1, the following corollary holds.

Corollary 7.5.2 *Let I be a prime ideal of S. Then C_I is a weakly prime fuzzy ideal of S.* ∎

Theorem 7.5.3 *Let S be a commutative semigroup and f be a fuzzy ideal of S. Then the following assertions hold.*

(1) f is completely prime if and only if f is prime.

(2) f is weakly completely prime if and only if f is weakly prime.

Proof. (1) Let f be a prime fuzzy ideal of S and $x_t, y_r \subseteq C_S$. If $x_t \circ y_r \subseteq f$, then, since S is commutative, we have $(x_t) \circ (y_r) = (t \wedge r)C_{(x)(y)} \subseteq (t \wedge r)C_{(xy)} = ((xy)_{t \wedge r}) = (x_t \circ y_r)$. Thus $(x_t) \circ (y_r) \subseteq (x_t \circ y_r) \subseteq f$. Hence $x_t \subseteq (x_t) \subseteq f$ or $y_r \subseteq (y_r) \subseteq f$. Therefore, f is completely prime. The proof of (2) is similar to that of (1). ∎

By Theorem 7.5.1, the reverse implications of (1) and (2) are clear.

7.6 Types of Prime Fuzzy Left Ideals

We now examine prime, quasi-prime and weakly quasi-prime fuzzy left ideals of a semigroup S. The results are from [152]. We characterize prime and quasi-prime fuzzy left ideals of S by fuzzy points of S. We also introduce the concept of fuzzy m-systems and prove that a fuzzy left ideal f of S is quasi-prime if and only if $1-f$ is a fuzzy m-system. We describe the largest fuzzy ideal $i(f)$ of S contained in a fuzzy left ideal f of S and the largest fuzzy subsemigroup $I(f)$ of S such that the fuzzy left ideal f is a fuzzy ideal of $I(f)$. We also characterize the strongly semisimple semigroups by quasi-prime fuzzy left ideals of S. We show that S is strongly semisimple if and only if each fuzzy left ideal of S is the intersection of all quasi-prime left ideals of S containing it.

S always denotes a semigroup. Recall that a nonempty subset I of S is called a left ideal of S if $SI \subseteq I$. Let L be a left ideal of S. L is called **prime** if for any two ideals A, B of S such that $AB \subseteq L$, either $A \subseteq L$ or $B \subseteq L$. L is called **quasi-prime** if for any two left ideals L_1 and L_2 of S $L_1L_2 \subseteq L$ implies $L_1 \subseteq L$ or $L_2 \subseteq L$. L is called **weakly quasi-prime** if for any two left ideals of L_1 and L_2 of S such that $L \subseteq L_1$, $L \subseteq L_2$ and $L_1L_2 \subseteq L$, $L_1 = L$ or $L_2 = L$.

Recall that a fuzzy subset f of S is called a fuzzy left ideal of S if $S \circ f \subseteq f$ and f is called a fuzzy right ideal of S if $f \circ S \subseteq f$. If f is both a fuzzy left ideal and a fuzzy right ideal of S, then f is called a fuzzy ideal of S. f is a fuzzy ideal of S if and only if $f(xy) \geq f(x) \vee f(y)$ $\forall x, y \in S$.

By definition of product of two fuzzy subsets of S, we have the following result.

Lemma 7.6.1 *Let f, g and h be fuzzy subsets of S. Then the following statements hold:*

(1) $f \circ (g \cup h) \subseteq f \circ g \cup f \circ h$.

(2) $f \circ (g \cap h) \subseteq f \circ g \cap f \circ h$.

(3) If f_1, f_2 are fuzzy subsets of S such that $f_1 \subseteq f$, $f_2 \subseteq g$, then $f_1 \circ f_2 \subseteq f \circ g$.

Proof. The proof is straightforward. ■

Lemma 7.6.2 [144, 150] *Let a_t be a fuzzy point of S. Then the following conditions hold:*

(1) The fuzzy left ideal generated by a_t, denoted by $L(a_t)$, is such that $\forall x \in S$,

$$L(a_t)(x) = \begin{cases} t & \text{if } x \in L(a) \\ 0 & \text{otherwise,} \end{cases}$$

where $L(a)$ is a left ideal of S generated by a.

(2) The fuzzy ideal generated by a_t, denoted by (a_t), is such that $\forall x \in S$,

$$(a_t)(x) = \begin{cases} t & \text{if } x \in (a) \\ 0 & \text{otherwise,} \end{cases}$$

where (a) is an ideal of S generated by a. ■

Lemma 7.6.3 *Let a_t be a fuzzy point of S. Then the following assertions hold:*

(1) For all $x \in S$

$$(S \circ a_t \circ S)(x) = \begin{cases} t & \text{if } x \in SaS \\ 0 & \text{otherwise.} \end{cases}$$

(2) $a_t \circ b_r = (ab)_{t \wedge r}$ for all fuzzy points b_r of S.
(3) $(a_t) = a_t \cup a_t \circ S \cup S \circ a_t \cup S \circ a_t \circ S$; $L(a_t) = a_t \cup S \circ a_t$.
(4) $(a_t)^3 \subseteq S \circ a_t \circ S$.

Proof. We only show that (4) holds. By (3),

$$(a_t)^3 = (a_t \cup a_t \circ S \cup S \circ a_t \cup S \circ a_t \circ S)^2 \circ (a_t).$$

Now

$$\begin{aligned} (a_t)^2 &= (a_t \cup a_t \circ S \cup S \circ a_t \cup S \circ a_t \circ S) \circ \\ &\quad (a_t \cup a_t \circ S \cup S \circ a_t \cup S \circ a_t \circ S) \\ &\subseteq S \circ (a_t \cup a_t \circ S \cup S \circ a_t \cup S \circ a_t \circ S) \\ &\subseteq S \circ a_t \cup S \circ a_t \circ S \qquad \text{(by Lemma 7.6.1).} \end{aligned}$$

Thus

$$\begin{aligned} (a_t)^3 &\subseteq (S \circ a_t \cup S \circ a_t \circ S) \circ (a_t \cup a_t \circ S \cup S \circ a_t \cup S \circ a_t \circ S) \\ &\subseteq (S \circ a_t \cup S \circ a_t \circ S) \\ &\subseteq S \circ a_t \circ S. \end{aligned}$$

■

Lemma 7.6.4 *Let A be a subset of S. Then for any $t \in [0,1]$ the following statements hold:*

(1) $tC_A \circ tC_B = tC_{AB}$.
(2) $tC_A \cap tC_B = tC_{A \cap B}$.
(3) $tC_A = \cup_{a \in A} a_t$.
(4) $S \circ tC_A = tC_{SA}$.
(5) If A is an ideal (right, left ideal) of S, then tC_A is a fuzzy ideal (fuzzy right, fuzzy left ideal) of S.

Proof. The proof follows easily. ■

7.7 Prime Fuzzy Left Ideals

In the following, when we say that the fuzzy left ideals of S form a chain, we mean with respect to the inclusion relation " $\subseteq$ ".

Definition 7.7.1 *A fuzzy left ideal f is called* **prime** *if for any two fuzzy ideals f_1 and f_2, $f_1 \circ f_2 \subseteq f$ implies that $f_1 \subseteq f$ or $f_2 \subseteq f$.*

Theorem 7.7.2 *A fuzzy left ideal f of S is prime if and only if for any two fuzzy points x_r, y_t of $S(rt > 0), x_r \circ S \circ y_t \circ S \subseteq f$ implies that $x_r \subseteq f$ or $y_t \subseteq f$.*

Proof. Let x_r and y_t be fuzzy points of S such that $x_r \circ S \circ y_t \circ S \subseteq f$. Then

$$(S \circ x_r \circ S) \circ (S \circ y_t \circ S) \subseteq f.$$

Since f is prime and $S \circ x_r \circ S$ and $S \circ y_t \circ S$ are fuzzy ideals of S, we have $S \circ x_r \circ S \subseteq f$ or $S \circ y_t \circ S \subseteq f$, say $S \circ x_r \circ S \subseteq f$. Then by Lemma 7.6.3(3),

$$(x_r)^3 \subseteq S \circ x_r \circ S \subseteq f.$$

Thus $x_r \in (x_r) \subseteq f$.

Conversely, let f_1, f_2 be fuzzy ideals of S such that $f_1 \circ f_2 \subseteq f$. If $f_1 \not\subseteq f$ and $f_2 \not\subseteq f$, then there exist $x, y \in S$ such that $f_1(x) > f(x), f_2(y) > f(y)$. Let $r = f_1(x), t = f_2(y)$. Then $rt > 0$, $x_r \subseteq f_1$, $y_t \subseteq f_2$ and by Lemma 7.6.1(3),

$$x_r \circ S \circ y_t \circ S \subseteq (x_r) \circ (y_t) \subseteq f_1 \circ f_2 \subseteq f.$$

By hypothesis, $x_r \subseteq f$ or $y_t \subseteq f$. Say $x_r \subseteq f$. Then $f(x) \geq r = f_1(x)$, a contradiction. ∎

Theorem 7.7.3 *A left ideal L of S is prime if and only if C_L is a prime fuzzy left ideal of S.*

Proof. Let L be a left ideal of S. By Lemma 7.6.4(5), C_L is a fuzzy left ideal of S. For any two fuzzy ideals, $f_1$and f_2 of S, if $f_1 \circ f_2 \subseteq C_L$, then $f_1 \subseteq C_L$,or $f_2 \subseteq C_L$. In fact, if $f_1 \not\subseteq C_L, f_2 \not\subseteq C_L$, then there exist $x, y \in S$ such that $f_1(x) > C_L(x), f_2(y) > C_L(y)$. Thus $f_1(x) > 0, f_2(y) = f_2(y) > 0$, and $C_L(x) = C_L(y) = 0$. Thus $x, y \notin L$. We now show that there exist $r_1, r_2 \in S$ such that $xr_1yr_2 \notin L$. If $xSyS \subseteq L$, then $(SxS)(SyS) \subseteq L$. Since SxS and SyS are ideals of S and L is a prime left ideal of S, we have $SxS \subseteq L$ or $SyS \subseteq L$. Say $SxS \subseteq L$. Then $(x)^3 \subseteq SxS \subseteq L$. Thus $x \in (x) \subseteq L$.

Let $a = xr_1yr_2$. Then $C_L(a) = 0$ and

$$\begin{aligned} f_1 \circ f_2(a) &= \vee_{a=cd}(f_1(c) \wedge f_2(d)) \\ &\geq f_1(xr_1) \wedge f_2(yr_2) \\ &\geq f_1(x) \wedge f_2(y) \\ &> 0, \end{aligned}$$

which contradicts the fact that $f_1 \circ f_2 \subseteq C_L$. Therefore, C_L is a prime fuzzy left ideal of S.

Conversely, let C_L be a prime fuzzy left ideal of S and A, B are ideals of S such that $AB \subseteq L$. Then $C_A \circ C_B \subseteq C_L$. In fact, for any $x \in S$, we consider the following two cases:

(1) If $C_A \circ C_B(x) = 0$, then $C_L(x) \geq C_A \circ C_B(x)$.

(2) If $C_A \circ C_B(x) \neq 0$, then there exists $a, b \in S$ such that $x = ab$ and $C_A(a) \wedge C_B(b) \neq 0$.

Thus $a \in A$ and $b \in B$, that is, $x \in AB \subseteq L$. Hence $C_L(x) = 1 \geq C_A \circ C_B(x)$. By hypothesis and Lemma 7.6.4(5), since C_L is prime, $C_A \subseteq C_L$ or $C_B \subseteq C_L$, i.e., $A \subseteq L$ or$B \subseteq L$. ■

7.8 Fuzzy m-systems and Quasi-prime Fuzzy Left Ideals

Definition 7.8.1 *A fuzzy left ideal f is called **quasi-prime** if for any two fuzzy left ideals f_1 and f_2, $f_1 \circ f_2 \subseteq f$ implies that $f_1 \subseteq f$ or $f_2 \subseteq f$. f is called **quasi-semiprime** if for any fuzzy left ideal g of $S, g^2 \subseteq f$ implies that $g \subseteq f$.*

Similar to Theorems 7.7.2 and 7.7.3, we have the following characterizations of quasi-prime left ideals of S. The proofs are easy modifications of Theorems 7.7.2 and 7.7.3, and so we omit them.

Theorem 7.8.2 *A fuzzy left ideal f of S is quasi-prime if and only if for any two fuzzy points x_r, y_t of S $(rt > 0)$, $x_r \circ S \circ y_t \subseteq f$ implies that $x_r \subseteq f$ or $y_t \subseteq f$.* ■

Theorem 7.8.3 *A left ideal L of S is quasi-prime if and only if C_L is a quasi-prime fuzzy left ideal of S.* ■

Definition 7.8.4 *A fuzzy subset f of S is called a **fuzzy m-system** if for any $t, s \in [0,1]$ and $a, b \in S$, $f(a) > t, f(b) > s$ imply that there exists an element $x \in S$ such that $f(axb) > t \vee s$.*

Theorem 7.8.5 *Let M be a subset of S. Then M is a m-system of S if and only if C_M is a fuzzy m-system.*

Proof. For any $t, s \in [0,1]$ and $a, b \in S$, if $C_M(a) > t$, $C_M(b) > s$, then $a, b \in M$. By hypothesis, there exists an element $x \in S$ such that $axb \in M$, that is, $C_M(axb) = 1$. Thus $C_M(axb) > t \vee s$.

Conversely, let $a, b \in M$. Then $C_M(a) = C_M(b) = 1$. Thus for any $t, s \in [0,1]$, $C_M(a) > t$, $C_M(b) > s$, which implies that there exists an element $x \in S$ such that $C_M(axb) > s \vee t$. Therefore, $axb \in M$. ■

Theorem 7.8.6 *Let f be a fuzzy left ideal of S. Then f is quasi-prime if and only if $1 - f$ is a fuzzy m-system.*

Proof. Suppose that f is quasi-prime. For any $t, s \in [0, 1], a, b \in S$, if $(1-f)(a) > t, (1-f)(b) > s$, then $f(a) < 1-t$, $f(b) < 1-s$, which implies that $a_{1-t} \not\subseteq f, b_{1-s} \not\subseteq f$. Since f is a quasi-prime fuzzy left ideal of S, by Theorem 7.8.2, there exists $x \in S$ such that

$$a_{1-t} \circ f_{\{x\}} \circ b_{1-s} = (axb)_{(1-t)\wedge(1-s)} \not\subseteq f.$$

Thus

$$f(axb) < (1-t) \wedge (1-s) = 1 - (t \vee s).$$

Therefore, $(1 - f)(axb) > t \vee s$.

Conversely, let $a_t, b_s (ts > 0) \in S$ be such that $a_t \circ S \circ b_s \subseteq f$. Suppose $a_t \not\subseteq f$. Then $f(a) < t$, $f(b) < s$. Thus $(1 - f)(a) > 1-t$ and $(1 - f)(b) > 1 - s$. By hypothesis, there exists $x \in S$ such that

$$(1 - f)(axb) > (1 - t) \vee (1 - s) = 1 - t \wedge s,$$

that is, $f(axb) < t \wedge s$. By Lemma 7.6.3(2), it follows that

$$a_t \circ f_{\{x\}} \circ b_s = (axb)_{t\wedge s} \not\subseteq f,$$

a contradiction. ■

7.9 Weakly Quasi-prime Fuzzy Left Ideals

Definition 7.9.1 *A fuzzy left ideal f is called* ***weakly quasi-prime*** *if for any two fuzzy left ideals f_1 and f_2, $f \subseteq f_1, f \subseteq f_2$, and $f_1 \circ f_2 \subseteq f$ implies $f_1 \subseteq f$ or $f_2 \subseteq f$.*

Theorem 7.9.2 *A left ideal L is weakly quasi-prime if and only if C_L is weakly quasi-prime.*

Proof. Let L be a weakly quasi-prime left ideal of S, f_1 and f_2 be fuzzy left ideals of S such that $C_L \subseteq f_1, C_L \subseteq f_2$ and $f_1 \circ f_2 \subseteq C_L$. Suppose $f_1 \not\subseteq C_L$ and $f_2 \not\subseteq C_L$. Then there exist $x, y \in S$ such that $f_1(x) > C_L(x), f_2(y) > C_L(y)$, so that $x, y \notin L$ and $f_1(x) > 0, f_2(y) > 0$. Furthermore, there exist $r_1, r_2 \in S$ such that

$$(\{r_1x\} \cup L)(\{r_2y\} \cup L) \not\subseteq L.$$

If $(Sx \cup L)(Sy \cup L) \subseteq L$, then since $Sx \cup L$ and $Sy \cup L$ are all left ideals of S containing L, by hypothesis, we have $Sx \cup L = L$ or $Sy \cup L = L$. Say $Sx \cup L = L$. Then $Sx \subseteq L$. Thus

$$(L \cup L(x))(L \cup L(x)) \subseteq L^2 \cup LL(x) \cup L(x)L(x) \subseteq L \cup Sx = L.$$

Hence $x \in L \cup L(x) = L$, which is impossible.

Since $(\{r_1x\} \cup L)(\{r_2y\} \cup L) \not\subseteq L$, we have $r_1xr_2y \notin L$ or $Lr_2y \not\subseteq L$.

(1) If $r_1xr_2y \notin L$, then

$$f_1 \circ f_2(r_1xr_2y) \geq f_1(r_1x) \wedge f_2(r_2y) \geq f_1(x) \wedge f_2(y) > 0.$$

However, $f_L(r_1xr_2) = 0$ and we have a contradiction.

(2) If $Lr_2y \not\subseteq L$, then there exists an element $l \in L$ such that $lr_2 \notin L$, that is, $C_L(lr_2y) = 0$. However,

$$f_1 \circ f_2(lr_2y) \geq f_1(l) \wedge f_2(r_2y) \geq C_L(l) \wedge f_2(y) = f_2(y) = f_2(y) > 0,$$

a contradiction. Therefore, $f_1 \subseteq C_L$ or $f_2 \subseteq C_L$.

Conversely, if C_L is weakly quasi-prime left ideal of S, then L is weakly quasi-prime by a simple modification of the proof of the converse of Theorem 7.7.3. ■

Theorem 7.9.3 *Let S be commutative and f be a fuzzy left ideal of S. Then the following statements are equivalent.*

(1) f is prime.

(2) f is quasi-prime.

(3) f is weakly quasi-prime.

Proof. That (1) $\Rightarrow$ (2) $\Rightarrow$ (3) is obvious.

(3) $\Rightarrow$ (1): Suppose $f_1 \circ f_2 \subseteq f$ for fuzzy ideals f_1 and f_2 of S. Since S is commutative, we have f is a fuzzy ideal of S and $(f_1 \cup f) \circ (f_2 \cup f) \subseteq f_1 \circ f_2 \cup f_1 \circ f \cup f \circ f_2 \cup f^2 \subseteq f$. By (3), $f_1 \cup f = f$ or $f_2 \cup f = f$, that is, $f_1 \subseteq f$ or $f_2 \subseteq f$. ■

Theorem 7.9.4 *Let f be a fuzzy left ideal of S. Then the following statements are equivalent.*

(1) f is weakly quasi-prime fuzzy left ideal of S.

(2) For any two fuzzy left ideals $f_1, f_2 \subseteq S$, if $(f_1 \cup f) \circ (f_2 \cup f) \subseteq f$, then $f_1 \subseteq f$ or $f_2 \subseteq f$.

(3) For any two fuzzy left ideals $f_1, f_2 \subseteq S$, if $f \subseteq f_1$ and $f_1 \circ f_2 \subseteq f$, then $f_1 = f$ or $f_2 \subseteq f$.

(4) For any two fuzzy left ideals $f_1, f_2 \subseteq S$, if $(f_1 \cup f) \circ f_2 \subseteq f$, then $f_1 \subseteq f$ or $f_2 \subseteq f$.

(5) For any two fuzzy points a_r, b_t of S ($rt > 0$), if $(a_r \cup f) \circ S \circ (b_t \cup f) \subseteq f$, then $a_r \subseteq f$, or $b_t \subseteq f$.

Proof. (1) $\Rightarrow$ (2): Since $f_1 \cup f$ and $f_2 \cup f$ are fuzzy left ideals of S such that $f \subseteq f_1 \cup f, f \subseteq f_2 \cup f$ and $(f_1 \cup f) \circ (f_2 \circ f) \subseteq f$, we have $f = f_1 \cup f$ or $f = f_2 \cup f$, that is, $f_1 \subseteq f$ or $f_2 \subseteq f$.

(2) $\Rightarrow$ (3): For two fuzzy left ideals f_1 and f_2 of S, if $f \subseteq f_1$ and $f_1 \circ f_2 \subseteq f$, then $(f \cup f_1) \circ (f_2 \cup f) \subseteq f_1 \circ (f \cup f_2) \subseteq f_1 \circ f \cup f_1 \circ f_2 \subseteq f$.

By (2), we have $f_1 \cup f = f$ or $f_2 \cup f = f$ so that $f_1 = f$ or $f_2 \subseteq f$.

(3) $\Rightarrow$ (4): Since $f \subseteq f \cup f_1$ and $(f \cup f_1) \circ f_2 \subseteq f$, we have $f_1 \cup f = f$ or $f_2 \subseteq f$ by (3). It follows that $f_1 \subseteq f$ or $f_2 \subseteq f$.

(4) $\Rightarrow$ (5): Let $a_r, b_t (rt > 0)$ be two fuzzy points of S such that $(a_r \cup f) \circ S \circ (b_t \cup f) \subseteq f$. Then $a_r \circ S \circ b_t \subseteq f$ and $f \circ S \circ b_t \subseteq f$. Thus

$$\begin{aligned}(a_r \cup S \circ a_r \cup f) \circ (S \circ b_t) &\subseteq a_r \circ S \circ b_t \cup S \circ a_r \circ S \circ b_t \cup f \circ S \circ b_t \\ &\subseteq f \cup S \circ f \\ &\subseteq f.\end{aligned}$$

Now $a_r \cup S \circ a_r = L(a_r)$ and $S \circ b_t$ are fuzzy left ideals of S. By (4), we have $L(a_r) \subseteq f$ or $S \circ b_t \subseteq f$. If $L(a_r) \subseteq f$, then $a_r \in L(a_r) \subseteq f$. If $S \circ b_t \subseteq f$, then

$$(L(b_t) \cup f) \circ L(b_t) \subseteq L(b_t)^2 \cup f \circ L(b_t) \subseteq S \circ b_t \subseteq f.$$

By (4), we have $b_t \in L(b_t) \subseteq f$.

(5) $\Rightarrow$ (1): Let f_1 and f_2 be two fuzzy left ideals of S such that $f \subseteq f_1, f \subseteq f_2$ and $f_1 \circ f_2 \subseteq f$. If $f_1 \neq f, f_2 \neq f$, then there exist $x, y \in S$ such that $f_1(x) > f(x), f_2(y) > f(y)$. Let $r = f_1(x), t = f_2(y)$. Then $rt > 0$ and

$$(x_r \cup f) \circ S \circ (y_t \cup f) \subseteq f_1 \circ S \circ f_2 \subseteq f_1 \circ f_2 \subseteq f.$$

By (5), we have $x_r \subseteq f$ or $y_t \subseteq f$, which is a contradiction. ∎

Let Φ denote the fuzzy subset of S such that $\Phi(x) = 0 \; \forall x \in S$.

Theorem 7.9.5 *Let f be fuzzy left ideal of S and h be a fuzzy subset of S satisfying the following conditions:*

(1) $f \cap h = \Phi$.

(2) For any $a_t, b_r \subseteq h, (a_t \cup f) \circ S \circ (b_r \cup f) \cap h \neq \Phi$.

If g is a maximal fuzzy left ideal of S with respect to containing f and $g \cap h = \Phi$, then g is a weakly quasi-prime fuzzy left ideal of S.

Proof. Let f_1 and f_2 be a fuzzy left ideal of S such that $g \subseteq f_1, f_2$ and $f_1 \circ f_2 \subseteq g$. If $g \subset f_1$ and $g \subset f_2$, then there exist $a_t \subseteq f \backslash g$, $b_r \subseteq f_2 \backslash g$ $(rt > 0)$. Thus $g \subset g \cup L(a_t) \subseteq f_1$, $g \subset g \cup L(b_r) \subseteq f_2$.

By hypothesis, we have

$$(g \cup L(a_t)) \cap h \neq \Phi, \qquad (g \cup L(b_r)) \cap h \neq \Phi.$$

Let

$$\begin{aligned} c_k &\subseteq (g \cup L(a_t)) \cap h \quad (k > 0), \\ d_l &\subseteq (g \cup L(b_r)) \cup h \quad (l > 0). \end{aligned}$$

Then

$$\begin{aligned}(c_k \cup f) \circ S \circ (d_l \cup f) &\subseteq (f_1 \cup f) \circ S \circ (f_2 \cup f) \\ &\subseteq f_1 \circ S \circ f_2 \cup f_1 \circ S \circ f \cup f \circ S \circ f_2 \cup f \circ S \circ f \\ &\subseteq f_1 \circ f_2 \\ &\subseteq g,\end{aligned}$$

which contradicts the fact that

$$(c_k \cup f) \circ S \circ (d_l \cup f) \cap h \neq \Phi.$$

■

7.10 Fuzzy Ideals $i(f)$ and $I(f)$

Let f be a fuzzy left ideal of S. We define two fuzzy subsets of S, denoted by $i(f)$ and $I(f)$, respectively, as follows: $\forall x \in S$,

$$i(f)(x) = \vee\{t \mid x_t \subseteq f,\ x_t \circ S \subseteq f, t \in [0,1]\},$$

$$I(f)(x) = \vee\{t \mid f \circ x_t \subseteq f, t \in [0,1]\}.$$

Theorem 7.10.1 *Let f be a fuzzy left ideal of S. Then $i(f)$ is the largest fuzzy ideal of S contained in f.*

Proof. For any $x, y \in S$, let

$$A(x) = \{t \mid x_t \subseteq f,\ x_t \circ S \subseteq f, t \in [0,1]\},$$

$$B(xy) = \{w \mid (xy)_w \subseteq f, (xy)_w \circ S \subseteq f, w \in [0,1]\},$$

$$C(y) = \{q \mid y_q \subseteq f,\ y_q \circ S \subseteq f,\ q \in [0,1]\}.$$

Then

$$i(f)(x) = \vee_{t \in A(x)} t,\ i(f)(xy) = \vee_{w \in B(xy)} w,$$

and

$$i(f)(y) = \vee_{q \in C(y)} q.$$

If $t \in A(x)$, then

$$x_t \circ y_t = (xy)_t \subseteq x_t \circ S \subseteq f,$$

$$(xy)_t \circ S = x_t \circ (y_t \circ S) \subseteq x_t \circ S \subseteq f.$$

Thus $t \in B(xy)$. It follows that $i(f)(xy) \geq i(f)(x)$. In a similar manner, it follows that $i(f)(xy) \geq i(f)(y)$. Therefore, $i(f)$ is a fuzzy ideal of S. Let $x \in S$ and let $t \in A(x)$. Then $x_t \subseteq f$, i.e., $t \leq f(x)$. Thus $i(f)(x) \leq f(x)$. Let g be a fuzzy ideal of S such that $g \subseteq f$ and $x_t \subseteq g \subseteq f$. Then $t \in A(x)$. Thus $g(x) = \vee_{x_t \subseteq g} t \leq i(f)(x)$. Therefore, $g \subseteq i(f)$. ■

Theorem 7.10.2 *Let S be a semigroup with an identity e and f a prime fuzzy left ideal of S. If $i(f) \neq 0$, then $i(f)$ is a quasi-prime fuzzy ideal of S.*

Proof. Let a_t and b_r $(tr > 0)$ be fuzzy points of S such that $a_t \circ S \circ b_r \subseteq i(f)$. Then

$$(S \circ a_t \circ S) \circ (S \circ b_r \circ S) \subseteq i(f) \subseteq f.$$

Since f is prime, we have $S \circ a_t \circ S \subseteq f$ or $S \circ b_r \circ S \subseteq f$. Say, $S \circ a_t \circ S \subseteq f$. By Theorem 7.10.1, we have $S \circ a_t \circ S \subseteq i(f)$. Since S has an identity e,

$$a_t = (eae)_t = e_t \circ a_t \circ e_t \subseteq S \circ a_t \circ S \subseteq i(f).$$

By Theorem 7.8.2, the desired result holds. ∎

Theorem 7.10.3 *Let f be a fuzzy left ideal of S. Then $I(f)$ is the largest fuzzy subsemigroup of S such that f is a fuzzy ideal of $I(f)$.*

Proof. For any $x, y \in S$, let

$$D(x) = \{t \mid f \circ x_t \subseteq f, t \in [0,1]\},$$

$$E(y) = \{r \mid f \circ y_r \subseteq f,\ r \in [0,1]\}.$$

(1) Let $x \in S$ and $f(x) = t_0$. Since $f \circ x_{t_0} \subseteq f \circ f \subseteq f$, $t_0 \in D(x)$ and it follows that

$$f(x) = t_0 \leq \vee_{t \subseteq D(x)} t = I(f)(x).$$

Hence $I(f)$ is a fuzzy subset of S containing f.

(2) $I(f)$ is a fuzzy subsemigroup of S. In fact,

$$I(f)(x) \wedge I(f)(y) = (\vee_{t \subseteq D(x)} t) \wedge (\vee_{r \subseteq E(y)} r) = \vee_{t \subseteq D(x), r \subseteq E(y)} (t \wedge r).$$

Since

$$f \circ x_t \circ y_r = (f \circ x_t) \circ y_r = f \circ (xy)_{t \wedge r} \subseteq f \circ y_r \subseteq f,$$

we have $t \wedge r \leq I(f)(xy)$. Thus $I(f)(x) \wedge I(f)(y) \leq I(f)(xy)$.

(3) We show that f is a fuzzy ideal of $I(f)$. Since f is a left ideal of S, we have $I(f) \circ f \subseteq S \circ f \subseteq f$. Furthermore, for any $x \in S$, if x can be expressed as $x = yz$, let

$$F(z) = \{q \mid f \circ z_q \subseteq f, q \in [0,1]\}.$$

Then $f \circ I(f)(x) = \vee_{x=yz}(f(y) \wedge \vee_{q \in F(z)} q) = \vee_{x=yx} \vee_{q \in F(z)} (f(y) \wedge q)$. Since for any $q \in F(z)$, $f \circ z_q \subseteq f$, we have $f(x) \geq f(y) \wedge z_q(z) \geq f(y) \wedge q$. Thus

$$f(x) \geq \vee_{x=yz} \vee_{q \in F(z)} (f(y) \wedge q) = f \circ I(f)(x).$$

If x is not expressible as $x = yz$, then clearly $f(x) \geq 0 = f \circ I(f)(x)$. Therefore, $f \circ I(f) \subseteq f$.

(4) Let g be a fuzzy subsemigroup of S such that f is a fuzzy ideal of g. For any $x \in S$, if $g(x) = t$, then $x_t \subseteq g$, and $f \circ x_t \subseteq f \circ g \subseteq f$. Therefore, $I(f)$ is the largest fuzzy subsemigroup of S such that f is a fuzzy ideal of $I(f)$. ∎

Theorem 7.10.4 *Let S be a semigroup with an identity e and f a fuzzy left ideal of S. Then the following statements are equivalent.*

(1) f is a weakly quasi-prime fuzzy left ideal of S.

(2) If f_1 is a fuzzy left ideal of S such that $f \circ f_1 \subseteq f$, then $f_1 \subseteq f$.

(3) For any fuzzy point $a_t \subseteq S$, if $f \circ (S \circ a_t) \subseteq f$, then $a_t \subseteq f$.

(4) f is the largest fuzzy left ideal of S contained in $I(f)$.

Proof. (1) ⇒ (2): Let f_1 be a fuzzy left ideal of S such that $f \circ f_1 \subseteq f$. Then $(f \circ S) \circ f_1 \subseteq f \circ f_1 \subseteq f$. Since S has an identity e, for any $x \in S$,

$$(f \circ S)(x) = \vee_{x=yz}(f(y) \wedge S(e)) \geq f(x) \wedge S(e) = f(x).$$

Then $f \subseteq f \circ S$. Since f is not a fuzzy ideal of S, $f \neq f \circ S$. By Theorem 7.9.4(3), it follows that $f_1 \subseteq f$.

(2) ⇒ (3): For any fuzzy point $a_t \subseteq S$, if $f \circ (S \circ a_t) \subseteq f$, by (2), $S \circ a_t \subseteq f$. Since S has an identity e, $S \circ a_t(a) \geq S(e) \wedge a_t(a) = t$. Thus $a_t \subseteq S \circ a_t \subseteq f$.

(3) ⇒ (4): By Theorem 7.10.3, $f \subseteq I(f)$. Let g be a fuzzy left ideal of S such that $g \subseteq I(f)$. We show that $g \subseteq f$. For any $a_t \subseteq g$, since

$$f \circ (S \circ a_t) \subseteq f \circ (S \circ g) \subseteq f \circ g \subseteq f \circ I(f) \subseteq f,$$

by (3), we have $a_t \subseteq f$. Thus $g = \vee_{a_t \subseteq g} a_t \subseteq f$.

(4) ⇒ (1): Let f_1, f_2 be fuzzy left ideals of S such that $f \subseteq f_1, f \subseteq f_2$ and $f_1 \circ f_2 \subseteq f$. Since $f \circ f_2 \subseteq f_1 \circ f_2 \subseteq f$, $f \circ x_t \subseteq f$ holds for any $x_t \subseteq f_2$. Thus $I(f) \geq t = x_t(x)$, which implies that $x_t \in I(f)$. Therefore, $f_2 = \vee_{x_t \in f_2} x_t \subseteq I(f)$. Since f is the largest fuzzy left ideal of S contained in $I(f)$, $f_2 \subseteq f$. Thus $f = f_2$. ∎

7.11 Strongly Semisimple Semigroups

A fuzzy left ideal f of a semigroup S is called **idempotent** if $f = f \circ f$, that is, $f = f^2$.

Definition 7.11.1 *A semigroup S is called* ***strongly semisimple*** *if every left ideal of S is idempotent.*

Lemma 7.11.2 *For a semigroup S, the following conditions are equivalent:*

(1) S is strongly semisimple.

(2) For any $a \in S, a \in SaSa$.

Proof. Suppose S is strongly semisimple. Then for any $a \in S$, we have $L(a) = L(a)^2$. Thus $L(a) = L(a)^4$. Since $L(a)^2 = (a \cup Sa)(a \cup Sa) \subseteq Sa \cup SaSa$, $a \in L(a) = L(a)^4 \subseteq (Sa \cup SaSa)(Sa \cup SaSa) \subseteq SaSa$.

Conversely, suppose (2) holds. Let L be any left ideal of S. For any $a \in L$, $a \in SaSa \subseteq L(a)L(a) \subseteq L^2$. Thus $L \subseteq L^2$. Clearly, $L^2 \subseteq L$. Therefore, $L^2 = L$. ■

The following theorems characterize the strongly semisimple semigroups by means of fuzzy quasi-prime left ideals of S.

Theorem 7.11.3 *Let S be a semigroup. Then the following statements are equivalent.*

(1) Every fuzzy left fuzzy ideal of S is idempotent.

(2) For any two fuzzy left ideals f_1 and f_2 of S, $f_1 \cap f_2 \subseteq f_1 \circ f_2$.

(3) For any fuzzy point $a_r \subseteq S, L(a_r) = L(a_r)^2$.

(4) For any fuzzy point $a_r \subseteq S, a_r \subseteq S \circ a_r \circ S \circ a_r$.

(5) Every fuzzy left ideal of S is a quasi-semiprime fuzzy left ideal of S.

(6) Every fuzzy left ideal of S is the intersection of all quasi-prime fuzzy left ideals of S containing it.

Proof. (1) $\Rightarrow$ (2): Let f_1 and f_2 be fuzzy left ideals of S. Then by Lemma 7.6.1,

$$S \circ (f_1 \cap f_2) \subseteq S \circ f_1 \cap S \circ f_2 \subseteq f_1 \cap f_2.$$

Thus $f_1 \cap f_2$ is a fuzzy left ideal of S. By (1) and Lemma 7.6.1(3)

$$f_1 \cap f_2 = (f_1 \cap f_2)^2 = (f_1 \cap f_2) \circ (f_1 \cap f_2) \subseteq f_1 \circ f_2.$$

(2) $\Rightarrow$ (3): For any fuzzy point a_r of S, let $f_1 = f_2 = L(a_r)^2$. By (2), $L(a_r) \subseteq L(a_r)^2$. On the other hand, by Lemma 7.6.1(3), $L(a_r) \subseteq S \circ L(a_r) \subseteq L(a_r)$. Therefore, $L(a_r) = L(a_r)^2$ holds.

(3) $\Rightarrow$ (4): For any fuzzy point a_r of S, by (3) and Lemma 7.6.1(3), we have $a_r \in L(a_r) = L(a_r)^2 = (a_r \cup S \circ a_r) \circ (a_r \cup S \circ a_r) \subseteq a_r^2 \cup a_r \circ S \circ a_r \cup S \circ a_r^2 \cup S \circ a_r \circ S \circ a_r$. Thus

$$\begin{aligned} a_r^2 &\subseteq (a_r^2 \cup a_r \circ S \circ a_r \cup S \circ a_r^2 \cup S \circ a_r \circ S \circ a_r) \circ a_r \\ &\subseteq a_r^3 \cup a_r \circ S \circ a_r^2 \cup S \circ a_r^3 \cup S \circ a_r \circ S \circ a_r^2 \\ &\subseteq S \circ a_r^2 \end{aligned}$$

and

$$\begin{aligned} a_r \circ S \circ a_r &\subseteq (a_r^2 \cup a_r \circ S \circ a_r \cup S \circ a_r^2 \cup S \circ a_r \circ S \circ a_r) \circ (S \circ a_r) \\ &\subseteq S \circ a_r \circ S \circ a_r. \end{aligned}$$

Therefore, $a_r \in S \circ a_r^2 \cup S \circ a_r \circ S \circ a_r$, which implies that

$$\begin{aligned} a_r &\in (S \circ a_r) \circ a_r \cup S \circ a_r \circ S \circ a_r \\ &\subseteq (S \circ a_r) \circ (S \circ a_r^2 \cup S \circ a_r \circ S \circ a_r) \cup S \circ a_r \circ S \circ a_r \\ &\subseteq S \circ a_r \circ S \circ a_r. \end{aligned}$$

(4) $\Rightarrow$ (5): Let g be a fuzzy left ideal of S such that $g^2 \subseteq f$. For any $a_r \in g$, by (4),

$$a_r \in S \circ a_r \circ S \circ a_r \subseteq S \circ g \circ S \circ g \subseteq g^2 \subseteq f.$$

Since $g = \cup_{a_r \subseteq g} a_r$, $g \subseteq f$.

(5) $\Rightarrow$ (1): Let f be any fuzzy left ideal of S. Then f^2 is also a fuzzy left ideal of S. By (5), since $f^2 \subseteq f^2$, we have $f \subseteq f^2$. Clearly, $f^2 \subseteq f$. It follows that $f^2 = f$.

(2) $\Rightarrow$ (6): Let f be a fuzzy left ideal of S, and

$$N = \{g \mid g \text{ is a quasi-prime left ideal of } S \text{ such that } f \subseteq g\}.$$

Then clearly, $f \subseteq \cap_{g \in N} g$

Conversely, for any $a_r \subseteq \cap_{g \in N} g$, if $a_r \not\subseteq f$, then $r > 0$ and $f(a) < r$. Let

$$B = \{h \mid h \text{ is a fuzzy left ideal of } S \text{ such that } f \subseteq h, f(a) = h(a)\}.$$

Since $f \in B$, then $B \neq \emptyset$. Thus $(B, \subseteq)$ is an ordered set. Let C be a chain in B. Then the set $\cup_{h_\beta \in C} h_\beta$ is a fuzzy left ideal of S and $f \subseteq \cup_{h \in C} h$. Since for any $h \in C, f(a) = h(a)$, we have $\cup_{h \in C} h(a) = f(a)$. Thus the fuzzy left ideal $\cup_{h_\beta \in C} h$ is an upper bound of C in B. By Zorn's Lemma, B has a maximal element, say $h_{\max}$. Then $a_r \not\subseteq h_{\max}$. We now prove that $h_{\max}$ is a quasi-prime fuzzy left ideal of S. Let f_1 and f_2 be two fuzzy left ideals of S such that $f_1 \circ f_2 \subseteq h_{\max}$. Then by (2), $f_1 \cap f_2 \subseteq f_1 \circ f_2 \subseteq h_{\max}$. Thus

$$\begin{aligned} h_{\max} &= h_{\max} \cup (f_1 \cap f_2) \\ &= (h_{\max} \cup f_1) \cap (h_{\max} \cup f_2). \end{aligned}$$

We claim that $h_{\max} = h_{\max} \cup f_1$ or $h_{\max} \cup f_2$, that is, $f_1 \subseteq h_{\max}$. In fact, if $h_{\max} \subset h_{\max} \cup f_1, h_{\max} \subset h_{\max} \cup f_2$, then since $h_{\max}$ is maximal with respect to the property that $f \subseteq h_{\max}$ and $h_{\max}(a) \neq f(a), h_{\max} \cup f_2(a) \neq f(a)$. Hence

$$h_{\max}(a) = ((h_{\max} \cup f_1)(a)) \cap ((h_{\max} \cup f_2)(a)) \neq f(a),$$

a contradiction.

(6) $\Rightarrow$ (1): Let f be any fuzzy left ideal of S. Then f^2 is also a fuzzy left ideal of S. By (6), we have $f^2 = \cap_{g \in M} g$, where M is the set of all quasi-prime fuzzy left ideals of S containing f^2.

For any $g \subseteq M$, $f \circ f \subseteq g$. Since g is quasi-prime, $f \subseteq g$ holds. Thus $f \subseteq \cap_{g \in M} g = f^2$. Therefore, $f^2 = f$. ■

By Theorem 7.11.3 and Lemma 7.11.2, we have the following result.

Theorem 7.11.4 *A semigroup S is strongly semisimple if and only if every fuzzy left ideal of S is idempotent.*

Theorem 7.11.5 *Let S be commutative. Then the fuzzy left ideals of S are quasi-prime if and only if they form a chain and S is strongly semisimple.*

Proof. Let g and h be fuzzy left ideals of S. Since $g \circ h$ is a fuzzy left ideal of S, we have $g \subseteq g \circ h \subseteq S \circ h \subseteq h$ or $h \subseteq g \circ h \subseteq g \circ S \subseteq g$. Thus the fuzzy left ideals of S form a chain. Moreover, for any fuzzy left ideal f of S, $f^2 \subseteq f$. Since $f \circ f \subseteq f^2$, we have $f \subseteq f^2$ so that $f^2 = f$.

Conversely, let f, g be two fuzzy left ideals of S and $f \circ g \subseteq h$. Since the fuzzy left ideals of S form a chain, we have $f^2 \subseteq f \circ g \subseteq h$ or $g^2 \subseteq f \circ g \subseteq h$. By hypothesis, $f \subseteq h$ or $g \subseteq h$ holds. ■

7.12 Fuzzy Multiplication Semigroups

The concept of a multiplication ring was introduced for the consideration of factorization of ideals. A parallel line of research work was initiated in the field of semigroups and the idea of a multiplication semigroup was introduced. A commutative semigroup S is called a **multiplication semigroup** [91] if for any two ideals A, B of S such that $A \subseteq B$, there exists an ideal C of S such that $A = BC$. In [91], Mannepalli has determined the structure of multiplication semigroups with identity with the help of the theory of ideals.

We now consider fuzzy proper fuzzy ideals as initiated in [21].

We state below some definitions and results concerning fuzzy ideals of semigroup analogous to those in the case of rings. The proofs of the results stated in the preliminaries are similar to those in the case of rings.

Let S and T be commutative semigroups with identity.

A nonzero fuzzy ideal f of S is called a **proper fuzzy ideal** if $f \neq C_S$.

Definition 7.12.1 *Let f be a fuzzy ideal of S. The* ***fuzzy radical*** *of f, denoted by rad f, is defined by*

$$radf(x) = \vee\{f(x^n) \mid n \in \mathbb{N}\}$$

for all $x \in S$.

It follows easily that $radf$ is a fuzzy ideal of S.

Definition 7.12.2 *A fuzzy ideal f of S is called* ***primary*** *if for any two fuzzy ideals g, h of S, $g \circ h \subseteq f$ implies $g \subseteq f$ or $h \subseteq radf$.*

Definition 7.12.3 *A fuzzy ideal f of S is called* ***maximal*** *if there does not exist any proper fuzzy ideal g of S such that $f \subset g$.*

Theorem 7.12.4 *A fuzzy ideal f of S is prime if for any two fuzzy points x_r, y_s, $x_r \circ y_s \subseteq f$ implies $x_r \subseteq f$ or $y_s \subseteq f$.* ■

Theorem 7.12.5 *A fuzzy ideal f of S is primary if for any two fuzzy points x_r, y_s, $x_r \circ y_s \subseteq f$ implies $x_r \subseteq f$ or $y_s^n \subseteq f$ for some $n > 0$.* ∎

Theorem 7.12.6 *Let f be a fuzzy ideal of S. Then f^n is also a fuzzy ideal of S for all $n > 0$ and $f^n \supseteq f^{n+1}$.*

Theorem 7.12.7 *If f is prime fuzzy ideal of S, then $f = radf$.* ∎

Theorem 7.12.8 *For any fuzzy ideal f of S, $radf = \cap\{g \mid g$ is a prime fuzzy ideal of S such that $g \supseteq f\}$.* ∎

Theorem 7.12.9 *If f is a primary fuzzy ideal of S, then $radf$ is a prime fuzzy ideal of S.* ∎

Theorem 7.12.10 *Let $\alpha : S \to T$ be an epimoprhism of semigroups and f, g be fuzzy ideals of S and T, respectively. Then the following properties hold.*

(1) $\alpha^{-1}(g)$ is a fuzzy ideal of S.
(2) $\alpha(f)$ is a fuzzy ideal of T. ∎

7.13 Properties of Fuzzy Multiplication Semigroups

Let S and T be commutative semigroups.

Definition 7.13.1 *S is called a **fuzzy multiplication semigroup** if for any two fuzzy ideals g and h of S such that $g \subseteq h$, there exists a fuzzy ideal f of S such that $g = h \circ f$.*

Theorem 7.13.2 *Let S be a fuzzy multiplication semigroup. If $\alpha : S \to T$ is an epimorphism, then T is a fuzzy multiplication semigroup.* ∎

Unless otherwise stated, S denotes a fuzzy multiplication semigroup with identity and f denotes a fuzzy ideal of S satisfying the sup property.

Theorem 7.13.3 *If h be a prime fuzzy ideal of S and g is any fuzzy ideal of S such that $h \subset g$, then $h = h \circ g$ and $h = g^\omega$ or $h = h \circ g^\omega$, where $g^\omega = \cap\{g^i \mid i \in \mathbb{N}\backslash\{0\}\}$.*

Proof. Since $h \subset g$ and S is a fuzzy multiplication semigroup, there exists a fuzzy ideal k of S such that $h = g \circ k$. Then since h is prime,

$h \supseteq k$. Now $h = g \circ k \subseteq k$. Thus $h = k$ and hence $h = g \circ h$. It now follows that $h = g^{\omega}$ or $h = h \circ g^{\omega}$. ■

Theorem 7.13.4 *If radf is a prime fuzzy ideal, then f is primary.*

Proof. Let $g = radf$. If $g = C_S$, then clearly f is primary. Assume $g \neq C_S$. Suppose that f is not primary. Then there exist fuzzy points x_r, y_t such that $x_r \circ y_t \subseteq f, x_r \in g$, but $x_r \not\subseteq f$ and $y_t^n \not\subseteq f$ for all $n > 0$. Let $k = f \cup g(x_r \circ C_S)$. Clearly, k is a fuzzy ideal of S. Suppose $x_r \subseteq k$. Then since $x_r \not\subseteq f$, $x_r \subseteq g(x_r \circ C_S)$. Thus $(g \circ (x_r \circ C_S))(x) \geq r$, or $\vee\{g(a) \wedge (x_r \circ C_S)(b) \mid x = ab\} \geq r$. Since f has the sup property, g also possesses the sup property. Hence there exists $z \in S$ such that $g(z) \geqslant r$ and $x = zxs = xzs$. Thus $f(z^n) \geqslant r$, for some $n > 0$. Now $x = xz^n s^n$ and since f is a fuzzy ideal, $f(x) = f(xz^n s^n) \geqslant f(z^n) \geqslant r$, i.e. $x_r \subseteq f$, a contradiction. Hence $x_r \not\subseteq k$. Now, $k \cup x_r \circ C_S \subseteq g$. Thus there exists a fuzzy ideal h of S such that $k \cup x_r \circ C_S = g \circ h$. Again since $y_t \not\subseteq g$, $g \subset g \cup y_t \circ C_S$. Then by Theorem 7.13.3, $g = g \circ (g \cup y_t \circ C_S)$. Now $k \cup x_r \circ C_S = g \circ h = g \circ (g \cup y_t \circ C_S) \circ h = g \circ h \circ (g \cup y_t \circ C_S)$ (since S is commutative) $= (k \cup x_r \circ C_S) \circ (g \cup y_t \circ C_S) \subseteq k$. Hence $x_r \subseteq k$, a contradiction. Therefore, f is primary. ■

Corollary 7.13.5 *Let f be prime. Then for all positive integers n, f^n is primary and its fuzzy radical is f.*

Proof. We first prove that $radf^n = f$ for all $n > 0$. If $n = 1$, the result is obvious. Let $n > 1$. Then $radf^n(x) = \vee\{f^n(x^m) \mid m > 0\} \geqslant f^n(x^n) \geqslant f(x)$ for all $x \in S$. Since f is prime, $f(x) = radf(x) = \vee\{f(x^m) \mid m > 0\} \geqslant \vee\{f^n(x^m) \mid m > 0\} = radf^n(x)$ for all $x \in S$. Hence $radf^n = f$. The desired result follows from Theorem 7.13.4. ■

Theorem 7.13.6 *Let f be a prime fuzzy ideal and $f^n \neq f^{n+1}$ for all $n > 0$. Then f^{ω} is a prime fuzzy ideal.*

Proof. Let x_l, y_m be fuzzy points such that $x_l \not\subseteq f^{\omega}$ and $y_m \not\subseteq f^{\omega}$. We show that $x_l \circ y_m \not\subseteq f^{\omega}$. If $x_l \not\subseteq f, y_m \not\subseteq f$, then since f is prime, $x_l \circ y_m \not\subseteq f$ and so $x_l \circ y_m \not\subseteq f^{\omega}$. Suppose $x_l \subseteq f$, $y_m \not\subseteq f$. Since $x_l \not\subseteq f^{\omega}$, there exists a positive integer p such that $x_l \in f^p, x_l \not\subseteq f^{p+1}$. Since by Corollary 7.13.5, f^{p+1} is a primary fuzzy ideal with fuzzy radical f, $x_l \circ y_m \not\subseteq f^{p+1}$ and so $x_l \circ y_m \not\subseteq f^{\omega}$. The case when $x_l \not\subseteq f$, $y_m \subseteq f$ is similar.

Finally, let $x_l, y_m \subseteq f$. Then there exist positive integers q, r such that $x_l \subseteq f^q, x_l \not\subseteq f^{q+1}$ and $y_m \subseteq f^r$, $y_m \not\subseteq f^{r+1}$. Then $x_l \circ C_S \subseteq f^q$, $y_m \circ C_S \subseteq f^r$. Since S is a fuzzy multiplication semigoup, there exist fuzzy ideals g, h of S such that $x_l \circ C_S \subseteq f^q \circ g$, $y_m \circ C_S = f^r \circ h$, $h, g \not\subseteq f$. Now, if $x_l \circ y_m \subseteq f^{q+r+1}$, then $f^{q+r} \circ h \circ g = (f^q \circ g)(f^r \circ h) = x_l \circ y_m \circ C_S \subseteq f^{q+r+1}$. Since f^{q+r+1} is a primary fuzzy ideal with fuzzy radical f and $h \circ g \not\subseteq f$ (since f is prime), $f^{q+r} \subseteq f^{q+r+1}$. Also $f^{q+r} \supseteq f^{q+r+1}$. Thus

$f^{q+r} = f^{q+r+1}$, a contradiction. Hence $x_l \circ y_m \not\subseteq f^{q+r+1}$, i.e., $x_l \circ y_m \not\subseteq f^{\omega}$. Thus f^{ω} is prime fuzzy ideal. ■

Theorem 7.13.7 *If f is a primary fuzzy ideal, then $f = g^n$ for some positive integer n, where $g = radf$.*

Proof. If $g = C_S$, then $f = C_S$. Assume $g \neq C_S$. Suppose $f \subseteq g^{\omega}$. Now g is a prime fuzzy ideal of S having the sup property. If $g^n \neq g^{n+1}$ for all $n > 0$, then by Theorem 7.13.6, g^{ω} is prime. Thus $g = radf \subseteq radg^{\omega} = g^{\omega}$, a contradiction. Thus either $f \subseteq g^{\omega}$ with $g^n = g^{n+1}$ for some $n > 0$, or $f \subseteq g^n$, but $g \not\subseteq g^{n+1}$ for some $n > 0$. In the first case, let $x_r \subseteq g^n$. Then $x_r \circ C_S \subseteq g^n$. Also, there exists fuzzy ideal h of S such that $x_r \circ C_S = g^n \circ h$. Thus $x_r \subseteq x_r \circ C_S = g^n \circ h = g^{n+1} \circ h = g \circ (x_r \circ C_S)$. Then as in Theorem 7.13.4, it can be shown that $x_r \subseteq f$. Hence $f = g^n$.

In the second case, there exists a fuzzy ideal k of S such that $f = g^n \circ k$, $k \not\subseteq g$. Since f is primary and $k \not\subseteq g, g^n \not\subseteq f$. Hence $f = g^n$. ■

Let f, g be two fuzzy ideals of S. Define the fuzzy subset $f : g$ of S as follows:

$$f : g = \cup\{h \mid h \text{ is a fuzzy ideal of } S \text{ such that } h \circ g \subseteq f\}.$$

It follows easily that $f : g$ is a fuzzy ideal of S.

Theorem 7.13.8 *If f is a proper prime fuzzy ideal and g is a fuzzy ideal of S such that $g \subseteq f^n$ and $g \not\subseteq f^{n+1}$ for some $n > 0$, then $f^n = g : (y_t \circ C_S)$, where $y_t \not\subseteq f$.*

Proof. Since $g \subseteq f^n$, there exists a fuzzy ideal h of S such that $g = f^n \circ h, h \not\subseteq f$. Let $y_t \subseteq h, y_t \not\subseteq f$. Then $y_t \circ C_S \subseteq h$ and $f^n(y_t \circ C_S) \subseteq f^n \circ h = g$. Thus $f^n \subseteq g : (y_t \circ C_S)$. Now let k be any fuzzy ideal of S such that $k \circ (y_t \circ C_S) \subseteq g$. Then $k \circ (y_t \circ C_S) \subseteq f^n$. Now since by Corollary 7.13.5, f^n is primary with fuzzy radical f and $y_t \circ C_S \not\subseteq f$, $k \subseteq f^n$. Therefore, $g : (y_t \circ C_S) \subseteq f^n$ and hence $f^n = g : (y_t \circ C_S)$. ■

Definition 7.13.9 *A fuzzy ideal g of S is said to be* ***idempotent*** *if $g = g^2$.*

The proof of the next result is straightforward.

Theorem 7.13.10 *If k is a nonidempotent prime fuzzy ideal of S, then there is at most one prime fuzzy ideal $g \subset k$ such that there is no prime fuzzy ideal between g and k.*

Definition 7.13.11 *A fuzzy ideal of the form $x_r \circ C_S$, where x_r is a fuzzy point of S called a* ***principal fuzzy ideal*** *of S.*

Theorem 7.13.12 *If prime fuzzy ideals of S are linearly ordered with respect to the inclusion relation $\subseteq$, then every nonidempotent prime fuzzy ideal g is principal.*

Proof. Since g is nonidempotent, $g \supset g^2$. Thus there exists a fuzzy point x_r such that $x_r \subseteq g$, $x_r \nsubseteq g^2$. Now $x_r \circ C_S \subseteq g$. Since g is prime, by Theorem 7.12.8, $rad(x_r \circ C_S) \subseteq g$. Let h be a prime fuzzy ideal of S such that $x_r \circ C_S \subseteq h \subset g$. There exists a fuzzy ideal k of S such that $x_r \circ C_S = g \circ k$, $k \nsubseteq g$. Hence $g \circ k \subseteq h$, $g \nsubseteq h$, $k \nsubseteq h$ which contradicts that h is a prime. Since prime fuzzy ideals of S are linearly ordered, g is the smallest prime fuzzy ideal of S containing $x_r \circ C_S$. Thus $g = rad(x_r \circ C_S)$. Hence by Theorems 7.13.4 and 7.13.7, $x_r \circ C_S$ is a primary fuzzy ideal and $x_r \circ C_S = g^n$ for some $n > 0$. Since $x_r \nsubseteq g^2$, it follows that $x_r \circ C_S = g$. ∎

Theorem 7.13.13 *Let m be the unique maximal fuzzy ideal of S having the sup property. Then for every fuzzy ideal g of S, either $g = m^n$ for some $n > 0$ or $g \subseteq m^\omega$.*

Proof. $g \subset m$ and m is prime. If $g \nsubseteq m^\omega$, there exists $n > 0$ such that $g \subseteq m^n$, $g \nsubseteq m^{n+1}$. Thus by the Theorem 7.13.8, $m^n = g : y_t \circ C_S$, where $y_l \notin \tau$. Since $y_l \nsubseteq m$, $y_t \circ C_S \nsubseteq m$. If $y_t \circ C_S$ is a proper fuzzy ideal of S, this contradicts the fact that m is the unique maximal fuzzy ideal of S. Hence $y_t \circ C_S = C_S$ and $m^n = g : C_S$. Now it can be easily verified that $g : C_S = g$. Thus $g = m^n$. ∎

Corollary 7.13.14 *Let S be a commutative semigroup with identity. Suppose that the unique maximal fuzzy ideal m satisfying the sup property and $m^\omega = 0$. Then S is a fuzzy multiplication semigroup if and only if the fuzzy ideals of S are powers of m.*

Definition 7.13.15 *A semigroup S is said to be a* ***fuzzy principal ideal semigroup*** *if every fuzzy ideal of S is a principal fuzzy ideal.*

Theorem 7.13.16 *Let S be a commutative semigroup with identity. If S is regular or a fuzzy principal ideal semigroup, then S is a fuzzy multiplication semigroup.*

Proof. Let S be regular. Let g and h be any two fuzzy ideals of S such that $g \subseteq h$. Then $g \circ h \subseteq g$. Next, let $x \in S$. Then there exists $a \in S$ such that $xax = x$. Thus $(g \circ h)(x) \geqslant g(x) \wedge h(ax) \geqslant g(x) \wedge h(x)$ (since h is a fuzzy ideal) $= g(x)$. Hence $g \circ h \supseteq g$ and thus $g \circ h = g$. Hence S is a fuzzy multiplication semigroup. The case when S is a fuzzy principal ideal semigroup is obvious. ∎

Definition 7.13.17 *A semigroup S is said to have* ***fuzzy dimension*** *n if there exists a chain of prime fuzzy ideals $g_0 \subset g_1 \subset \ldots \subset g_n$ of length n and there does not exist any chain of prime fuzzy ideals of length greater than n.*

Theorem 7.13.18 *Let S be a fuzzy multiplication semigroup with identity such that every fuzzy ideal has the sup property, there exists a unique* *maximal fuzzy ideal m of S, and S has finite fuzzy dimension n. Then the following conditions are equivalent.*

(1) S is a fuzzy principal ideal semigroup.

(2) Every idempotent prime fuzzy ideal of S is principal.

(3) The prime fuzzy ideals of S are linearly ordered.

Proof. (1) $\Rightarrow$ (2): The proof here is clear.

(2) $\Rightarrow$ (3): Since the fuzzy dimension of S is n, there exists a chain $g_0 \subset g_1 \subset \ldots \subset g_n = m$ of prime fuzzy ideals of length n of S. Clearly, there is no prime fuzzy ideal between g_{i-1} and g_i for every i. If g_i is nonidempotent, then by Theorem 7.13.10, g_{i-1} is the only prime fuzzy ideal such that $g_{i-1} \subset g_i$ with no prime fuzzy ideal between g_{i-1} and g_i.

Suppose that g_i is idempotent. Then by (2), $g_i = x_r \circ C_S$, where x_r is a fuzzy point of S. Let $h(\neq g_{i-1})$ be a prime fuzzy ideal such that $h \subset g_i$ with no prime fuzzy ideal between h and g_i. Then $h \cup g_{i-1}$ is also a prime fuzzy ideal and $h \cup g_{i-1} = g_i = x_r \circ C_S$. From this it follows that $x_r \circ C_S \subseteq h$ or $x_r \circ C_S \subseteq g_{i-1}$, a contradiction. Thus g_{i-1} is the only prime fuzzy ideal such that $g_{i-1} \subset g_i$ with no prime fuzzy ideal between g_{i-1} and g_i. Hence there is no prime fuzzy ideal of S except those present in the chain. Thus the prime fuzzy ideals of S are linearly ordered.

(3)$\Rightarrow$(1): By Theorem 7.12.8, it follows that the fuzzy radical of every fuzzy ideal of S is prime. Hence by Theorems 7.13.4 and 7.13.7, every fuzzy ideal of S is some power of a prime fuzzy ideal of S. Thus it suffices to show that every prime fuzzy ideal of S is a principal.

Let g be any prime fuzzy ideal of S. If g be non-idempotent, then by Theorem 7.13.12, g is principal. Suppose that g is idempotent. Since fuzzy dimension of S is finite, g contains properly at most a finite number of prime fuzzy ideals of S. Let $x_r \subseteq g$. If g does not contain properly any prime fuzzy ideal, clearly $rad(x_r \circ C_S) = g$. Then by Theorems 7.13.4 and 7.13.7, $x_r \circ C_S = g^n = g$ since g is idempotent.

Next, let $g_1, \ldots, g_m$ be the prime fuzzy ideals properly contained in g. Clearly, $\cup\{g_i \mid i = 1, \ldots, m\} \subset g$. Let, $x_r \subseteq g$, but $x_r \nsubseteq \cup\{g_i \mid i = 1, \ldots, m\}$. Then a similar argument as before yields $x_r \circ C_S = g$. ■

Theorem 7.13.19 *Let S be a commutative semigroup with identity such that every fuzzy ideal of S has the sup property, S has only two proper prime fuzzy ideals, and there exists a unique fuzzy maximal ideal m of S. Then S is a fuzzy multiplication semigroup with identity if and only if*

(1) every fuzzy ideal having prime fuzzy radical is primary,

(2) every primary fuzzy ideal is a power of its fuzzy radical.

Proof. The necessity of (1) and (2) follow from Theorems 7.13.4 and 7.13.7.

Conversely, suppose the conditions (1) and (2) hold. Let g and m be the two proper prime fuzzy ideals of S. Clearly $g \subset m$. Let k and h be any two fuzzy ideals of S such that $k \subseteq h$. If $k = h$, then $hC_S = k$. Again if

$h = C_S$, then $h \circ k = k$. Suppose that $k \subset h$ and $h \neq C_S$. Now if k and h both have the same fuzzy radical, then the result follows immediately from the given conditions. Suppose now that $radk = g$, $radh = m$. Then by the given conditions, $k = g^k$ and $h = m^l$ for some positive integers j, l. Now, $g^j \circ m^l \subseteq g$. Thus the fuzzy radical of $g^j \circ m^l$ is g and by the given conditions, $g^j \circ m^l = g^n$ for some positive integer n. Since g^n is a primary fuzzy ideal having the fuzzy radical g and since $m^l \not\subseteq g$, $g^j \subseteq g^n$. Hence $g^j = g^n$. Thus $k \circ h = k$. ■

A problem remaining is to find a sufficient condition for a fuzzy multiplication semigroup with identity to be a regular semigroup

7.14 Fuzzy Ideal Extensions

In the remaining sections of this chapter, we introduce the concepts of the extension of a fuzzy ideal of S and a 3-prime fuzzy ideal of S. We discuss some properties of the extension of a fuzzy ideal of S and study the relationship between l-prime fuzzy ideals of S and 3-prime fuzzy ideals of S by means of the extensions of fuzzy ideals. If a commutative semigroup S has an identity element, then the 3-prime fuzzy ideals and the l-prime fuzzy ideals of S coincide. If f is a semiprime fuzzy ideal of S, then f is the infimum of all extensions $\langle x, f\rangle$, $x \in S$. Moreover, if f is 3-prime, then f can be expressed as the infimum of all l-prime fuzzy ideals of S containing it. As a consequence, if S is a semilattice, then a 3-prime fuzzy ideal of S is the infimum of l-prime fuzzy ideals of S containing it. The results are from [153].

Let f be a fuzzy subset of S. f is called **l-prime** if

$$f(x_1x_2) = f(x_1) \vee f(x_2) \; \forall x_1, x_2 \in S.$$

A fuzzy subset f of S is called **semiprime** if

$$f(a) \geq f(a^2) \; \forall a \in S.$$

Lemma 7.14.1 *Let S be a commutative semigroup and f a fuzzy subset of S. Then f is l-prime if and only if for any $t \in (0,1]$ the level subset f_t, if not empty, is a prime ideal of S.*

Proof. Let f be l-prime. Then f is a fuzzy ideal of S. If $f_t \neq \emptyset$ for $t \in [0,1]$, then by Proposition 2.3.6, f_t is an ideal of S. Let $x, y \in S$ be such that $xy \in f_t$. Then $f(yx) \geq t$. Since f is l-prime, we have $f(xy) = f(x) \vee f(y)$. Thus $f(x) \geq t$ or $f(y) \geq t$, i.e. $x \in f_t$ or $y \in f_t$.

Conversely, let f be a fuzzy subset of S. Then by Proposition 2.3.6 and the hypothesis, f is a fuzzy ideal of S. Let $x, y \in S$ and $f(xy) = t$. Then $t \in [0,1]$ and $f_t \neq \emptyset$. Thus by hypothesis, f_t is prime. Since $xy \in f_t$, we have $x \in f_t$ or $y \in f_t$. It implies that $f(x) \geq t$ or $f(y) \geq t$. Consequently,

$$f(xy) = t \leq f(x) \vee f(y) \leq f(xy).$$

Therefore, $f(xy) = f(x) \vee f(y)$. ■

Lemma 7.14.2 *Let S be a semigroup and $\emptyset \neq I \subseteq S$. Then I is a prime ideal if and only if C_I is an l-prime fuzzy ideal of S.*

Definition 7.14.3 *Let f a fuzzy subset of S and $x \in S$. The fuzzy subset $\langle x, f\rangle$ of S defined by $\forall y \in S$,*

$$\langle x, f\rangle(y) = f(xy),$$

is called the ***extension*** *of f by x.*

Proposition 7.14.4 *Let S be a commutative semigroup. If f is a fuzzy ideal of S and $x \in S$, then the extension of f by x is a fuzzy ideal of S.*

Proof. Clearly, $\langle x, f\rangle$ is a fuzzy subset of S. Let $y, z \in S$. We have

$$\langle x, f\rangle(yz) = f(xyz) \geq f(xy) = \langle x, f\rangle(y).$$

Since S is commutative and f a fuzzy ideal of S, we have

$$\langle x, f\rangle(yz) = f(xyz) = f(yxz) \geq f(xz) = \langle x, f\rangle(z).$$

■

Proposition 7.14.5 *Let S be a semigroup, f be a fuzzy ideal of S, and $x \in S$. Then the following assertions hold:*
(1) $f \subseteq \langle x, f\rangle$.
(2) $\langle x^n, f\rangle \subseteq \langle x^{n+1}, f\rangle$ for every $n \in \mathbb{N}$.
(3) If $f(x) > 0$, then $Supp\langle x, f\rangle = S$.

Proof. (1) Let $y \in S$. Since f is a fuzzy ideal of S, we have

$$\langle x, f\rangle(y) = f(xy) \geq f(y).$$

Thus $f \subseteq \langle x, f\rangle$.
(2) Let $n \in \mathbb{N}$ and $y \in S$. Since f is a fuzzy ideal of S, we have

$$\langle x^{n+1}, f\rangle(y) = f(x^{n+1}y) = f(xx^n y) \geq f(x^n y) = \langle x^n, f\rangle(y).$$

Hence $\langle x^n, f\rangle \subseteq \langle x^{n+1}, f\rangle$.
(3) Clearly, $Supp\langle x, f\rangle \subseteq S$.
Let $y \in S$. Since f is a fuzzy ideal of S, we have

$$\langle x, f\rangle(y) = f(xy) \geq f(x) > 0.$$

Thus $\langle x, f\rangle(y) > 0$, and $y \in Supp\langle x, f\rangle$. ■

Remark 7.14.6 *Let S be a semigroup. If f is an l-prime fuzzy subset of S, then*

$$\langle x, f\rangle = \langle x^2, f\rangle \ \forall x \in S.$$

In fact, let $x, y \in S$. Then $\langle x, f\rangle(y) = f(xy)$ and $\langle x^2, f\rangle(y) = f(x^2y)$. Since f is an l-prime fuzzy subset of S, $f(x) = f(x^2)$ and

$$f(xy) = f(x) \vee f(y) = f(x^2) \vee f(y) = f(x^2y).$$

If $A \subseteq S$ and $x \in S$, we define $\langle x, A\rangle = \{y \in S \mid xy \in A\}$.

Proposition 7.14.7 *Suppose that A is a nonempty subset of S. Then*

$$\langle x, f_A\rangle = f_{\langle x,A\rangle}$$

for every $x \in S$.

Proof. Let $x \in S$, $y \in S$. Then

$$\langle x, f_A\rangle(y) = f_A(xy). \tag{7.1}$$

(1) If $xy \in A$, then $f_A(xy) = 1$.
Since $xy \in A$, we have $y \in \langle x, A\rangle$. Thus $f_{\langle x,A\rangle}(y) = 1$. By (7.1), $\langle x, f_A\rangle(y) = f_{\langle x,A\rangle}(y)$.
(2) Let $xy \notin A$. Then $f_A(xy) = 0$.
Since $xy \notin A$, $y \notin \langle x, A\rangle$. Hence $f_{\langle x,A\rangle}(y) = 0$. By (7.1), $\langle x, f_A\rangle(y) = f_{\langle x,A\rangle}(y)$. ■

Theorem 7.14.8 *If f is an l-prime fuzzy subset of S and $x \in S$ is such that $f(x) = \wedge_{y\in S} f(y)$, then $\langle x, f\rangle = f$. Conversely, let f be a fuzzy ideal of S. Suppose that for any $y \in S$ such that $f(y)$ is not maximal in $f(S)$, we have $\langle y, f\rangle = f$. Then f is l-prime.*

Proof. Let f be an l-prime fuzzy subset of S and $x \in S$ be such that $f(x) = \wedge_{y\in S} f(y)$.
We notice first that the element $\wedge_{y\in S} f(y)$ exists in $[0, 1]$.
Let $y \in S$. Then

$$\langle x, f\rangle(y) = f(xy). \tag{7.2}$$

Since f is l-prime, we have $f(xy) = f(x) \vee f(y)$. Thus

$$f(xy) = f(x) \text{ or } f(xy) = f(y). \tag{7.3}$$

Let $f(xy) \neq f(y)$. Then by (7.3), we have

$$f(xy) = f(x). \tag{7.4}$$

Since f is a fuzzy ideal of S, we have $f \subseteq \langle x, f \rangle$ by Proposition 7.14.5(1). Thus by (7.4),

$$f(y) \leq \langle x, f \rangle(y) = f(xy) = f(x).$$

Since $f(x) = \wedge_{y \in S} f(y)$, we have $f(x) \leq f(y)$. Hence $f(x) = f(y)$, and $f(xy) = f(y)$ by (7.4), a contradiction. Thus $f(xy) = f(y)$. Hence by (7.1), $\langle x, f \rangle(y) = f(y)$.

We now consider the converse. Let $x_1, x_2 \in S$. Then

$$f(x_1 x_2) = f(x_1) \vee f(x_2).$$

In fact, since f is a fuzzy ideal of S, we have

$$f(x_1 x_2) \geq f(x_1) \text{ and } f(x_1 x_2) \geq f(x_2). \tag{7.5}$$

Suppose $f(x_1 x_2) = f(x_1)$. Then by (7.5), $f(x_1) \geq f(x_2)$. Thus

$$f(x_1) \vee f(x_2) = f(x_1) = f(x_1 x_2).$$

Suppose $f(x_1 x_2) \neq f(x_1)$. Then the element $f(x_1)$ can not be a maximal element of $f(S)$.

If $f(x_1)$ is a maximal element of $f(S)$, then by (7.5), we have $f(x_1 x_2) = f(x_1)$, a contradiction. By hypothesis, we have $\langle x_1, f \rangle = f$. Thus $\langle x_1, f \rangle(x_2) = f(x_2)$, and $f(x_1 x_2) = f(x_2)$. Then by (7.5), $f(x_2) \geq f(x_1)$. Hence $f(x_1) \vee f(x_2) = f(x_2)$, and $f(x_1 x_2) = f(x_1) \vee f(x_2)$. ■

Corollary 7.14.9 *Let I be an ideal of S. Then I is prime if and only if for $\forall\, x \in S$ such that $x \notin I$, $\langle x, C_I \rangle = C_I$.*

Proof. Let I be a prime ideal of S. Then by Lemma 7.14.2, C_I is a prime fuzzy ideal of S. For $x \in S$ such that $x \notin I$, we have

$$C_I(x) = 0 = \wedge_{y \in S} C_I(y).$$

By Theorem 7.14.8, $\langle x, C_I \rangle = C_I$ holds.

Conversely, let I be an ideal of S. By Lemma 2.3.1, C_I is a fuzzy ideal of S. Let $x \in S$ be such that $C_I(x)$ is not maximal in $C_I(S)$. Since $C_I(y) = 0$ or $C_I(y) = 1$ for every $y \in S$, we have $C_I(x) = 0$ else $C_I(x) = 1$ which is the maximum in $f(S)$. Thus $x \notin I$. By hypothesis, we have $\langle x, C_I \rangle = C_I$. By Theorem 7.14.8, f_I is prime. Therefore, by Lemma 7.14.2, I is prime. ■

Theorem 7.14.10 *Let S be a commutative semigroup and f a fuzzy subset of S such that $\langle x, f \rangle = f$ for every $x \in S$. Then f is constant.*

Proof. Let $x, y \in S$. By hypothesis, we have

$$f(x) = \langle y, f \rangle(x) = f(xy) = \langle x, f \rangle(y) = f(y).$$

Thus f is constant. ■

Proposition 7.14.11 *Let f be an l-prime fuzzy subset of S. Then $\langle x, f\rangle$ is an l-prime fuzzy subset of S for every $x \in S$.*

Proof. Let f be an l-prime fuzzy subset of S. Clearly, $\langle x, f\rangle$ is a fuzzy subset of S. For every $x \in S$ and $y, z \in S$, we have

$$\begin{aligned}\langle x, f\rangle(yz) &= f(xyz)\\ &= f(xy) \vee f(z)\\ &= f(x) \vee f(y) \vee f(z)\\ &= f(xy) \vee f(xz)\\ &= \langle x, f\rangle(y) \vee \langle x, f\rangle(z).\end{aligned}$$

Thus $\langle x, f\rangle$ is l-prime. ∎

Proposition 7.14.12 *Let S be a commutative semigroup and f a semiprime fuzzy ideal of S. Then $\langle x, f\rangle$ is a semiprime fuzzy ideal of S for every $x \in S$.*

Proof. Let f be a semiprime fuzzy ideal of S and $x \in S$. Then by Proposition 7.14.4, $\langle x, f\rangle$ is a fuzzy ideal of S. For all $a \in S$, since f is semiprime, we have

$$\langle x, f\rangle(a^2) = f(xa^2) \leq f((xa)^2) \leq f(xa) = \langle x, f\rangle(a).$$

Thus $\langle x, f\rangle$ is semiprime. ∎

Corollary 7.14.13 *Let S be a commutative semigroup, $\{f_i\}_{i\in I}$ a nonempty family of a semiprime fuzzy ideals of S and let $f = \cap_{i\in I} f_i$. Then for all $x \in S$, $\langle x, f\rangle$ is a semiprime fuzzy ideal of S.*

Proof. Let $x, y \in S$. Since $f_i(i \in I)$ is a fuzzy ideal of S, we have $f(xy) = (\cap_{i\in I} f_i)(xy) = \wedge\{f_i(xy) \mid i \in I\} \geq \wedge\{f_i(x) \vee f_i(y) \mid i \in I\} \geq (\wedge\{f_i(x) \mid i \in I\}) \vee (\wedge\{f_i(y) \mid i \in I\}) = f(x) \vee f(y)$. Thus f is a fuzzy ideal of S. Let $a \in S$. Since f_i is semiprime $\forall i \in I$, we have

$$\begin{aligned}f(a) &= (\cap_{i\in I} f_i)(a)\\ &= \wedge\{f_i(a) \mid i \in I\} \geq \wedge\{f_i(a^2) \mid i \in I\}\\ &= \wedge\{f_i \mid i \in I\}(a^2) = f(a^2).\end{aligned}$$

Thus f is semiprime. By Proposition 7.14.12, $\langle x, f\rangle$ is semiprime fuzzy ideal of S $\forall x \in S$. ∎

Corollary 7.14.14 *Let S be a commutative semigroup, $\{P_i\}_{i\in I}$ a nonempty family of semiprime ideals of S, and $A = \cap_{i\in I} P_i \neq \emptyset$. Then $\langle x, C_A\rangle$ is a semiprime ideal of S for every $x \in S$.*

Proof. Clearly, A is a semiprime ideal of S. Thus C_A is a semiprime fuzzy ideal of S. Hence by Proposition 7.14.12, $\langle x, C_A\rangle$ is a semiprime fuzzy ideal of S. ∎

Corollary 7.14.15 *Let S be a commutative semigroup and f an l-prime fuzzy subset of S. If f is not constant, then f is not a maximal l-prime fuzzy subset of S.*

Proof. Let f be a l-prime fuzzy subset of S. Then by Proposition 7.14.11 and hypothesis, for each $x \in S$, $\langle x, f\rangle$ is a l-prime fuzzy subset of S. Further, there exists $x \in S$ such that $f \subset \langle x, f\rangle$, else, $f = \langle x, f\rangle \; \forall x \in S$. By Theorem 7.14.10, f is constant, a contradiction. ■

If f is a fuzzy ideal of S, we denote by f_ρ the equivalence relation on S defined by

$$f_\rho = \{(x, y) \mid \langle x, f\rangle = \langle y, f\rangle\}.$$

Proposition 7.14.16 *Let S be a commutative semigroup and f a fuzzy ideal of S. Then the following assertions hold:*

(1) f_ρ is a congruence on S.

(2) If f is semiprime, then f_ρ is a semilattice congruence on S.

Proof. (1) To show that f_ρ is a congruence on S, we need only prove that f_ρ is compatible with the operation on S. Let $(x, y) \in f_\rho$, $c \in S$. Then $(xc, yc) \in f_\rho$. Indeed $\forall z \in S$,

$$\begin{aligned}\langle xc, f\rangle(z) &= f(xcz)\\ &= \langle x, f\rangle(cz)\\ &= \langle y, f\rangle(cz) \text{ (since } \langle x, f\rangle = \langle y, f\rangle)\\ &= f(ycz)\\ &= \langle yc, f\rangle(z).\end{aligned}$$

Similarly, we have $(cx, cy) \in f_\rho$.

(2) Let f be semiprime. Then for any $z \in S$,

$$\begin{aligned}\langle x^2, f\rangle(z) &= f(x^2 z)\\ &\le f((xz)^2)\\ &\le f(xz)\\ &= \langle x, f\rangle(z).\end{aligned}$$

Hence $\langle x^2, f\rangle \subseteq \langle x, f\rangle$. By Proposition 7.14.5, $\langle x, f\rangle \subseteq \langle x^2, f\rangle$. Thus $\langle x, f\rangle = \langle x^2, f\rangle$, that is, $(x, x^2) \in f_\rho$. ■

7.15 Prime Fuzzy Ideals

Definition 7.15.1 *A fuzzy subset f of S is called **3-prime** if for any $x_1, x_2, x_3 \in S$,*

$$\begin{aligned}f(x_1x_2x_3) &= f(x_1x_2) \vee f(x_1x_3)\\ &= f(x_2x_3) \vee f(x_2x_1)\\ &= f(x_3x_1) \vee f(x_3x_2).\end{aligned}$$

Proposition 7.15.2 *Let S be a semigroup and f a fuzzy subset of S. If f is l-prime, then f is 3-prime.*

Proof. Let f be a prime fuzzy subset of S. Then for any $x_1, x_2, x_3 \in S$, we have

$$\begin{aligned} f(x_1x_2x_3) &= f(x_1) \vee f(x_2x_3) \\ &\leq f(x_1x_2) \vee f(x_2x_3) \\ &\leq f(x_1x_2x_3). \end{aligned}$$

Thus $f(x_1x_2x_3) = f(x_1x_2) \vee f(x_2x_3)$. Since

$$\begin{aligned} f(x_1x_2x_3) &= f(x_2x_1) \vee f(x_3) \\ &= f(x_1x_2) \vee f(x_3) \\ &= f(x_1x_2x_3), \end{aligned}$$

we have

$$\begin{aligned} f(x_1x_2x_3) &= f(x_1x_2) \vee f(x_3) \\ &\leq f(x_1x_2) \vee f(x_1x_3) \\ &= f(x_2x_1) \vee f(x_1x_3) \\ &\leq f(x_2x_1x_3) \vee f(x_2x_1x_3) \\ &= f(x_2x_1x_3) \\ &= f(x_1x_2x_3). \end{aligned}$$

Hence

$$f(x_1x_2x_3) = f(x_2x_1) \vee f(x_1x_3).$$

In a similar manner, we have

$$\begin{aligned} f(x_1x_2x_3) &= f(x_1) \vee f(x_2x_3) \\ &\leq f(x_3x_1) \vee f(x_2x_3) \\ &= f(x_1x_3) \vee f(x_2x_3) \\ &\leq f(x_2x_1x_3) \vee f(x_1x_2x_3) \\ &= f(x_1x_2x_3). \end{aligned}$$

Thus

$$f(x_1x_2x_3) = f(x_3x_1) \vee f(x_2x_3).$$

■

In general, a 3-prime fuzzy ideal is not l-prime as shown in the following example.

Example 7.15.3 *Let S be a semigroup defined by the following multiplication table:*

$\cdot$	a	b	c	d
a	b	b	d	d
b	b	b	d	d
c	d	d	c	d
d	d	d	d	d

Let f be a fuzzy subset of S defined as follows:

$$f(d) = 0.5, \ f(a) = f(b) = f(c) = 0.$$

Then by Proposition 2.3.6, f is a fuzzy ideal of S. Since

$$f(bc) = f(d) = 0.5 \neq f(b) \vee f(c),$$

f is not prime. But f is a 3-prime fuzzy ideal. In fact, for any $x_1, x_2, x_3 \in S$, we have

$$x_1x_2x_3 \in \{a^3, a^2b, a^2c, a^2d, b^3, b^2a, b^2c, b^2d, c^3, c^2a, c^2b, c^2d, abc, acd, bcd, abd\}.$$

Let one of x_1, x_2 and x_3 be d. Since S is commutative and $xd = d$ for every $x \in S$, we have

$$\begin{aligned} f(x_1x_2x_3) &= f(d) \\ &= f(x_1x_2) \vee f(x_1x_3) \\ &= f(x_2x_3) \vee f(x_2x_1) \\ &= f(x_3x_1) \vee f(x_3x_2). \end{aligned}$$

Furthermore, let x_1, x_2 and x_3 be different from d. Then $f(x_1x_2x_3)$ has the following cases:

$$\begin{aligned} &f(a^3) = f(b) = f(a^2) \vee f(a^2); \\ &f(a^2b) = f(b) = f(a^2) \vee f(ab) = f(ab) \vee f(ab); \\ &f(a^2c) = f(d) = f(a^2) \vee f(ac) = f(ac) \vee f(ac); \\ &f(b^3) = f(b) = f(b^2) \vee f(b^2); \\ &f(b^2a) = f(b) = f(b^2) \vee f(ab) = f(ab) \vee f(ab); \\ &f(b^2c) = f(d) = f(b^2) \vee f(cb) = f(cb) \vee f(cb); \\ &f(c^3) = f(c) = f(c^2) \vee f(c^2); \\ &f(c^2a) = f(d) = f(c^2) \vee f(ac) = f(ac) \vee f(ac); \\ &f(c^2b) = f((d)) = f(c^2) \vee f(ac) = f(cb) \vee f(cb); \\ &f(abc) = f(d) = f(ab) \vee f(ac) = f(ab) \vee f(cb) = f(ac) \vee f(cb). \end{aligned}$$

Theorem 7.15.4 *Let f be a fuzzy subset of S. If f is 3-prime, then any extension of f is l-prime. In particular, if S is commutative, then f is 3-prime if and only if any extension of f is l-prime.*

Proof. Let f be 3-prime. Then for any $x, y, z \in S$,

$$\begin{aligned} \langle x, f\rangle(yz) &= f(xyz) \\ &= f(xy) \vee f((xz) \\ &= \langle x, f\rangle(y) \vee \langle x, f\rangle(z). \end{aligned}$$

Thus f is l-prime. Conversely, let S be a commutative semigroup and f a fuzzy subset of S. Then by Proposition 7.14.4 and the hypothesis, any extension of f is an l-prime fuzzy subset of S. Thus $\forall x, y, z \in S$,

$$\begin{aligned} f((xyz) &= \langle x, f\rangle(yz) \\ &= f(xy) \vee f(xz) \\ &= \langle y, f\rangle(xz) \\ &= f(yx) \vee f(zy) \\ &= \langle z, f\rangle(xy) \\ &= f(zx) \vee f(zy). \end{aligned}$$

Therefore, f is 3-prime. ■

Corollary 7.15.5 *Let S be a semigroup with an identity element e and f a fuzzy subset of S. If f is 3-prime, then f is l-prime.*

Proof. Since e is the identity of S, $\langle e, f\rangle = f$. Since f is 3-prime, by Theorem 7.15.4, $\langle e, f\rangle$ is l-prime. Thus f is l-prime. ■

Now, by Proposition 7.15.2 and Corollary 7.15.5, in semigroups which have an identity element, the l-prime fuzzy subsets and the 3-prime fuzzy subsets coincide.

By the definition of a semiprime fuzzy subset of S, we have the following result.

Theorem 7.15.6 *If S is a semigroup and f a semiprime fuzzy ideal of S, then*

$$f = \cap_{x \in S}\langle x, f\rangle.$$

Proof. Let f be a fuzzy ideal of S. Then by Proposition 7.14.5 (1), we have $f \subseteq \langle x, f\rangle$ $\forall x \in S$.

Let g be a fuzzy subset of S such that $g \subseteq \langle x, f\rangle$ $\forall x \in S$, and let $y \in S$. Since f is semiprime, we have

$$g(y) \leq \langle y, f\rangle(y) = f(y^2) \leq f(y).$$

Thus $g \subseteq f$. ■

Corollary 7.15.7 *Let f be a fuzzy ideal of S. If f is 3-prime and semiprime, then f is the infimum of all l-prime fuzzy ideals of S containing f.*

Proof. The result follows by Theorems 7.15.6 and 7.15.4. ■

If S is a semilattice, then any fuzzy subset of S is semiprime. Thus we have the following result.

Corollary 7.15.8 *Let S be a semilattice. Then any 3-prime fuzzy ideal f is expressible as the infimum of all l-prime fuzzy ideals of S containing f.*

Corollary 7.15.7 is not true in general if f is not semiprime. This can be seen by the following example.

Example 7.15.9 *Let S be the semigroup defined by the following multiplication table:*

$\cdot$	a	b	c	d
a	a	a	a	a
b	a	a	a	a
c	a	a	a	a
d	a	a	a	d

Let f be the fuzzy subset of S defined as follows:

$$f(a) = 0.5,\ f(b) = f(c) = f(d) = 0.$$

By Proposition 2.3.6, *f is a fuzzy ideal of S. As in Example* 7.15.3, *it can be shown that f is* 3*-prime . However, f is not semiprime since*

$$f(b^2) = f(a) = 0.5 > f(b).$$

Let A be the set of all fuzzy subsets of S containing f. Then A is nonempty. We have $g(b) \geq 0.5$ *for every* $g \in A$. *Indeed, let* $g \in A$. *Since* $a = b^2$, *we have*

$$0.5 = f(a) \leq g(a) = g(b^2) = g(b) \vee g(b) = g(b).$$

Hence

$$(\cap_{h \in A} h)(b) = \wedge\{h(b) \mid h \in A\} \geq 0.5.$$

Since $f(b) = 0$, $(\cap_{h \in S} h)(b) \neq f(b)$. *Thus* $f \neq \cap_{h \in A} h$.

8

Fuzzy Codes on Free Monoids

8.1 Fuzzy Codes

We introduce the concept of a fuzzy base of a fuzzy monoid. We also introduce the concepts of a fuzzy code, fuzzy prefix code, and maximal fuzzy prefix code, and examine some of their algebraic properties. The results of Section 8.1-8.9 are essentially from [133].

In the following, we assume that X is an alphabet with $1 \leq |X| < \infty$ and $X^*(X^+)$ is the free monoid (semigroup) generated by X with the operation of concatenation. In this chapter, we let e denote the empty string of X^*. Recall that $\mathcal{FP}(X)$ denotes the set of all fuzzy subsets of X. $A \in \mathcal{FP}(X^*)$ is called a **fuzzy language** on the free monoid X^*. Let $A, B \in \mathcal{FP}(X^*)$. Define the fuzzy subset AB of X^* by $\forall x \in X^*$,

$$AB(x) = \vee\{A(y) \wedge B(z) \mid y, z \in X^*,\ x = yz\}.$$

It follows easily that $(AB)C = A(BC)\ \forall A, B, C \in \mathcal{FP}(X^*)$. $\forall n \in \mathbb{N}$, define $A^{n+1} = AA^n$, i.e.,

$$A^{n+1}(x) = \vee\{A(y) \wedge A^n(z) \mid y, z \in X^*,\ x = yz\},$$

$\forall x \in X^*$. $A^* = C_{\{e\}} \cup A \cup A^2 \ldots$ is the Kleene's closure of A, where the empty string e is the identity of X^*.

Let Φ denote the fuzzy subset of X^* such that $\Phi(x) = 0\ \forall x \in X^*$. Let $A, B \in \mathcal{FP}(X^*)$. Define the fuzzy subset $A \backslash B$ of X^* by $\forall x \in X^*$, $(A \backslash B)(x) = A(x)$ if $x \notin SuppB$ and $(A \backslash B)(x) = 0$ otherwise.

Let $x, y \in X^*$. We define $[x = y] = 1$ if $x = y$ and $[x = y] = 0$ if $x \neq y$. Similarly, $[x \neq y] = 1$ if $x \neq y$ and $[x \neq y] = 0$ if $x = y$.

For all $a, b \in [0,1]$, let $a \propto b = 1$ if $a \leq b$ and $a \propto b = b/a$ if $a > b$. Let $x_i, y_j \in X^*$, $i = 1, \ldots, n$; $j = 1, \ldots, m$. In keeping with the notation in [133], we let

$$\wedge([n = m], [x_1 = y_1], \ldots, [x_n = y_m]) = \begin{cases} 1 & \text{if } m = n,\ x_i = y_i, \\ & \qquad i = 1, \ldots, n; \\ 0 & \text{if } m = n,\ x_i \neq y_i, \\ & \qquad \text{for some } i \in \{1, \ldots, n\}; \\ 0 & \text{if } m \neq n. \end{cases}$$

Definition 8.1.1 *Let A be a fuzzy submonoid of X^* and let $B \in \mathcal{FP}(X^*)$ be such that $B \subseteq A$. Then B is called a* ***fuzzy base*** *of A if $B(e) = 0$ and*

(B1) for all $x \in Supp(A) \backslash \{e\}$, $B^*(x) \geq A(x)$;

(B2) for all $x \in Supp(A) \backslash \{e\}$, $x_i, y_j \in X^*$, $i = 1, \ldots, n$; $j = 1, \ldots, m$ and $x = x_1 \ldots x_n = y_1 \ldots y_m$, $\wedge\{B(x_1), \ldots, B(x_n), B(y_1), \ldots, B(y_n)\} \propto \wedge\{[m = n], [x_1 = y_1], \ldots, [x_n = y_n]\} \geq A(x)$.

Definition 8.1.2 *A fuzzy submonoid A of X^* is called a* ***fuzzy free submonoid*** *if there exists a fuzzy base B of A such that $B^* = A$.*

Definition 8.1.3 *Let $A \in \mathcal{FP}(X^+)$ be such that $A \neq \Phi$. Then A is called a* ***fuzzy code*** *on X if A is a fuzzy base of A^*.*

Theorem 8.1.4 *Let $A \in \mathcal{FP}(X^+)$ be a fuzzy code. Then A^* is free.*

Proof. The proof is straightforward. ■

Let S be a semigroup and $\widehat{S} = S \cup \{e\}$, the monoid with identity e. Let $A \in \mathcal{FP}(S)$ and $f \in \widehat{S}$. Define the fuzzy subsets Af and fA of S as follows: $\forall x \in S$,

$$\begin{aligned} Af(x) &= \vee\{A(v) \mid v \in S,\ vf = x\} \\ fA(x) &= \vee\{A(u) \mid v \in S,\ fu = x\}. \end{aligned}$$

Theorem 8.1.5 *For any $A \in \mathcal{FP}(X^+)$, $A \neq \Phi$, the following conditions are equivalent $\forall f \in X^*$.*

(1) If $A^*f \cap A^* \neq \Phi$ and $fA^* \cap A^* \neq \Phi$, then $A^*(f) > 0$.

(2) If $A^*(f) = 0$, then $fA^* \cap A^* \cap A^*f = \Phi$.

Proof. (1) $\Rightarrow$ (2): Assume $A^*(f) = 0$. Suppose $fA^* \cap A^* \cap A^*f \neq \Phi$. Then there exists $x \in X^+$ such that $(fA^*)(x) \wedge A^*(x) \wedge (A^*f)(x) > 0$. It follows that $A^*f \cap A^* \neq \Phi$ and $fA^* \cap A^* \neq \Phi$. By (1), we have $A^*(f) > 0$, which is a contradiction. Thus (2) holds.

(2) $\Rightarrow$ (1): Suppose $A^*f \cap A^* \neq \Phi$ and $fA^* \cap A^* \neq \Phi$. Then there exists $x, y \in X^+$ such that $(A^*f \cap A^*)(x) > 0$ and $(fA^* \cap A^*)(y) > 0$, i.e., $(A^*f)(x) > 0$, $A^*(x) > 0$, $(fA^*)(y) > 0$ and $A^*(y) > 0$. Thus since

$(A^*f)(x) = \vee\{A^*(v) \mid v \in X^+, vf = x\} > 0$, there exists $v \in X^+$ such that $vf = x$ and $A^*(v) > 0$, and since $(fA^*)(y) = \vee\{A^*(u) \mid u \in X^+, fu = y\} > 0$, there exists $u \in X^+$ such that $fu = y$ and $A^*(u) > 0$. Hence

$$A^*(yx) \geq A^*(y) \wedge A^*(x) > 0,$$

$$\begin{aligned} (A^*f)(yx) &= (A^*f)(yvf) \\ &= \vee\{A^*(z) \mid z \in X^+, zf = yvf\} \\ &= \vee\{A^*(z) \mid z \in X^+, z = yv\} \\ &\geq A^*(y) \wedge A^*(v) > 0, \end{aligned}$$

$$\begin{aligned} (fA^*)(yx) &= (fA^*)(fux) \\ &= \vee\{A^*(z) \mid z \in X^+, fz = fux\} \\ &= \vee\{A^*(z) \mid z \in X^+, z = ux\} \\ &\geq A^*(z) \wedge A^*(x) \\ &> 0. \end{aligned}$$

Therefore, $fA^* \cap A^* \cap A^*f \neq \Phi$. By (2), $A^*(f) \neq 0$, i.e., $A^*(f) > 0$. Thus (1) holds. ∎

Theorem 8.1.6 *Let $A \in \mathcal{FP}(X^+)$ be such that $A \neq \Phi$. If A satisfies one of the conditions in Theorem* 8.1.5*, then the generating set $A^0 = A_1^* \backslash (A_1^*)^2$ is a fuzzy code, where $A_1^*(x) = A^*(x)$ if $x \neq e$; $A_1^*(x) = 0$ if $x = e$. Conversely, if A is a fuzzy code, then the conditions in Theorem* 8.1.5 *hold.*

Proof. If A satisfies the condition (1), then for any $f \in X^+$, $A^*f \cap A^* \neq \Phi$ and $fA^* \cap A^* \neq \Phi$ imply $A^*(f) > 0$. If A^0 is not a fuzzy code, then there exists $x \in SuppA^0 \backslash \{e\}$ and $x_1, \ldots, x_n, y_1, \ldots, y_m \in X^+$ such that $x = x_1 \ldots x_n = y_1 \ldots y_m$, and $\wedge\{A^0(x_1), \ldots, A^0(x_n), A^0(y_1), \ldots, A^0(y_m)\}$ $\propto \wedge\{[n = m], [x_1 = y_1], \ldots, [x_n = y_n]\} < A^{0*}(x)$. Thus

$$\wedge\{[n = m], [x_1 = y_1], \ldots, [x_n = y_n]\} = 0$$

and $A^0(x_i) > 0$, $i = 1, \ldots, n$, $A^0(y_j) > 0$, $j = 1, \ldots, m$. Without loss of generality, we assume that $x_1 \neq y_1$. We also assume that $x_1 = y_1 h$, $h \in X^+$. Then since $A^0(x_1) > 0$ and $A^0(y_1) > 0$, we have $A^0(x_1) = A^*(x_1) > 0$ and $A^*(y_1) > 0$. Hence

$$(A^*h)(x_1) = (A^*h)(y_1 h) = \vee\{A^*(z) \mid z \in X, zh = y_1 h\} \geq A^*(y_1) > 0.$$

Thus $(A^*h \cap A^*)(x) > 0$ and $A^*h \cap A^* \neq \Phi$. Since $y_1 h x_2 \ldots x_n = y_1 y_2 \ldots y_m$, we have $hx_2 \ldots x_n = y_2 \ldots y_m$. Set $u = x_2 \ldots x_n$, $v = y_2 \ldots y_m$. Then $hu = v$. Similarly, we also have $hA^* \cap A^* \neq \Phi$. From our assumption, it follows that $A^*(h) > 0$. Hence

$$(A_1^*)^2(x) = (A_1^*)^2(y_1 h) \geq A_1^*(y_1) \wedge A_1^*(h) = A^*(y_1) \wedge A^*(h) > 0.$$

Thus $A^0(x_1) = (A_1^* \backslash (A_1^*)^2)(x_1) = 0$, which is a contradiction. Hence A^0 is a fuzzy code.

Conversely, assume that A does not satisfy condition (1) of Theorem 8.1.5. Then there exists $f \in X^*$ such that $A^*f \cap A^* \neq \Phi$, $fA^* \cap A^* \neq \Phi$ and $A^*(f) = 0$. Since $A^*f \cap A^* \neq \Phi$, there exists $u \in X^+$ such that $A^*(u) > 0$ and

$$(A^*f)(u) = \vee \{A^*(v) \mid vf = u\} > 0.$$

Thus there exists $v \in X^+$ such that $vf = u$ and $A^*(v) > 0$. Hence there exists $a_1, \ldots, a_m, b_1, \ldots, b_n \in X^+$ such that $u = b_1 \ldots b_n$, $v = a_1 \ldots a_m$ and $A(b_i) > 0$, $i = 1, \ldots, n$, $A(a_j) > 0$, $j = 1, \ldots, m$. Therefore, $a_1 a_2 \ldots a_m f = b_1 b_2 \ldots b_n$ with $a_1 \neq b_1$. Also since $fA^* \cap A^* \neq \Phi$, we have that there exists $d_i, c_j \in X^+$, $(i = 1, \ldots, p;\ j = 1, \ldots, q)$, such that $fd_1 d_2 \ldots d_p = c_1 c_2 \ldots c_q$, $A^*(d_1 \ldots d_p) > 0$, $A^*(c_1 \ldots c_q) > 0$, $A(d_i) > 0$, $i = 1, \ldots, p$ and $A(c_j) > 0$, $j = 1, \ldots, q$. That is, there exists $b_1, \ldots, b_n$, $d_1, \ldots, d_p$, $a_1, \ldots, a_m$, $c_1, \ldots, c_q \in X^+$ and

$$b_1 \ldots b_n d_1 \ldots d_p = a_1 \ldots a_m c_1 \ldots c_q = a_1 \ldots a_m f d_1 \ldots d_p$$

such that $\wedge\{A(b_1), \ldots, A(b_n), \ldots, A(d_p), A(a_1), \ldots, A(a_m), A(c_1), \ldots, A(c_q)\} > 0$ with $[a_1 = b_1] = 0$. Thus A^0 does not satisfy the condition (B2), i.e., A^0 is not a fuzzy code. ■

8.2 Prefix Codes

Let X be a nonempty set and $A \in \mathcal{FP}(X^+)$. Define the fuzzy subset AX^+ of X^+ by $\forall u \in X^+$,

$$AX^+(u) = \vee\{A(x) \mid x, y \in X^+, xy = u\}.$$

Definition 8.2.1 *Let $A \in \mathcal{FP}(X^+)$. Then A is called a* ***fuzzy prefix code*** *if $A \neq \Phi$ and $A \cap AX^+ = \Phi$.*

Theorem 8.2.2 *A is a fuzzy prefix code if and only if*

$$\wedge\{A(xy) \propto 0 \mid y \in X^+\} \geq A(x)$$

for all $x \in X^+$.

Proof. Suppose A is a fuzzy prefix code and there exists $x \in X^+$ such that

$$\wedge\{A(xy) \propto 0 \mid y \in X^+\} < A(x).$$

Then there exists $y \in X^+$ such that $A(xy) \propto 0 < A(x)$. Thus $A(x) > 0$, $A(xy) > 0$. Set $u = xy$. Then

$$(AX^+)(u) = \vee\{A(x) \mid x, y \in X^+, xy = u\} = A(x) > 0.$$

Hence $(AX^+ \cap A)(u) > 0$, i.e., A is not a fuzzy prefix code, a contradiction.

Conversely, suppose $A \cap AX^+ \neq \Phi$. Then there exists $x \in X^+$ such that $A(x) > 0$ and

$$(AX^+)(x) = \vee\{A(u) \mid u, v \in X^+, uv = x\} > 0.$$

Thus there exists $u, v \in X^+$ such that $uv = x$ and $A(u) > 0$. Hence

$$\wedge\{A(uv) \propto 0 \mid v \in X^+\} \leq A(x) \propto 0 = 0 < A(u).$$

■

Theorem 8.2.3 *Let $A \in \mathcal{FP}(X^+)$. Then A is a fuzzy prefix code if and only if*

$$\wedge\{A(x) \propto [xy \neq u] \mid x, y \in X^+\} \geq A(u)$$

for any $u \in X^+$.

Proof. Suppose A is a fuzzy prefix code, $A(u) > 0$ and

$$\wedge\{A(x) \propto [xy \neq u] \mid x, y \in X^+\} < A(u).$$

Then there exists $x_0, y \in X^+$ such that $[x_0 y \neq u] = 0$, i.e., $x_0 y = u$, and $A(x_0) > 0$. Thus

$$(AX^+)(u) = \vee\{A(x) \mid x, y \in X^+, xy = u\} \geq A(x_0) > 0$$

and $(A \cap AX^+)(u) > 0$. However, this contradicts the assumption that A is a fuzzy prefix code by Theorem 8.2.2. Thus for all $u \in X^+$, we have

$$\wedge\{A(x) \propto [xy \neq u] \mid x, y \in X^+\} \geq A(u).$$

Conversely, assume that A is not a fuzzy prefix code. Then there exists $u \in X^+$ such that $(A \cap AX^+)(u) > 0$, i.e., $A(u) > 0$ and

$$(AX^+)(u) = \vee\{A(x) \mid x, y \in X^+, xy = u\} > 0.$$

Thus there exists $x, y \in X$ such that $xy = u$ and $A(x) > 0$. Hence $[x_0 y \neq u] = 0$ and $A(x) \propto [x_0 y \neq u] = 0$. Thus

$$\wedge\{A(x) \propto [xy \neq u] \mid x, y \in X\} = 0 < A(x).$$

Therefore, A does not satisfy the condition of the theorem. ■

Theorem 8.2.4 *Let $A \in \mathcal{FP}(X^+)$. Then A is a fuzzy prefix code if and only if*

$$\vee\{A(u) \wedge A(v) \mid u, v, x \in X^+,\ u = vx\} = 0.$$

Proof. Suppose A is a fuzzy prefix code and

$$\vee\{A(u) \wedge A(v) \mid u, v, x \in X^+,\ u = vx\} > 0.$$

Then there exists $u_0, v_0, x_0 \in X^+$ such that $u_0 = v_0 x_0$ and $A(u_0) > 0$, $A(v_0) > 0$. Thus

$$(AX^+)(u_0) = \vee\{A(v) \mid v, x \in X^+, vx = u_0\} \geq A(v_0) > 0.$$

Hence $(AX^+ \cap A)(u_0) > 0$, which is a contradiction.

Conversely, assume that A is not a fuzzy prefix code. Then by Theorem 8.2.3, there exists $u_0 \in X^+$ such that

$$\wedge\{A(v) \propto [vx \neq u_0] \mid v, x \in X^+\} < A(u_0)$$

and $A(u_0) > 0$. Hence there exists $v_0, x_0 \in X^+$ such that $v_0 x_0 = u_0$ and $A(v_0) > 0$. Thus we have

$$\vee\{A(u) \wedge A(v) \mid u, v, x \in X^+, u = vx\} \geq A(u_0) \wedge A(v_0) > 0,$$

i.e., A does not satisfy the condition in the theorem. ■

Theorem 8.2.5 *Let $A, B \in \mathcal{FP}(X^+)$. If $\Phi \neq B \subseteq A$ and A is a fuzzy prefix code, then B is also a fuzzy prefix code.*

Proof. The theorem follows immediately from Theorem 8.2.4. ■

Theorem 8.2.6 *A fuzzy prefix code is a fuzzy code.*

Proof. Suppose A is a fuzzy prefix code and is not a fuzzy code. Then there exists $x \in SuppA$ and $x_i, y_j \in A$, $i = 1, \ldots, n$, $j = 1, \ldots, m$, $m \geq n \geq 1$, such that $x_1 \ldots x_n = y_1 \ldots y_m = x$ and $\wedge\{A(x_1), \ldots, A(x_n), A(y_1), \ldots, A(y_m)\} \propto \wedge\{[m = n], [x_1 = y_1], \ldots, [x_n = y_n]\} < A^*(x)$. Thus $A^*(x) > 0$,

$$\wedge\{[m = n], [x_1 = y_1], \ \ldots, \ [x_n = y_n]\} = 0$$

and $A(x_i) > 0$, $A(y_j) > 0$, $i = 1, \ldots, n$, $j = 1, \ldots, m$. We may assume $x_1 \neq y_1$ and $|x_1| < |y_1|$. Then $x_1 = y_1 h$ for some $h \in X^+$. Since $A(x_1) > 0$ and

$$(AX^+)(x_1) = \vee\{A(y) \mid y, h \in X^+, yh = x_1\} \geq A(y_1) > 0,$$

we have $A \cap AX^+ \neq \Phi$, which contradicts the assumption that A is a fuzzy prefix code. Thus $x_1 = y_1$. Since cancellation holds in the free monoid X^*, we have $x_2 x_3 \ldots x_n = y_2 y_3 \ldots y_m$. Similarly, we can show that $x_2 = y_2, \ldots, x_n = y_m$ and $m = n$. However,

$$\wedge\{[m = n], [x_1 = y_1], \ldots, [x_n = y_n]\} = 0,$$

which is a contradiction. Hence A is a fuzzy code. ■

Theorem 8.2.7 *A is a fuzzy prefix code if and only if* $A(B \cap C) = AB \cap BC$ *for all* $B, C \in \mathcal{FP}(X^*)$.

Proof. Suppose A is a fuzzy prefix code. Clearly, $A(B \cap C) \subseteq AB \cap AC$. On the other hand, for any $x \in X^+$, let

$$\begin{aligned}(AB \cap AC)(x) &= (\vee\{A(a_1) \wedge B(b) \mid a_1, b \in X^*, a_1 b = x\}) \\ &\quad \wedge(\vee\{A(a_2) \wedge C(c) \mid a_2, c \in X^*, a_2 c = x\}) \\ &> t,\end{aligned}$$

where $t \in [0,1)$. Then there exists $a_1, a_2, b, c \in X^*$ such that $a_1 b = a_2 c = x$ and $A(a_1)$, $A(a_2)$, $B(b)$, $C(c) > t$. Since $A \in \mathcal{FP}(X^+)$, we have $a_1, a_2 \in X^+$. If $b = c = e$, then $a_1 = a_2$. If b (or c) $\in X^+$ and $a_1 \neq a_2$, we may assume $|a_1| < |a_2|$. Then $a_1 = a_2 h$, for some $h \in X^+$. Thus

$$(A \cap AX^+)(a_1) = A(a_1) \wedge AX^+(a_1) \geq A(a_1) \wedge A(a_2) > t,$$

which contradicts the assumption that A is a fuzzy prefix code. Hence $a_1 = a_2$ and $b = c$. Therefore,

$$\begin{aligned}(A(B \cap C))(x) &= \vee\{A(a_1') \wedge B(b') \wedge C(c) \mid a_1', b' \in X^*,\ a_1' b' = x\} \\ &\geq A(a_1) \wedge B(b) \wedge C(c) \\ &> t.\end{aligned}$$

Hence $A(B \cap C) \supseteq AB \cap AC$.

For the converse, it suffices to show that $A \cap AX^+ = \Phi$. In fact, since $A(B \cap C) = AB \cap BC$ for all $B, C \in \mathcal{FP}(X^*)$, we have $A \cap AX^+ = A(C_{\{e\}} \cap X^+) = A\Phi = \Phi$. ■

Theorem 8.2.8 *If A and B are fuzzy prefix codes, then AB is also a fuzzy prefix code.*

Proof. By Theorem 8.2.7,

$$AB \cap ABX^+ = A(B \cap BX^+) = A\Phi = \Phi.$$

■

Let $A \in \mathcal{FP}(X^+)$. Define the fuzzy subset $A \backslash X^n$ of X^+ as follows: $\forall x \in X^+$, $(A \backslash X^n)(x) = A(x)$ if $x \notin X^n$ and $(A \backslash X^n)(x) = 0$ if $x \in X^n$.

Theorem 8.2.9 *Let A be a fuzzy prefix code. Then* $A \backslash X^n$ *is also a fuzzy prefix code.*

Proof. Suppose $(A \backslash X^n) \cap (A \backslash X^n) X^+ \neq \Phi$. Then there exists $x \in X^+$ such that $(A \backslash X^n)(x) > 0$ and

$$((A \backslash X^n) X^+)(x) = \vee\{(A \backslash X^n)(y) \mid y, z \in X^+, yz = x\} > 0.$$

Thus $x \in X^n$, $A(x) > 0$ and there exists $y, z \in X^+$ such that $yz = x$, $(A \backslash X^n)(y) > 0$. Hence $y \in X^n$, $A(y) > 0$ and $n = |x| = |yz| > n$, which is a contradiction. Therefore, $(A \backslash X^n) \cap (A \backslash X^n) X^+ = \Phi$ and so $A \backslash X^n$ is a fuzzy prefix code. ■

Corollary 8.2.10 *For any $n \in \mathbb{N}$, $\Phi \neq A \in \mathcal{FP}(X^n)$ is a fuzzy prefix code.*

We usually call such a code a fuzzy uniform code.

8.3 Maximal Fuzzy Prefix Codes

Definition 8.3.1 *For $A, B \in \mathcal{FP}(X^+)$, define $A \dot{\subseteq} B$ if $A \subseteq B$ and there exists $x \in X$ such that $B(x) = 1, A(x) = 0$.*

Definition 8.3.2 *A fuzzy prefix code $A \in \mathcal{FP}(X^+)$ is called a **maximal fuzzy prefix code** if for any $B \in \mathcal{FP}(X^+)$ such that $A \dot{\subseteq} B$, B is not a fuzzy prefix code.*

Recall that for $x \in X^+$, $C_{\{x\}}$ denotes the characteristic function of $\{x\}$ in X^+.

Theorem 8.3.3 *A fuzzy prefix code $A \in \mathcal{FP}(X^+)$ is a maximal fuzzy prefix code if and only if for any $x \notin SuppA$, $A \cup C_{\{x\}}$ is not a fuzzy prefix code.*

Proof. Suppose A is a maximal fuzzy prefix code. Then for any $x \notin SuppA$, $A \subseteq A \cup C_{\{x\}}$, $A(x) = 0$ and $A \cup C_{\{x\}}(x) = 1$. Thus $A \cup C_{\{x\}}$ is not a fuzzy prefix code.

For the converse, suppose for any $x \notin SuppA$, $A \cup C_{\{x\}}$ is not a fuzzy prefix code. For any B such that $A \dot{\subseteq} B$, we have $A \subseteq B$ and there exists $x \in X$ such that $B(x) = 1$, $A(x) = 0$. Hence for all $x \in SuppB \backslash SuppA$, we have $A \subseteq A \cup C_{\{x\}} \subseteq B$. By hypothesis, $A \cup C_{\{x\}}$ is not a fuzzy prefix code. By Theorem 8.2.5, B is not a fuzzy prefix code. Thus A is a maximal fuzzy prefix code. ■

Corollary 8.3.4 *Let A be a fuzzy prefix code. If $\forall B \in \mathcal{FP}(X^+)$ such that $B \supset A$, B is not a fuzzy prefix code, then A is a maximal fuzzy prefix code.*

Proposition 8.3.5 *Let $a \in (0, 1]$. Let $A \in \mathcal{FP}(X^n)$ and $A(x) = a$ $\forall x \in X^n$. Then A is a maximal fuzzy prefix code.*

Proof. By Corollary 8.2.10, A is a fuzzy prefix code. For any $x \in SuppA = X^n$, we have $(A \cup C_{\{x\}}) \cap (A \cup C_{\{x\}})X^+ = (A \cap AX^+) \cup (C_{\{x\}} \cap C_{\{x\}}X^+) \cup (A \cap C_{\{x\}}X^+) \cup (C_{\{x\}} \cap AX^+)$. Since A is a fuzzy prefix code, we have $A \cap AX^+ = \Phi$. $C_{\{x\}} \cap C_{\{x\}}X^+ = \Phi$ is obvious. Let $|x| = m$, $m \neq n$. Then $x = x_1 x_2 \ldots x_m$, $x_i \in X$, $i = 1, \ldots, m$. If $m \leq n$, then for any $y = y_1 y_2 \ldots y_{n-m}$, $y \in X^+$, $i = 1, \ldots, n - m$, we have $xy \in X^n$ and $A(xy) = a > 0$. Thus

$$(C_{\{x\}}X^+)(xy) = \vee \{C_{\{x\}}(s) \mid s, t \in X^+, st = xy\} \geq C_{\{x\}}(x) = 1.$$

Therefore, $(A \cap C_{\{x\}}X^+)(xy) = a > 0$ and $(A \cup C_{\{x\}}) \cap (A \cup C_{\{x\}})X^+ \neq \Phi$. If $m > n$, then set $y = x_1 \ldots x_n$, $z = x_{n+1} \ldots x_m$. We have

$$(AX^+)(x) = \vee\{A(u) \mid uv = x\} \geq A(y) = a > 0.$$

Thus $(C_{\{x\}} \cap AX^+)(x) \geq a > 0$ and

$$(A \cup \{x\}) \cap (A \cup \{x\})X^+ \neq \Phi.$$

Therefore, $A \cup \{x\}$ is not a fuzzy prefix code. By Theorem 8.3.3, A is maximal fuzzy prefix code. ■

For any $L \subseteq X^*$, we let $\|L\| = \vee\{|x| \mid x \in L\}$ if it exists, otherwise $\|L\| = \infty$.

Theorem 8.3.6 *Let $A \in \mathcal{FP}(X^+)$ be a fuzzy prefix code and $\|SuppA\| = n$. Then there exists maximal fuzzy prefix code $M \in \mathcal{FP}(X^+)$ such that $A \subseteq M$ and $\|SuppM\| = n$.*

Proof. Let $M = A \cup (C_{X^n} \backslash AX^*)$. We want to show that M is a required maximal fuzzy prefix code, i.e., (1) $A \subseteq M$; (2) $\|SuppM\| = n$; (3) M is a fuzzy prefix code and (4) M is a maximal fuzzy prefix code.

(1) Clearly, $A \subseteq M$.

(2) For any $x \in SuppM$, we have $M(x) = (A \cup (C_{X^n} \backslash AX^*))(x) > 0$. If $A(x) > 0$, then $x \in SuppA$ and $|x| \leq n$. If $(C_{X^n} \backslash AX^*)(x) > 0$, then $(AX^*)(x) = 0$ and $C_{X^n}(x) > 0$, i.e., $x \in X^n$. Then $|x| \leq n$ and $\|SuppM\| \leq n$. On the other hand, since $M \supseteq A$, we have $SuppM \supseteq SuppA$. Hence $\|SuppM\| \geq \|SuppA\| = n$. Thus $\|SuppM\| = n$.

(3)

$$\begin{aligned}
& M \cap MX^+ \\
= \; & (A \cup (C_{X^n} \backslash AX^*)) \cap (A \cup (C_{X^n} \backslash AX^*))X^+ \\
= \; & (A \cup (C_{X^n} \backslash AX^*)) \cap (AX^+ \cup (C_{X^n} \backslash AX^*)X^+) \\
= \; & (A \cap AX^+) \cup (A \cap (C_{X^n} \backslash AX^*)X^+) \cup \\
& ((C_{X^n} \backslash AX^*) \cap AX^+) \cup ((C_{X^n} \backslash AX^*) \cap (C_{X^n} \backslash AX^*)X^+).
\end{aligned}$$

Since A is a fuzzy prefix code, we have $A \cap AX^+ = \Phi$. Since $C_{X^n} \backslash AX^* \in \mathcal{FP}(X^n)$ is a fuzzy uniform code, we obtain $(C_{X^n} \backslash AX^*) \cap (C_{X^n} \backslash AX^*)X^+ = \Phi$. For any $x \in SuppA$, we have $|x| \leq n$. If $((C_{X^n} \backslash AX^*)X^+)(x) > 0$, then there exists $y, z \in X^+$ such that $yz = x$ and $(C_{X^n} \backslash AX^*)(y) > 0$. Thus $C_{X^n}(y) > 0$ and $|y| = n$. Therefore, $|x| = |y| + |z| > n$, which is a contradiction. Hence $A \cap (C_{X^n} \backslash AX^*)X^+ = \Phi$. Similarly, $(C_{X^n} \backslash AX^*) \cap AX^+ = \Phi$. Consequently, $M \cap MX^+ = \Phi$ and M is a fuzzy prefix code.

(4) It suffices to show that $M \cup C_{\{y\}}$ is not a fuzzy prefix code for any $y \in SuppM$.

Case (a): $|y| > n$. Let $y = x_1 \ldots x_n x_{n+1} \ldots x_m = st$, where $s = x_1 \ldots x_n \in X^n$, $t = x_{n+1} \ldots x_m \in X^+$, $|s| = n$. Then $(M \cup C_{\{y\}})(y) \geq C_{\{y\}}(y) = 1$,

$$\begin{aligned}
((M \cup C_{\{y\}})X^+)(y) &= \vee\{(M \cup C_{\{y\}})(s) \mid s, t \in X^+, st = y\} \\
&= \vee\{M(s) | s, t \in X^+\} \geq M(s).
\end{aligned}$$

If $A(s) = 0$ and $(AX^*)(s) = 0$ then $M(s) \geq (C_{X^n} \backslash AX^*)(s) = C_{X^n}(s) = 1$. Hence $(M \cup C_{\{y\}}) \cap (M \cup C_{\{y\}})X^+ \neq \Phi$ and $M \cup C_{\{y\}}$ is not a fuzzy prefix code. If $A(s) = 0$ and $(AX^*)(s) > 0$, then $\vee\{A(u) \mid u, v \in X^*, uv = s\} > 0$. So there exists $u, v \in X^*$ such that $uv = s$ and $A(u) > 0$. Thus $y = st = u(vt)$, $vt \in X^+$, $A(u) > 0$. Therefore, $M(u) > 0$ and

$$\begin{aligned} \left((M \cup C_{\{y\}})X^+\right)(y) &= \vee\{(M \cup C_{\{y\}})(s) \mid s, t \in X^+, st = y\} \\ &\geq (M \cup C_{\{y\}})(u) \\ &= M(u) > 0. \end{aligned}$$

Hence $M \cup C_{\{y\}}$ also is not a fuzzy prefix code.

Case (b): $|y| = n$. Since $M(y) = 0$, we have $A(y) = 0$ and

$$(C_{X^n} \backslash AX^*)(y) = 0.$$

Thus $(AX^*)(y) > 0$, i.e., $\vee\{A(u) \mid u, v \in X^*, uv = y\} > 0$. Therefore, there exists $u, v \in X^*$ such that $uv = y$ and $A(u) > 0$. Hence $M(u) > 0$, $v \in X^+$ and

$$\begin{aligned} (((M \cup C_{\{y\}})X^+)(y) &= \vee\{(M \cup C_{\{y\}})(u) \mid u, v \in X^+, uv = y\} \\ &\geq M(u) \\ &> 0. \end{aligned}$$

Therefore, $(M \cup C_{\{y\}}) \cap (M \cup C_{\{y\}})X^* \neq \Phi$ and $M \cup C_{\{y\}}$ is not a fuzzy prefix code.

Case (c): $|y| < n$. Since the length of y is less than n, there exists $m \in X^n$ and $x \in X^+$ such that $m = yx$. If $A(m) > 0$, then $M(m) > 0$. Thus $(M \cup C_{\{y\}})(m) \geq M(m) > 0$ and

$$\begin{aligned} ((M \cup C_{\{y\}})X^+)(m) &= \vee\{(M \cup C_{\{y\}})(u) \mid u, v \in X^+, uv = m\} \\ &\geq (M \cup C_{\{y\}})(y) \\ &= 1. \end{aligned}$$

Hence $M \cup C_{\{y\}}$ is not a fuzzy prefix code. If $A(m) = 0$ and $(AX^*)(m) = 0$, then $(C_{X^n} \backslash AX^*)(m) = C_{X^n}(m) = 1 > 0$ and $M(m) > 0$. Therefore, $M \cup C_{\{y\}}$ is not a fuzzy prefix code. If $A(m) = 0$ and $(AX^*)(m) > 0$, then $\vee\{A(u) \mid u, v \in X^*, uv = m\} > 0$. Thus there exists $u, v \in X^*$ such that $uv = m$, $A(u) > 0$, $v \in X^+$ and $uv = yx$. This implies that either $y = u.u'$ or $u = yy'$ for some $u', y' \in X^+$. Hence we have

$$((M \cup C_{\{y\}}) \cap (M \cup C_{\{y\}})X^+)(y) > 0$$

or

$$((M \cup C_{\{y\}}) \cap (M \cup C_{\{y\}})X^+)(u) > 0.$$

But both cases imply that $M \cup C_{\{y\}}$ is not fuzzy prefix code. Thus case (c) follows. ■

8.4 Algebraic Properties of Fuzzy Prefix Codes on a Free Monoid

We now examine algebraic properties of fuzzy prefix codes and maximal fuzzy prefix codes. In particular, we examine the relationship between fuzzy prefix codes and fuzzy factor theorems. We consider results from [97].

Theorem 8.4.1 *Let $A, B \in \mathcal{FP}(X^+)$ be such that $a, ya \in SuppA$, $b, by \in SuppB$ for some $a, b, y \in X^+$. Then AB is not a fuzzy code.*

Proof. Now $(AB)(ab) = \vee\{A(y) \wedge B(z) \mid yz = ab,\ y, z \in X^*\} \geq A(a) \wedge B(b) > 0$. Similarly, $(AB)(aby) > 0$, $(AB)(yab) > 0$, $(AB)(yaby) > 0$. Since AB is a fuzzy monoid, $(AB)(abyab) \geq (AB)(aby) \wedge (AB)(ab) > 0$, and $(aby)(ab) = (ab)(yab)$, $\wedge((AB)(aby), (AB)(ab), (AB)(yab)) \propto \wedge([2] = [2], [aby] = [ab], [ab] = [yab]) = 0 < (AB)(abyab)$. Hence AB is not a fuzzy code. ■

Let $Y \subseteq X^+$ and let $B \in C_f(X)$. Define the fuzzy subset YB of X^+ by $\forall w \in X^+$,

$$YB(w) = \vee\{B(u) \mid yu = w,\ y \in Y,\ u \in X^*\}.$$

For any nonempty alphabet X, let $C_f(X)$ denote the class of all fuzzy codes over X. We also consider $C_{\{e\}}$ as a fuzzy code. We now define a subclass $P_f(X)$ of $C_f(X)$ as follows:

(i) If $x = \{a\}$, then $P_f(X) = \{A \mid SuppA = \{a^n\},\ n \geq 0\}$.

(ii) If X contains more than one letter, then for every $x \in X$ and $y \in H_x$, where $H_x = X^+ \backslash xX^*$, let $M_y^x = \{x, yx\}$ and

$$C_f M_y^x = \{B \in C_f(X) \mid \left(M_y^x\right) B \in C_f(X)\}.$$

We then define $P_f(X) = \cap_{x \in X,\ y \in H_x} C_f M_y^x$. It is clear that $P_f(X)$ is a nonempty subclass of $C_f(X)$. If X consists of only one letter, then $P_f(X) = C_f(X)$ and $P_f(X) \subset C_f(X)$ when X contains more than one letter.

Theorem 8.4.2 *Let X be an alphabet. Then a fuzzy code A over X is a fuzzy prefix code if and only if A is in $P_f(X)$.*

Proof. The theorem is trivial if X consists of only one letter. Now assume that X contains more than one letter.

Assume A is a fuzzy prefix code. Then $A \cap AX^+ = \Phi$. Let $M_y^x = \{x, yx\}$, $x \in X$, $y \in H_x$. Suppose $(M_y^x)A \cap (M_y^x)AX^+ \neq \Phi$. Then there exists $u \in X^+$ such that

$$\left[(M_y^x)A \cap (M_y^x)AX^+\right](u) = ((M_y^x)A)(u) \wedge ((M_y^x)AX^+)(u) > 0,$$

i.e., $((M_y^x)A)(u) = \vee\{A(a) \mid m \in M_y^x, a \in X^+, ma = u\} > 0$ and

$$(M_y^x AX^+)(u) = \vee\{A(a') \mid m' \in M_y^x, a', v \in X^+, m'a'v = u\} > 0.$$

Then there exists $m, m' \in M_y^x$, $a', a, v \in X^+$ such that $A(a) > 0$, $A(a') > 0$, $ma = u = m'a'v$. It follows that either $m = m'd'$ or $m' = md$ for some $d, d' \in X^*$. Since $M_y^x = \{x, yx\}$, we have in either case $d' = 1$ or $d = 0$. Therefore, $a = a'v$ holds. However, this is impossible since $A \cap AX^+ = \Phi$. Thus $M_y^x A \cap M_y^x AX^+ = \Phi$ must hold for any $x \in X$ and $y \in H_x$. By Definition 8.2.1 and Theorem 8.2.6, we thus have that $M_y^x A$ is a fuzzy code and so $A \in C_f M_y^x$ for every $x \in X$, $y \in H_x$. Hence $A \in P_f(X)$.

Assume $A \in P_f(X)$. Suppose $A \cap AX^+ \neq \Phi$. Then there exists u such that $(A \cap AX^+)(u) = A(u) \wedge (\vee \{A(y) \mid y, v \in X^*, yv = u\}) > 0$, i.e., there exists $y, v \in X^+$ such that $yv = u$, $A(u) > 0$, $A(y) > 0$. Thus $y, yv \in SuppA$. Since X contains at least two elements, we can select an $x \in X$ such that $v \neq xz$ for every $z \in X^*$. Consider $M_y^x = \{x, ux\}$. Then by Theorem 8.4.1, $M_v^x A$ is not a fuzzy code. Therefore, $A \notin C_f M_v^x$ and so $A \notin P_f(X)$, which is a contradiction. Hence $A \cap AX^+ = \Phi$ must hold. ■

Corollary 8.4.3 *Let $A \in \mathcal{FP}(X^+)$. Then A is a fuzzy prefix code if and only if for every $x \in X$, $y \in H_x$, $M_y^x A$ is a fuzzy prefix code.*

The set of all maximal fuzzy prefix codes over X is denoted by $M_f(X)$. For $Y, Z \subseteq X^*$, let

$$YZ = \{yz \mid y \in Y,\ z \in Z\}$$

and $yZ = \{y\}Z$, where $y \in Y$.

Theorem 8.4.4 *Suppose $C_{\{e\}} \neq M \in M_f(X)$. Then the following statements hold:*

(1) If $y \in SuppM$, then $(M \backslash C_{\{y\}}) \cup C_{yX} \in M_f(X)$;

(2) If $C_{yX} \subseteq M$, then $(M \backslash C_{yX}) \cup C_{\{y\}} \in M_f(X)$.

Proof. (1) Now

$$\begin{aligned}
& \left[(M \backslash C_{\{y\}}) \cup C_{yX}\right] \cap \left[(M \backslash C_{\{y\}}) \cup C_{yX}\right] X^+ \\
= & \left[(M \backslash C_{\{y\}}) \cup C_{yX}\right] \cap \left[(M \backslash C_{\{y\}}) X^+ \cup C_{yX} X^+\right] \\
= & \left[(M \backslash C_{\{y\}}) \cap (M \backslash C_{\{y\}}) X^+\right] \cup \\
& \left[(M \backslash C_{\{y\}}) \cap C_{yX} X^+\right] \cup \\
& \left[C_{yX} \cap (M \backslash C_{\{y\}}) X^+\right] \cup \left[C_{yX} \cap C_{yX} X^+\right]
\end{aligned}$$

Clearly, $C_{yX} \cap C_{yX}X^+ = \emptyset$. Since $M \in M_f(X)$, $M \cap MX^+ = \Phi$ and so $(M \backslash C_{\{y\}}) \cap (M \backslash C_{\{y\}}X^+) \subseteq M \cap MX^+ = \Phi$. Suppose $(M \backslash C_{\{y\}}) \cap C_{yX}X^+ = \Phi$. Then there exists $u \in X^+$, such that $(M \backslash C_{\{y\}})(u) > 0$ and $(C_{yX}X^+)(u) > 0$, i.e., there exists $z \in X$, $v \in X^+$ such that $yzv = u$ and $M(u) > 0$, which contradicts that $M \in M_f(X)$ and $M \cap MX^+ = \Phi$. (Since $y \in SuppM$, $yzv = u \in SuppM$ and $yzu \in SuppMX^+$, so $M \cap MX^+ \neq \Phi$.) Suppose $C_{yX} \cap (M \backslash C_{\{y\}})X^+ \neq \Phi$. Then there exists $u \in X^+$ such that $C_{yX}(u) > 0$ and

$$\left((M \backslash C_{\{y\}})X^+\right)(u) = Supp\{(M \backslash C_{\{y\}})(a) \mid av \in X^*, av = u\},$$

i.e., there exists $z \in X, a, v \in X^+$ such that $yz = u = av$, $a \neq y$ and $M(a) > 0$. It follows that either $y = ad$ or $a = yd'$, and both contradict that $M \cap MX^+ = \Phi$. Therefore, $\left[(M\backslash C_{\{y\}}) \cup C_{yX}\right] \cap \left[(M\backslash C_{\{y\}}) \cup C_{yX}\right] X^+ = \Phi$, i.e., $(M\backslash C_{\{y\}}) \cup C_{yX} \in P_f(X)$. Now, let $(M\backslash C_{\{y\}}) \cup C_{yX} = M_1$, and suppose $M_1 \notin M_f(X)$. Then there exists $f \in Supp(X^*\backslash M_1)$ such that $M_1 \cup C_{\{f\}}$ is a fuzzy prefix code, i.e., $M_1 \cup C_{\{f\}} \in P_f(X)$. If $f = y$ then for any $a \in X$, we have $fa = ya \in Supp\,(yX) \subseteq SuppM_1$. (By Definition 8.2.1, if $f \in Supp(M_1 \cup C_{\{f\}})$, then $fa \notin Supp(M_1 \cup C_{\{f\}})$, which is not the case. Hence we may assume that $f \neq y$ and $f \in Supp\,(X^*\backslash M_1)$. By assumption, $M \in M_f(X)$, and so we have that $M \cup C_{\{f\}} \notin P_f(X)$, i.e., $(M \cup C_{\{f\}}) \cap (M \cap C_{\{f\}})X^+ \neq \Phi$. Then there exists u, such that $(M \cup C_{\{f\}})(u) = M(u) \vee C_{\{f\}}(u) > 0$, and

$$((M \cup C_{\{f\}})X^+)(u) = \vee\{M(z) \vee C_{\{f\}}(z) \mid z \in X^*, x \in X^+, zx = u\} > 0,$$

i.e., there exists $m \in X^*$, $x \in X^+$ such that $mx = u$, $M(m) \vee C_{\{f\}}(m) > 0$. If $u = f$, then $m \neq f$, and $M(m) \vee C_{\{f\}}(m) = M(m) > 0$. If $u \neq f$, then $M(u) = M(u) \vee C_{\{f\}}(u) > 0$ and $C_{\{f\}}(m) = M(m) \vee C_{\{f\}}(m) > 0$ (if $m \in SuppM$, $mx = u \in SuppM$, then $M \cap MX^+ \neq \Phi$, a contradiction). Hence either (i) there exists $m_1 \in SuppM$ and $x_1 \in X^+$ such that $m_1x_1 = f$ or (ii) there exists $m_2 \in SuppM$ and $x_2 \in X^+$ such that $m_2 = fx_2$. In the former case, if $m_1 \neq y$, since $f \in Supp(M_1 \cup C_{\{f\}})$,

$$\begin{aligned} & ((M_1 \cup C_{\{f\}})X^+)(f) \\ = \; & \vee\{(M_1 \cup C_{\{f\}})(m) \mid m_1, x_1 \in X^*, m_1x_1 = f \geq (M_1 \cup C_{\{f\}})(m_1)\} \\ \geq \; & M_1(m_1) \\ = \; & \left[(M\backslash C_{\{y\}} \cup C_{\{yx\}})\right](m_1) \\ \geq \; & M \end{aligned}$$

$$C_{\{y\}}(m_1) = M(m_1) > 0,$$

i.e.,

$$(M_1 \cup C_{\{f\}}) \cap (M_1 \cup C_{\{f\}})X^+ \neq \Phi.$$

Thus $M_1 \cup C_{\{f\}} \notin P_f(X)$, a contradiction. On the other hand, if $m_1 = y$, let $x_1 = a_1a_2..a_l$, where $a_1, a_2, \ldots, a_l \in X$. Since $l = 1$, $f = m_1x_1 = ya_1 \in Supp\,(C_{yX}) \subseteq SuppM_1$, which contradicts the fact that $f \in Supp(X^*\backslash M_1)$. Since $l \geq 2$,

$$\begin{aligned} (ya_1)a_2a_3\ldots a_l \; &= \; f \in Supp[(M_1 \cup C_{\{f\}})], \\ & \quad (M_1 \cup C_{\{f\}}X^+)(f)) \\ &= \; \vee\{(M_1 \cup C_{\{f\}})(ya_1) \mid y, a_1, \ldots, a_l \in X^*, \\ & \qquad ya_1a_2\ldots a_l = f\} \\ &\geq \; (M_1 \cup C_{\{f\}})(ya_1) \\ &\geq \; M_1(ya_1) \\ &\geq \; (C_{yX})(ya_1) \\ &= \; 1, \end{aligned}$$

i.e.,

$$(M_1 \cup C_{\{f\}}) \cap (M_1 \cup C_{\{f\}})X^+ \neq \Phi.$$

Hence $M_1 \cup C_{\{f\}} \notin P_f(X)$, a contradiction. Now for the latter case, if $m_2 = y$, then for any $a \in X$, $ya = m_2a = fx_2a$. Since $f \in Supp(M_1 \cup C_{\{f\}})$, $fx_2a = ya \in Supp\{yX\} \subseteq SuppM_1$, which contradicts. $M_1 \cup C_{\{f\}} \in P_f(X)$. If $m_2 \neq y$, then

$$M_1(m_2) = ((M \backslash C_{\{y\}}) \cup C_{yX})(m_2) \geq (M \backslash C_{\{y\}})(m_2) = M(m_2) > 0,$$

i.e., $m_2 \in SuppM_1 \subseteq Supp(M_1 \cup C_{\{f\}})$, $m_2x_2 = f \in Supp(M_1 \cup C_{\{f\}})$ again, in contradiction to $M_1 \cup C_{\{f\}} \in P_f(X)$. This completes the proof of (1).

(2) Now

$$\begin{aligned} & \left[(M \backslash C_{yX}) \cup C_{\{y\}}\right] \cap [(M \backslash C_{yX}) \cup C_{\{y\}}]X^+ \\ = & \left[(M \backslash C_{yX}) \cup C_{\{y\}}\right] \cap \left[(M \backslash C_{yX})X^+ \cup C_{yX^+}\right] \\ = & \left[(M \backslash C_{yX}) \cap (M \backslash C_{yX})X^+\right] \cup \left[C_{\{y\}} \cap (M \backslash C_{yX})X^+\right] \cup \\ & \left[(M \backslash C_{yX}) \cap C_{yX^+}\right] \cup \left[C_{\{y\}} \cap C_{yX^+}\right]. \end{aligned}$$

Clearly, $C_{\{y\}} \cap C_{yX^+} = \Phi$. Since $M \in M_f(X)$,

$$(M \backslash C_{yX}) \cap (M \backslash C_{yX})X^+ \subseteq M \cap MX^+ = \Phi.$$

Suppose $C_{\{y\}} \cap (M \backslash C_{yX})X^+ \neq \Phi$. Then there exists $z, u \in X^+$ such that $zu = y = u, M(z) > 0$, $C_{yX}(z) = 0$. For any $a \in X$, $zua = ya \in Supp\,(C_{yX}) \subseteq SuppM$,

$$(MX^+)\,(ya) = \vee\{M(z) \mid z, v, a \in X^*, zva = ya\} \geq M(z) > 0.$$

Thus $M \cap MX^+ \neq \Phi$, which contradicts that $M \in M_f(X)$. Suppose $(M \backslash C_{yX}) \cap C_{yX^+} \neq \Phi$. Then there exists $u \in X^+$ such that $M(u) > 0$, $C_{yX}(u) = 0$, $C_{yX^+}(u) > 0$. i.e., there exists $z_1 \in X$, $z_2 \in X^+$ such that $u = yz_1z_2$, $M(u) > 0$ since $C_{yX} \subseteq M$. Hence

$$\begin{aligned} (MX^+)(u) &= \vee\{M(yz_1) \mid y, z_1, z_2 \in X^*, yz_1z_2 = u\} \\ &\geq M(yz_1) \\ &\geq C_{yX}(yz_1) \\ &= 1. \end{aligned}$$

Thus $M \cap AX^+ \neq \Phi$, a contradiction. Hence

$$\left[(M \backslash C_{yX}) \cup C_{\{y\}}\right] \cap \left[(M \backslash C_{yX}) \cup C_{\{y\}}\right] X^+ = \Phi$$

must hold and $(M \backslash C_{yX}) \cup C_{\{y\}} \in P_f(X)$.

Now let $(M \backslash C_{yX}) \cup C_{\{y\}} = M_2$. Suppose $M_2 \notin M_f(X)$. Then there exists $g \in Supp\,(C_{X^*} \backslash M_2)$ such that $M_2 \cup C_{\{g\}}$ is a fuzzy prefix code, i.e.,

$M_2 \cup C_{\{g\}} \in P_f(X)$. This condition implies that for every $a \in X, g \neq ya$. Since $ya = g \in Supp(M_2 \cup C_{\{g\}})$,

$$\begin{aligned}[(M_2 \cup C_{\{g\}})X^+](g) &= \vee\{(M_2 \cup C_{\{g\}})(y) \mid y, a \in X^*, ya = g\} \\ &\geq (M_2 \cup C_{\{g\}})(y) \\ &= M_2(y) \\ &\geq 1\end{aligned}$$

which contradicts $M_2 \cup C_{\{g\}} \in P_f(X)$. By analogy with case (1), either (i) $m_3x_3 = g$ with $x_3 \in X^+$ and $m_3 \in SuppM$ or (ii) $m_4 = gx_4$ with $x_4 \in X^+$ and $m_4 \in SuppM$. In case (i), if $m_3 = ya$ for some $a \in X$, then $m_3x_3 = y\,(ax_3) = g \in Supp(M_2 \cup C_{\{g\}})$ and

$$\begin{aligned}[(M_2 \cup C_{\{g\}})X^+](g) &= \vee\{(M_2 \cup C_{\{g\}})(y) \mid y, a, x_3 \in X^*, yax_3 = g\} \\ &\geq (M_2 \cup C_{\{g\}})(y) \\ &\geq M_2(y) \\ &\geq C_{\{y\}}(g) \\ &= 1\end{aligned}$$

in contradiction to $M_2 \cup C_{\{g\}} \in P_f(X)$. If $m_3 \neq ya$ for any $a \in X$, then $m_3 \in Supp(M_2 \cup C_{\{g\}})$ and

$$\begin{aligned}[(M_2 \cup C_{\{g\}})X^+](g) &= \vee\{(M_2 \cup C_{\{g\}})(m_3) \mid m_3, x_3 \in X^*, m_3x_3 = g\} \\ &\geq M_2\,(m_3) \\ &> 0,\end{aligned}$$

again in contradiction to $M_2 \cup C_{\{g\}} \in P_f(X)$. For case (ii), if $m_4 = ya$ for some $a \in X$, then, since $m_4 = ya = gx_4 = gx_5a$, where $x_5a = x_4$, we have $y = gx_5 \in SuppM_2$ and

$$\begin{aligned}[(M_2 \cup C_{\{g\}})X^+](y) &= \vee\{(M_2 \cup C_{\{g\}})(g) \mid g, x_5 \in X^*, gx_5 = y\} \\ &\geq (M_2 \cup C_{\{g\}})(g) \\ &\geq C_{\{g\}}(g) \\ &= 1,\end{aligned}$$

which contradicts $M_2 \cup C_{\{g\}} \in P_f(X)$. If $m_4 \neq ya$ for any $a \in X$, then

$$\begin{aligned}&m_4 \in SuppM_2, [(M_2 \cup C_{\{g\}})X^+](m_4) \\ =\ &\vee\{(M_2 \cup C_{\{g\}})(g) \mid g, x_4 \in X^*, gx_4 = m_4\} \\ \geq\ &(M_2 \cup C_{\{g\}})(g) \\ \geq\ &C_{\{g\}}(g) \\ =\ &1,\end{aligned}$$

a contradiction. Hence (2) holds. ∎

Let $L \subseteq X^*$. Recall that $\|L\| = \vee\{|x| \mid x \in L\}$ if the supremum exists. Let $\lceil L \rceil = \wedge\{|x| \mid x \in L\}$.

Theorem 8.4.5 *Let $A \in P_f(X)$ be such that $\|SuppA\| = n$ and $x \notin SuppA$. Then there exists $M \in M_f(X)$ such that $\|SuppM\| \leq n+1$, $A \subseteq M$ and $x \notin SuppM$.*

Proof. The theorem is trivial if $A \in M_f(X)$. Now suppose that $A \notin M_f(X)$. Then, by Theorem 8.3.6, there exists $M \in M_f(X)$ such that $A \subseteq M$ and $\|SuppM\| = n$. If $x \notin SuppM$, then M is the required maximal fuzzy prefix code. On the contrary, if $x \in SuppM$, then, by Theorem 8.4.4, $(M \backslash C_{\{x\}}) \cup C_{xX} \in M_f(X)$, where $A \subseteq (M \backslash C_{\{x\}}) \cup C_{xX}$ and

$$\left\| Supp\left[(M \backslash C_{\{x\}} \cup C_{xX})\right] \right\| \leq n+1.$$

■

Theorem 8.4.6 *Let A be a fuzzy prefix code over X with $\|SuppA\| = n$. Then there exists an index set T and maximal fuzzy prefix codes $M_i, i \in T$, such that $A = \cap_{i \in T} M_i$, where $\|SuppM_i\| \leq n+1, i \in T$.*

Proof. Let $C = \{M \in M_f(X) \mid A \subseteq M, \|SuppM\| \leq n+1\}$. It is clear that the set C is nonempty set by the Theorem 8.4.5. Let

$$C = \{M_i \mid M_i \in M_f, A \subseteq M_i, i \in T, \|SuppM\| \leq n+1\}.$$

Then $A \subseteq \cap_{i \in T} M_i$. If $x \in SuppA$, then, by Theorem 8.4.5, there is an $M'_{i_0} \in M_f(X)$ such that $A \subseteq M'_{i_0}$, $x \in SuppM'_{i_0}$ and $\left\|SuppM'_{i_0}\right\| \leq n+1$. But this M'_{i_0} is in C and so $x \notin Supp\left(\cap_{i \in T} M_i\right)$ since $\left(\cap_{i \in T} M_i\right)(x) = \wedge_{i \in T}\left(M_i(x)\right) \leq M'_{i_0}(x) = 0$. Therefore, $SuppA = Supp\left(\cap_{i \in T} M_i\right)$ for any $x_0 \in SuppA$. If there exists $i_0 \in T$ such that $A(x_0) < M_{i_0}(x_0)$, then we define M'_{i_0} as follows:

$$M'_{i_0}(x) = \begin{cases} A(x_0) & \text{if } x = x_0, \\ M_{i_0}(x) & \text{otherwise.} \end{cases}$$

Since $M'_{i_0} \leq M_{i_0}$ is obvious, $M'_{i_0} \cap M'_{i_0} X^+ \subseteq M_{i_0} \cap M_{i_0} X^+ = \Phi$, i.e., M'_{i_0} is a fuzzy prefix code. For any $f \in SuppM'_{i_0}$, $M'_{i_0}(f) = 0$ and so $f \neq x_0$, $f \notin SuppM_{i_0}$. Therefore, $M_{i_0} \cup C_{\{f\}} \in M_f(x)$, and there exists u such that

$$(M_{i_0} \cap C_{\{f\}})(u) > 0, [(M_{i_0} \cup C_{\{f\}})X^+](u) > 0.$$

If $u \neq x_0$ and $u \neq x_0 z$, $z \in X^+$, then

$$\begin{aligned} (M'_{i_0} \cup C_{\{f\}})(u) &= M'_{i_0}(u) \vee C_{\{f\}}(u) \\ &= M_{i_0}(u) \vee C_{\{f\}}(u) \\ &= (M_{i_0} \cup C_{\{f\}})(u) \\ &> 0 \end{aligned}$$

and

$$\begin{aligned} [(M'_{i_0} \cup C_{\{f\}})X^+](u) &= \vee\left\{(M'_{i_0} \cup C_{\{f\}})(y) \mid y, z \in X^*, yz = u\right\} \\ &= \vee\left\{(M_{i_0}(y) \vee C_{\{f\}}(y)) \mid yz \in X^*, yz = u\right\} \\ &= (M_{i_0} \cup C_{\{f\}})(u) \\ &> 0. \end{aligned}$$

If $u \neq x_0$ and $u = x_0 z$, $z \in X^+$, then $(M'_{i_0} \cup C_{\{f\}})(u) > 0$,

$$\begin{aligned}[(M'_{i_0} \cup C_{\{f\}})X^+](u) &= \vee\left\{M'_{i_0}(x) \vee C_{\{f\}}(x) \mid x, z \in X^+, xz = u\right\} \\ &\geq M'_{i_0}(x_0) \vee C_{\{f\}}(x_0) \\ &\geq M'_{i_0}(x_0) \\ &= A(x_0) \\ &> 0.\end{aligned}$$

If $u = x_0$, then

$$(M'_{i_0} \cup C_{\{f\}})(u) = M'_{i_0}(x_0) \vee C_{\{f\}}(x_0) = M'_{i_0}(x_0) = A(x_0) > 0,$$

$$\begin{aligned}&\vee\{M_{i_0}(y) \vee C_{\{f\}}(y) \mid y \in X^*, z \in X^+, yz = u\} \\ =\ & [(M_{i_0} \cup C_{\{f\}})X^+](u) \\ >\ & 0,\end{aligned}$$

and

$$\begin{aligned}&[(M'_{i_0} \cup C_{\{f\}})X^+](u) \\ =\ & \vee\left\{M'_{i_0}(y) \vee C_{\{f\}}(y) \mid y \in X^*, z \in X^+, yz = u = x_0\right\} \\ =\ & \vee\left\{M_{i_0}(y) \vee C_{\{f\}}(y) \mid y \in X^*, z \in X^+, yz = u\right\} \\ =\ & [(M_{i_0} \cup C_{\{f\}})X^+](u) \\ >\ & 0.\end{aligned}$$

Thus $M'_{i_0} \cup \{f\} \in M_f(X)$, and $(M'_{i_0} \cup \{f\})(x_0) = A(x_0)$, $A \subseteq M'_{i_0}$. Hence M'_{i_0} is in C, and $A(x_0) \leq (\cap_{i \in T} M_i)(x_0) \leq M'_{i_0}(x_0) = A(x_0)$. Therefore, $A = \cap_{i \in T} M_i$. ∎

Definition 8.4.7 *A fuzzy prefix code A is called a* ***synchronizing fuzzy prefix code*** *if $X^* x \subseteq A^*$ for some $x \in SuppA^*$.*

Theorem 8.4.8 *Let A be a synchronizing fuzzy prefix code over X with $\|SuppA\| = n$. Then there exists an index set T and synchronizing maximal fuzzy prefix codes M_i, $i \in T$, such that $A = \cap_{i \in T} M_i$, where $\|SuppM_i\| \leq n + 1$ for all $i \in T$.*

Proof. By Theorem 8.4.6, we have $A = \cap_{i \in T} M_i$, where $M_i \in M_f(X)$ and $\|SuppM_i\| \leq n + 1$ for all $i \in T$. Since for any $x \in X^*$,

$$A^*(x) = \vee_{j=0}^{\infty} A^j(x) = \vee_{j=0}^{\infty}\{(\cap_{i \in T} M_i)^j(x)\}$$

and

$$\begin{aligned}(\cap_{i \in T} M_i)^j(x) &= \vee_{x_1, x_2, \ldots, x_j \in X^*, x_1 x_2 \ldots x_j = x}\{(\cap_{i \in T} M_i)(x_1) \wedge \\ &\quad (\cap_{i \in T} M_i)(x_2) \wedge \ldots \wedge (\cap_{i \in T} M_i)(x_j)\} \\ &= \vee_{x_1, x_2, \ldots, x_j \in X^*, x_1 x_2 \ldots x_j = x}\{(\wedge_{i \in T}(M_i(x_1)) \wedge \\ &\quad (\wedge_{i \in T}(M_1(x_2)) \wedge \ldots \wedge (\wedge_{i \in T}(M_i(x_j))\} \\ &= \vee_{x_1, x_2, \ldots, x_j \in X^*, x_1 x_2 \ldots x_j = x}\{\wedge_{i \in T}[\wedge_{k=1}^{j}(M_i(x_k))]\} \\ &\leq \wedge_{i \in T}\{\vee_{x_1, x_2, \ldots, x_j \in X^*, x_1 x_2 \ldots x_j = x}[\wedge_{k=1}^{j}(M_i(x_k))]\} \\ &= \wedge_{i \in T}(M_i^j(x)) \\ &= (\cap_{i \in T} M_i^j)(x),\end{aligned}$$

$A^*(x) \leq \vee_{j=0}^{\infty}\{\wedge_{i\in T}(M_i^J(x))\} \leq \wedge_{i\in T}(\vee_{j=0}^{\infty} M_i^j(x)) = \wedge_{i\in T}(M_i^*(x)) = (\cap_{i\in T} M_i^*)(x)$, i.e., $A^* \subseteq \cap_{i\in T} M_i^*$. Since A is a synchronizing fuzzy prefix code, then there exists $x \in SuppA^*$ such that $X^*x \subseteq A^*$. Thus for every $i \in T$, since $A^* \subseteq M_i^*$, $x \in SuppA^* \subseteq SuppM_i^*$, $X^*x \subseteq A^* \in M_i^*$. Hence each M_i $(i \in T)$ is a synchronizing fuzzy prefix code. ■

8.5 Fuzzy Prefix Codes Related to Fuzzy Factor Theorems

Theorem 8.5.1 *Let $A_i, i = 1, 2, \ldots, n$, be nonempty fuzzy languages over X. Then the following assertion hold.*

(1) If $A_1, A_2, \ldots, A_n$ are fuzzy prefix codes, then $A_2A_3\ldots A_n$, $A_3A_4\ldots A_n$, $\ldots, A_{n-1}A_n$, A_n are fuzzy prefix codes.

(2) If $A_1A_2\ldots A_n$ is a fuzzy suffix code, then $A_1, A_1A_2, \ldots, A_1A_2\ldots A_{n-1}$ are fuzzy suffix codes.

Proof. We prove (1) for the case $n = 2$ and the result follows by induction. Assume A_1A_2 is a fuzzy prefix code. If A_2 is not a fuzzy prefix code, then $A_2 \cap A_2X^+ \neq \Phi$, i.e. there exists $v \in X^+$ such that $(A_2 \cap A_2X^+)(v) > 0$. Since $A_1 \neq \Phi$, there exists $u \in X^+$ such that $A_1(u) > 0$. Now

$$\begin{aligned} & uv \in X^+, [A_1(A_2 \cap X^+)(uv)] \\ = \ & \vee\{A_1(u') \wedge (A_2 \cap A_2X^+)(v) \mid u', v' \in X^*,\ u'v' = uv\} \\ \geq \ & A_1(u) \wedge (A_2 \cap A_2X^+)(u) \\ > \ & 0. \end{aligned}$$

It follows that $\Phi \neq A_1(A_2 \cap A_2X^+) \subseteq A_1A_2 \cap A_1A_2X^+$, which contradicts the fact that A_1A_2 is a fuzzy prefix code. Hence A_2 is a fuzzy prefix code.

A similar proof works for (2). ■

Definition 8.5.2 *Let $A, B \in \mathcal{FP}(X^+)$. Then A and B are called* ***separable*** *over X if for all $x \in Supp(AB)$, $x = a_1b_1 = a_2b_2$ implies $a_1 = a_2$ and $b_1 = b_2$, where $a_1, a_2 \in SuppA$ and $b_1, b_2 \in SuppB$.*

Theorem 8.5.3 *Let $A \in \mathcal{FP}(X^+)$. Then the following assertions hold:*

(1) A is a fuzzy prefix code if and only if A and B are separable for all $B \in \mathcal{FP}(X^+)$.

(2) D is a fuzzy suffix code if and only if C and D are separable for all $C \in \mathcal{FP}(X^+)$.

Proof. (1) Assume that there is a fuzzy language $B \in \mathcal{FP}(X^+)$ such that A and B are not separable. This then implies that there is a word $x \in Supp(AB)$ such that $x = x_1x_2 = y_1y_2$, $x_1, y_1 \in SuppA$, $x_2, y_2 \in SuppB$ and $x_1 \neq y_1$. Now either $x_1 = y_1z_1$ or $y_1 = x_1z_2$ for some $z_1, z_2 \in X^+$. Hence $A(x_1) > 0$ and $(AX^+)(x_1) = \vee\{A(y) \mid y, z \in X^*, yz = x_1\} \geq A(y_1) > 0$ or

$A(y_1) > 0$ and $(AX^+)(y_1) = \vee\{A(x') \mid x', z \in X^*, x'z = y_1\} \geq A(x_1) > 0$. This contradicts the fact that A is a fuzzy prefix code. Conversely, suppose for any $B \in \mathcal{FP}(X^+)$, A and B are separable. If A is not a fuzzy prefix code, then there exists $x \in X^+$ such that $(A \cap AX^+)(x) > 0$, i.e., $A(x) > 0$, $(AX^+)(x) = \vee\{A(a) \mid a, y \in X^*, ay = x\} > 0$. Therefore, there exists a, $y \in X^+$ such that $A(a) > 0$ and $A(ay) = A(x) > 0$, i.e., a, $ay \in SuppA$. Take $B = C_{\{b,yb\}}$, where $b \in X^+$. Then we have $(ay)b = a(yb)$. This implies that A and B are not separable and so (1) holds.

(2) We can prove (2) in a similar way. ∎

For any $G \in \mathcal{FP}(X^+)$, define the fuzzy subset $\bar{G}$ of X^+ as follows:

$$\bar{G}(x) = \begin{cases} G(x) & \text{if } x \in X^+ \text{ and } |x| \leq |y| \text{ for all } y \in SuppG \\ 0 & \text{otherwise.} \end{cases}$$

Theorem 8.5.4 *Let $A, B, C, D, E \in \mathcal{FP}(X^+)$ and let $A = BC = DE$. Then the following assertions hold:*

(1) $\lceil SuppB \rceil + \lceil SuppC \rceil = \lceil SuppD \rceil + \lceil SuppE \rceil = \lceil SuppA \rceil$;

(2) $\bar{B}\bar{C} = \bar{D}\bar{E} = \bar{A}$;

(3) If $\lceil SuppB \rceil = \lceil SuppD \rceil$, then $Supp\bar{B} = Supp\bar{D}$ and $Supp\bar{C} = Supp\bar{E}$;

(4) If B and D are fuzzy prefix codes and $\lceil SuppB \rceil = \lceil SuppD \rceil$, then $SuppB = SuppD$ and $SuppC = SuppE$;

(5) If $\lceil SuppB \rceil = \lceil SuppD \rceil$ and the pairs (B, C), (D, E), (B, E), (D, C) are separable, then $SuppB = SuppD$ and $SuppC = SuppE$.

Proof. (1) Immediate.

(2) For any $x \in Supp(BC)$,

$$\begin{aligned}
(\bar{B}\bar{C})(x) &= \vee\{\bar{B}(y) \wedge \bar{C}(z) \mid yz \in X^*, yz = x\} \\
&= \vee\{(B(y) \wedge C(z)) \mid y, z \in X^*, yz = x, |y| \leq |y'| \\
&\qquad \text{and } |z| \leq |z'| \text{ for all } y' \in SuppB \text{ and } z' \in SuppC\} \\
&= \vee\{\{B(y) \wedge C(z)\} \mid yz = x, \\
&\qquad |yz| \leq |y'z'| \text{ for all } y'z' \in SuppBC\} \\
&= \bar{BC}(x) \\
&= \bar{A}(x).
\end{aligned}$$

This then implies $\bar{B}\bar{C} \subseteq \bar{A}$. Conversely, suppose $\bar{A} \not\subseteq \bar{B}\bar{C}$. Then there exists $y \in Supp(\bar{A})$ such that $\bar{A}(y) > (\bar{B}\bar{C})(y)$. If $y \in Supp(\bar{B}\bar{C})$, by the previous proof, we have $(\bar{B}\bar{C})(y) = \bar{A}(y)$, a contradiction. If $y \notin Supp(\bar{B}\bar{C})$, then

$$(\bar{B}\bar{C})(y) = \vee\{\bar{B}(u) \wedge \bar{C}(u) \mid u, v \in X^*, uv = y\} = 0.$$

Thus for any $u, v \in X^*$ such that $uv = y$, $\bar{B}(u) = 0$ or $\bar{C}(v) = 0$, i.e. $y \neq uv$ for all $u \in Supp\bar{B}$ and $v \in Supp\bar{C}$. Since $y \in SuppA = Supp(BC)$, $y = u'v'$ for some $u' \in SuppB$ and $v' \in SuppC$. It follows that

$$|y| = \lceil SuppA \rceil > \lceil SuppB \rceil + \lceil SuppC \rceil,$$

contradiction to (1). Thus $\bar{B}\bar{C} \supseteq \bar{A}$ and therefore $\bar{B}\bar{C} = \bar{D}\bar{E} = \bar{A}$.

(3) Assume, without loss of generality, that $Supp\bar{B} \neq Supp\bar{D}$ and that there is a word $x \in X^+$ such that $x \in Supp\bar{B}$, $x \notin Supp\bar{D}$. Then for all $y \in Supp\bar{C}$,

$$\begin{aligned} \bar{D}\bar{E}(xy) &= \bar{B}\bar{C}(xy) \\ &= \vee\{\bar{B}(x') \wedge \bar{C}(y') \mid x', y' \in X^*, x'y' = xy\} \\ &\geq \bar{B}(x) \wedge \bar{C}(y) \\ &> 0. \end{aligned}$$

Thus there exists $u \in Supp\bar{D}$ and $v \in Supp\bar{E}$ such that $uv = xy$. Since $\lceil Supp\bar{B} \rceil = \lceil Supp\bar{D} \rceil$, we have $u = x$, a contradiction. Hence $Supp\bar{B} = Supp\bar{D}$ and $Supp\bar{C} = Supp\bar{E}$.

(4) Suppose $SuppB \neq SuppD$. Then $B \backslash D \neq \Phi$ or $D \backslash B \neq \Phi$. Let $G = (B \backslash D) \cup (D \backslash B)$ and let $g \in SuppG$. We consider the following two cases: (i) $g \in Supp(B \backslash D)$ and (ii) $g \in Supp(D \backslash B)$.

Case (i): Since $Supp\bar{B} = Supp\bar{D}$, we have $|g| > \lceil Supp\bar{B} \rceil$. Let $c_0 \in Supp\bar{C} = Supp\bar{E}$. Then $gc_0 = df$ for some $d \in SuppD$ and $f \in SuppE$. Since $|c_0| \leq |f|$, there exists $h \in X^*$ such that $f = hc_0$. It is immediate that $h \neq e$. Indeed, if $h = e$, then $f = c_0$ and $g = d \in SuppD$, a contradiction. Thus $g = dh$ and $|g| > |d|$. If $d \notin SuppB$, then $d \in Supp(D \backslash B)$ and this contradicts the choice of g. On the other hand, if $d \in SuppB$, then $dh = g \in SuppB$, and

$$(BX^+)(g) \in \vee\{B(d) \mid d, h \in X^*, dh = g\} \geq B(d) > 0,$$

i.e., $g \in Supp(B \cap BX^+)$, which contradicts the fact that B is a fuzzy prefix code. Thus $D \backslash B = \Phi$.

Case (ii): Similarly, we can show that $D \backslash B = \Phi$. Therefore, $SuppB = SuppD$ holds.

Let $b_0 \in SuppB$ and $c \in SuppC$. Then $b_0 c = bq$ for some $b \in SuppD = SuppB$ and $q \in SuppE$. If $b_0 \neq b$, then either $b_0 = ba_1$ or $b = b_0 a_2$ for some $a_1, a_2 \in X^+$. But this contradicts the fact that B is a fuzzy prefix code. So we have $b_0 = b$ and hence $c = q$, that is $SuppC = SuppE$. Similarly, we can show that $SuppE \subseteq SuppC$ and hence $SuppC = SuppE$ holds.

(5) Assume $SuppB \neq SuppD$. Then there exists at least one word $x \in SuppA$ such that $x = x_1x_2 = y_1y_2, x_1 \neq y_1$, where $x_1 \in SuppB$, $x_2 \in SuppC$, $y_1 \in SuppD$ and $y_2 \in SuppE$. Choose a smallest such x, denoted by $\hat{x}$, and let the decompositions be $\hat{x}_1\hat{x}_2 = \hat{y}_1\hat{y}_2 = \hat{x}$.

Now $Supp\bar{B} = Supp\bar{D}$ and $Supp\bar{C} = Supp\bar{E}$ by (3). Denote the elements of $Supp\bar{G}$ by $\bar{g}$. We will show that $\hat{x}_1$ is in $SuppD$ and $\hat{x}_2$ is in $SuppE$.

Consider any word $\hat{x}_1 x_2 \in SuppBC = SuppDE$. Since $|\hat{x}_1\bar{x}_2| \leq |\hat{x}_1\hat{x}_2|$,

$$\lceil SuppC \rceil \leq |\hat{x}_2|.$$

It follows that $\hat{x}_1 \in SuppD$ by the following arguments: Case (i) $\lceil SuppC \rceil < |\hat{x}_2|$. Assume $\hat{x}_1 \notin SuppD$. Let $\hat{x}_1 = z_1z_2z_2 \neq e$ be such that $z_1 \in SuppD$

and $z_2\bar{x}_2 \in SuppE$. Thus $|\hat{x}_1\bar{x}_2| < |\widehat{x}|$, which contradicts that $\hat{x}$ was a shortest word with two decompositions. Case (ii) $\lceil SuppC \rceil = |x_2|$. Assume $\hat{x}_1 \notin SuppD$. Let $\hat{x}_1 = z_1z_2, z_2 \neq e$ be such that $z_1 \in SuppD$ and $z_2\bar{x}_2 \in SuppE$. It follows that $z_1 \in SuppB$ since $z_1\bar{x}_2 \in Supp(DE)$ and $|z_1\bar{x}_2| < |\widehat{x}_2|$. Since $z_1 \in SuppB$, $z_1z_2 \in SuppB$ and $z_2\bar{x}_2 \in SuppE$, $\bar{x}_2 \in SuppE$, B and E are not separable. In both cases, we arrive at a contradiction. Therefore, $\hat{x}_1 \in SuppD$. Similarly, we can show that $\bar{x}_1\hat{x}_2 \in SuppBC$ implies $\hat{x}_2 \in SuppE$. This part of the proof uses the fact that D and C are separable.

We have shown that $\hat{x}_1 \in SuppD$ and $\hat{x}_2 \in SuppE$. Since $\hat{x}_1\hat{x}_2 = \hat{y}_1\hat{y}_2$, $\hat{x}_1 \neq \hat{y}_1$, $\hat{x}_1$, $\hat{y}_1 \in SuppD$ and $\hat{x}_2, \hat{y}_2 \in SuppE$, D and E are not separable which is a contradiction. ■

Definition 8.5.5 *A nonempty $G \in \mathcal{FP}(X^*)$ is called a* ***bounded fuzzy language*** *if there exists a positive integer k such that $|\omega| \leq k$ for all $\omega \in SuppG$.*

If G is a bounded fuzzy language, then $\|SuppG\|$ exists. Let $\widehat{G}$ denote the set of all words in $SuppG$ with longest length if G is bounded.

Theorem 8.5.6 *Let $A, B, C, D, E \in \mathcal{FP}(X^*)$ and let A be a bounded fuzzy language such that $A = BC = DE$. Then the following assertions hold.*

(1) $\|SuppB\| + \|SuppC\| = \|SuppD\| + \|SuppE\| = \|SuppA\|$.

(2) $\widehat{B}\widehat{C} = \widehat{D}\widehat{E} = \widehat{A}$.

(3) If $\|SuppB\| = \|SuppD\|$, then $Supp\widehat{B} = Supp\widehat{D}$ and $Supp\widehat{C} = Supp\widehat{E}$.

(4) If $\|SuppB\| = \|SuppD\|$ and the pairs (B,C), (B,E), $((D,C)$ and (D,E) are separable, then $SuppB = SuppD$ and $SuppC = SuppE$.

Proof. The proofs of (1), (2) and (4) are analogous to the proofs of Theorem 8.5.4 (1), (2) and (5), respectively. ■

Theorem 8.5.7 *Let A, B, C, $D \in P_f(X)$, where none of $SuppA$, $SuppB$, $SuppC$, $SuppD$ is $\{e\}$. Let $AB = CD$ and let*

$$\lceil SuppA \rceil \geq \lceil SuppC \rceil.$$

If $c_0 \in SuppC$, $|c_0| = \lceil SuppC \rceil$, and $U \in \mathcal{FP}(X^+)$ is such that

$$SuppU = \{u \in X^* \mid c_0u \in SuppA\},$$

then either $SuppU = \{e\}$ or U is a fuzzy prefix code.

Proof. Suppose $\lceil SuppA \rceil = \lceil SuppC \rceil$. Then by (4) of Theorem 8.5.4 $SuppA = SuppC$, $SuppB = SuppD$. If $SuppU \neq \{e\}$, then there exists $u \in X^+$ such that $c_0u \in SuppA$. Since $c_0 \in SuppC$, $SuppA = SuppC$, so $c_0u \in SuppC$ and

$$(CX^+)(c_0u) = \vee \left\{C(y) \mid y, z \in X^+, yz = c_0u\right\} \geq C(c_0) > 0,$$

i.e., $c_0u \in Supp(C \cap CX^+)$, which contradicts that C is a fuzzy prefix code. Hence $SuppU = \{e\}$ in this case.

Now suppose $\lceil SuppA \rceil > \lceil SuppC \rceil$. Then $U \neq \Phi$ and $SuppU \neq \{e\}$, and U is a fuzzy prefix code for if $U \cap UX^+ \neq \Phi$, then there exists u such that $U(u) > 0$ and $(UX^+)(u) = \vee\{U(x) \mid x, y \in X^*, xy = u\} > 0$, i.e., there exists $x \in X^*$, $y \in X^+$ such that $xy \in SuppU$ and $x \in SuppU$. Thus both c_0xy, $c_0x \in SuppA$ and

$$\begin{aligned}(A \cap AX^+)(c_0xy) &= A(c_0xy) \wedge (\vee\{A(z) \mid z, u \in X^*, zu = c_0xy\}) \\ &\geq A(c_0xy) \wedge A(c_0x) \\ &> 0.\end{aligned}$$

This contradicts the fact that A is a fuzzy prefix code. ∎

Theorem 8.5.8 *$\mathcal{FP}(X^*)$ is a monoid with the binary operation defined in Section 8.1.*

Proof. Clearly, $C_{\{e\}} \in \mathcal{FP}(X^*)$ is the identity. ∎

Theorem 8.5.9 *The family of fuzzy prefix codes $P_f(X)$ is a submonoid of the monoid $\mathcal{FP}(X^*)$.*

Proof. The proof of the theorem is immediate by Theorems 8.5.8 and 8.2.8. ∎

Theorem 8.5.10 *Let $P_M = \{SuppA \mid A \in P_f(X)\}$. Then P_M is a free monoid.*

Proof. It is obvious that P_M is a monoid. Now we prove the P_M is a free monoid by using Levi's characterization for free monoids.

(1) Let h be the function from P_M to $\mathbb{N} \cup \{0\}$ defined by that $h(SuppA) = \lceil SuppA \rceil$ for all $SuppA \in P_M$. It is immediate that h is a homomorphism from P_M to $\mathbb{N} \cup \{0\}$ and $h^{-1}_{(0)} = \{e\}$.

(2) Suppose A, B, C, $D \in P_f(X)$ are such that

$$(SuppA)(SuppB) = (SuppC)(SuppD)$$

and $h(SuppA) \geq h(SuppC)$. If one of $SuppA$, $SuppB$, $SuppC$, $SuppD$ is $\{e\}$, then Levi's conditions are trivially satisfied. Let $c_0 \in SuppC$ be such that $|c_0| = \lceil SuppC \rceil$ and let $U = \{u \in X^* \mid c_0u \in SuppA\}$. Then C_U is a fuzzy prefix code for if $x, xy \in U$ for some $y \in X^+$, then $c_0x, c_0xy \in SuppA$, which contradicts the fact that $A \in P_f(X)$. Since

$$\{c_0\} \cup SuppB \subseteq (SuppA)(SuppB) = (SuppC)(SuppD),$$

for any $uv \in USuppB$, we have that $c_0uv \in c_0uSuppB$ and $c_0uv = c_1d$, where $u \in U$, $v \in SuppB$, $c_1 \in SuppC$ and $d \in SuppD$. Since $|c_0| =$

$\lceil SuppC \rceil$, $c_1 = c_0 y$ for some $y \in X^*$. If $y \neq e$, then $y \in X^+$ and

$$\begin{aligned}(C \cap CX^+)(c_1) &= C(c_1) \wedge (\vee\{C(z) \mid z, y' \in X^*, zy' = c_1\} \\ &\geq C(c_1) \wedge C(c_0) \\ &> 0.\end{aligned}$$

This contradicts the fact that $C \in P_f(X)$. Hence $y = e$, $c_1 = c_0$ and $uv = d$, i.e., $SuppB \subseteq SuppD$. Moreover, for any $d \in SuppD$, $c_0 d \in (SuppA)(SuppB)$. Thus there exists $a \in SuppA$, $b \in SuppB$ such that $c_0 d = ab$. Since $\lceil SuppA \rceil \geq \lceil SuppC \rceil$, $a = c_0 a_1$ for some $a_1 \in X^*$ and so $a_1 \in U$, $d = a_1 b \in USuppB$. Hence $SuppD \subseteq USuppB$ and $SuppD = USuppB$. It follows that

$$(SuppA)(SuppB) = (SuppC) \cup (SuppB).$$

For any $a \in SuppA$, we have $ab_0 = c_1 ub$ for some $c_1 \in SuppC$, $u \in U$, $b \in SuppB$. Since $|b_0| = \|SuppB\|$, $b = b_1 b_0$ for some $b_1 \in X^*$ and $a = c_1 u b_1 \in (SuppC)UX^*$, i.e., $SuppA \subseteq (SuppC)UX^*$. Similarly, $(SuppC)Ub_0 \subseteq (SuppA)(SuppB)$ and so $(SuppC)U \subseteq (SuppA)X^*$. For any $a \in SuppA$, $a = c_1 u x_1$, where $c_1 \in SuppC$, $u \in U$, $x_1 \in X^*$. If $x \neq e$, then $x_1 \in X^+$. Since $c_1 u \in (SuppC)U \subseteq (SuppA)X^*$, there exists $a_1 \in SuppA$, $x_2 \in X^*$ such that $c_1 u = a_1 x_2$. Hence $a = a_1 x_2 x_1 = a_1 x$, where $x = x_2 x_1 \in X^+$ and

$$\begin{aligned}(A \cap AX^+)(a) &= A(a) \wedge (\vee\{A(y) \mid y, z \in X^+, yz = a\}) \\ &\geq A(a) \wedge A(a_1) \\ &> 0,\end{aligned}$$

which contradicts the fact that $A \in P_f(X)$. Thus $a = c_1 u$ and $SuppA \subseteq (SuppC)U$. On the other hand, for any $v \in (SuppC)U$, there exists $a_2 \in SuppA$, $x_3 \in X^*$ such that $v = a_2 x_3$. If $x_3 \neq e$ then $x_3 \in X^+$. Since $SuppA \subseteq (SuppC) \cup X^*$, there exists $v' \in (SuppC)$, $x_4 \in X^*$ such that $a_2 = v' x_4$. Hence $v = v' x_4 x_3 = v' x'$, where $x' = x_4 x_3 \in X^+$. Also

$$\begin{aligned}(CU \cap CUX^+)(v) &= (CU)(v) \wedge (\vee\{(CU)(y) \mid y, z \in X^*, yz = v\}) \\ &\geq (CU)(v) \wedge (CU)(v') \\ &> 0,\end{aligned}$$

which contradicts that CU is a fuzzy prefix code by Theorem 8.2.8. Thus $x_3 = e$ and $c = a_2$, i.e., $(SuppC)U \subseteq SuppA$. Therefore, $SuppA \subseteq (SuppC)U$. ■

8.6 Equivalent Depictions of Fuzzy Codes

We now consider [99]. Let X be an alphabet. A subset A of the free monoid X^* is a code over X if for all $n, m \geq 1$ and $x_1, \ldots, x_n$, $x'_1, \ldots, x'_m \in A$, $x_1 x_2 \ldots x_n = x'_1 x'_2 \ldots x'_m$ implies $n = m$ and $x_i = x'_i$ for $i = 1, \ldots, n$.

(1) A code never contains the empty word e.

(2) Any subset of a code is a code. In particular, the empty set is a code.

Let A be a subset of X^*. Then A is a prefix (suffix) set if no element of A is a proper left (right) factor of another element in A. A is called a biprefix set if it is both prefix and suffix.

A prefix (suffix, biprefix) code is a prefix (suffix, biprefix) set which is a code, that is distinct from $\{e\}$.

Theorem 8.6.1 *Let $A \in \mathcal{FP}(X^*)$ be a fuzzy subset of the free monoid X^*. Then A is a fuzzy code if and only if (1) if for any $n, m \geq 1$ and $x_1, \ldots, x_n$, $x'_1, \ldots, x'_m \in X^*$, the condition $x_1 x_2 \ldots x_n = x'_1 x'_2, \ldots, x'_m = x$ implies $\wedge\{A(x_1), \ldots, A(x_n),\ A(x'_1), \ldots,\ A(x'_m)\} \propto \wedge\{[m = n],\ [x_1 = x'_1],\ \ldots,$ $[x_n = x'_n]\} \geq A^+(x)$.*

Proof. Assume the condition holds. Since $ee = e$,

$$\wedge\{A(e), A(e), A(e)\} \propto \wedge\{[2 = 1],\ [e = e]\} \geq A^+(e).$$

Now $A(e) \propto 0 \geq A^+(e)$ and so $A(e) = 0$. From the hypothesis, we can obtain the following condition easily, for any $x \in SuppA \setminus \{e\}$, $x_i, y_j \in X^*$, $i = 1, 2, \ldots, n$, $j = 1, 2, \ldots, m$, $x = x_1 x_2 \ldots x_n = y_1 y_2 \ldots y_m$ implies $\wedge\{A(x_1), \ldots, A(x_n),\ A(y_1), \ldots,\ A(y_m)\} \propto \wedge\{[m = n],\ [x_1 = y_1],\ \ldots,$ $[x_n = y_n]\} \geq A^+(x)$ and so A is a fuzzy code.

Suppose A is a fuzzy code. For any $n, m > 1$ and $x_1 \ldots x_n, x'_1, \ldots, x'_m \in X^*$, $x = x_1 x_2 \ldots x_n = x'_1 \ldots x'_m$. If $A(x_i) > 0$, $i = 1, 2, \ldots, n$, $A(x'_j) > 0$, $j = 1, \ldots, m$, and $x \neq e$, then $A'(x) > 0$, $x \in SuppA^* \setminus \{e\}$, by Definition 8.1.3 we have that $\wedge\{A(x_1),\ \ldots, A(x_n),\ A(x'_1), \ldots,\ A(x'_m)\} \propto \wedge\{[m = n],$ $[x_1 = x'_1],\ \ldots,\ [x_n = x'_n]\} \geq A^+(x)$. If $x = e$, clearly $x_i = e$, $i = 1, \ldots, n$, $x'_j = e$, $j = 1, \ldots, m$ since $A(e) = 0$. Thus

$$\wedge\{(A(x_1), \ldots, A(x_n), A(x'_1), \ldots, A(x'_m))\} = 0$$

and

$$\wedge\{A(x_1), \ldots, A(x_n), A(x'_1), \ldots, A(x'_m)\} = 0,$$

and $\wedge\{A(x_1), \ldots,\ A(x_n),\ A(x'_1), \ldots,\ A(x'_m)\} \propto \wedge\{[m = n],\ [x_1 = x'_1], \ldots,$ $[x_n = x'_n]\} \geq A^+(x)$.

If there exist x_{i_n} or x'_{j_m} such that $A(x_{i_n}) = 0, 1 \leq i_0 \leq n$ or $A(x'_{j_m}) = 0, 1 \leq j_0 \leq m$, then

$$\wedge\{A(x_1), \ldots, A(x_n), A(x'_1), \ldots, A(x'_m)\} = 0$$

and $\wedge\{A(x_1), \ldots,\ A(x_n),\ A(x'_1), \ldots,\ A(x'_m)\} \propto \wedge\{[m = n],\ [x_1 = x'_1], \ldots,$ $[x_n = x'_n]\} \geq A^+(x)$. ■

Proposition 8.6.2 *Let $A, B \in \mathcal{FP}(X^+)$ be such that $B \subseteq A$. If A is a fuzzy code, then B is a fuzzy code.*

Proof. For all $n, m \geq 1$ and $x_1, \ldots, x_n, x'_1, \ldots, x'_m \in X^*$, $x_1 \ldots x_n = x'_1 \ldots x'_m = x$, since A is a fuzzy code, we have $\wedge\{A(x_1), \ldots, A(x_n), A(x'_1), \ldots, A(x'_m)\} \propto \wedge\{[n = m], [x_1 = x'_1], \ldots, [x_n = x'_n]\} \geq A^+(x)$.

If $A^+(x) > 0$, then $\wedge\{[n = m], [x_1 = x'_1], \ldots, [x_n = x'_n]\} = \wedge\{A(x_1), \ldots, A(x_n), A(x'_1), \ldots, A(x'_m)\} = 0$. Since $B \subseteq A$,

$$B(x_i) \leq A(x_i), B(x'_j) \leq A(x'_j),$$

$i = 1, \ldots, n; j = 1, \ldots, m$. Thus $\wedge\{B(x_1), \ldots, B(x_n), B(x'_1), \ldots, B(x'_m)\} \propto \wedge\{[n = m], [x_1 = x'_1], \ldots, [x_n = x'_n]\} \geq B^+(x)$. If $A^+(x) = 0$, then $B(x) = 0$ and the result is obvious. By Theorem 8.6.1, B is a fuzzy code. ■

Theorem 8.6.3 *$A \in \mathcal{FP}(X^*)$ is a fuzzy code if and only if $A_t \subseteq X^*$ is a code for any $t \in (0, 1]$.*

Proof. Suppose A is a fuzzy code. Let $x \in A_t$. For any $n, m \geq 1$ and $x_1 \ldots x_n, x'_1 \ldots x'_m \in A_t$ such that $x_1 \ldots x_n = x'_1 \ldots x'_m = x$, we have since A is a fuzzy code that $\wedge\{A(x_1), \ldots, A(x_n), A(x'_1), \ldots, A(x'_m)\} \propto \wedge\{[n = m], [x_1 = x'_1], \ldots, [x_n = x'_m]\} \geq A^+(x) \geq A(x) \geq t > 0$. Thus

$$\wedge\{[n = m], [x_1 = x'_1], \ldots, [x_n = x'_n]\} = 1,$$

i.e., $n = m$, $x_i = x'_i$ for $i = 1, \ldots, n$. Therefore, A_t is a code for $t \in (0, 1]$.

Suppose A_t is a code for all $t \in (0, 1]$. If A is not a fuzzy code, then there exists $w \in X^*$, $w = x_1 \ldots x_n = y_1 \ldots y_m$ such that $\wedge\{A(x_1), \ldots, A(x_n), A(y_1), \ldots, A(y_m)\} \propto \wedge\{[n = m], [x_1 = y_1], \ldots, [x_n = y_n]\} < A^+(w)$.

Thus

$$t = \wedge\{A(x_1), \ldots, A(x_n), A(y_1), \ldots, A(y_m)\} > 0$$

and

$$\wedge\{[n = m], [x_1 = y_1], \ldots, [x_n = y_m]\} = 0,$$

i.e., $x_1, \ldots, x_n, y_1, \ldots, y_m \in A_t$, and $m \neq n$ or $x_i \neq y_i$ for $1 \leq i \leq n$. Hence A_t is not a code, a contradiction. Thus A is a fuzzy code. ■

Theorem 8.6.4 *If $A \in \mathcal{FP}(X^*)$ is a fuzzy code, then any homomorphism $\beta : Y^* \to X^*$ which induces a bijection of some alphabet Y onto $SuppA$ is injective. Conversely, if there exists an injective homomorphism $\beta : Y^* \to X^*$ such that $SuppA = \beta(Y)$, then A is a fuzzy code.*

Proof. Let $\beta : Y^* \to X^*$ be a homomorphism such that β is a bijection of Y onto *SuppA*. Let $u, v \in Y^*$ be words such that $\beta(u) = \beta(v)$. If $u = e$, then $v = e$; indeed $\beta(y) \neq e$ for each letter $y \in Y$, since A is a fuzzy code. If $u \neq e$ and $v \neq e$, set $u = y_1 y_2 \ldots y_n$, $v = y'_1, y'_2 \ldots y'_m$ with $n, m \geq 1$, $y_1, \ldots, y_n, y'_1, \ldots, y'_m \in Y$. Since β is a homomorphism, we have

$\beta(y_1)\ldots\beta(y_n) = \beta(y'_1)\ldots\beta(y'_m)$. But A is a fuzzy code and $\beta(y_1), \beta(y'_1) \in SuppA$, so $\wedge\{A(\beta(y_1)), \ldots, A(\beta(y_n)), A(\beta(y'_1)), \ldots, A(\beta(y'_m))\} \propto \wedge\{[m = n], [\beta(y_1) = \beta(y'_1)], \ldots, [\beta(y_n) = \beta(y'_n)]\} \geq A'(\beta)(u) > 0$,

$$\wedge\{[m = n], [\beta(y_1) = \beta(y'_1)], \ldots, [\beta(y_n) = \beta(y'_n)]\} > 0.$$

Thus $n = m$ and $\beta(y_i) = \beta(y'_i)$ for $i = 1, \ldots, n$. Now β is injective on Y. Thus $y_i = y'_i$ for $i = 1, \ldots, n$, and $u = v$. This shows that β is injective.

Conversely, suppose $\beta : Y^* \to X^*$ is an injective homomorphism. For any $n, m \geq 1$, $x_1, \ldots, x_n, x'_1, \ldots, x'_m \in X^*$ and $x_1 \ldots x_n = x'_1 \ldots x'_m = x$ and $x_1, .., x_n, x'_1, \ldots, x'_m \in SuppA = \beta(y)$, let $y_i, y'_i \in Y$ be such that $\beta(y_i) = x_i$, $\beta(y'_j) = x'_j$, $i = 1, \ldots, n$, $j = 1, \ldots, m$. Since β is an injective homomorphism, $x_1 \ldots x_n = x'_1 \ldots x'_m \Rightarrow \beta(y_1)\ldots\beta(y_n) = \beta(y'_1)\ldots\beta(y'_m) \Rightarrow \beta(y_1 \ldots y_n) = \beta(y'_1 \ldots y'_m) \Rightarrow y_1 \ldots y_n = y'_1 \ldots y'_m$. Thus $n = m$ and $y_i = y'_i$, since Y is a code over itself, and so $x_i = x_i$ for $i = 1, \ldots, n$. Hence $\wedge\{A(x_1), \ldots, A(x_n), A(x'_1), \ldots, A(x'_m)\} \propto \wedge\{[m = n], [x_1 = x'_1], \ldots, [x_n = x'_n]\} = \wedge\{A(x_1), \ldots, A(x_n)\} \propto 1 \geq A^+(x)$.

If there exists x_i or $x'_i \notin SuppA$, then $\wedge\{A(x_1), \ldots, A(x_n), A(x'_1), \ldots, A(x'_m)\} \propto \wedge\{[n = m], [x_1 = x'_1], \ldots, [x_n = x'_n]\} = 0 \propto \wedge\{[n = m], [x_1 = x'_1], \ldots, [x_n = x'_n]\} = 1 \geq A^+(x)$. Therefore, A is a fuzzy code over X. ■

Definition 8.6.5 *A homomorphism $\beta : Y^* \to X^*$ which is injective and such that $SuppA = \beta(Y)$ is called a* ***fuzzy coding homomorphism*** *for A, where A is a fuzzy code.*

Corollary 8.6.6 *Let $\alpha : X^* \to Y^*$ be an injective homomorphism. If A is a fuzzy code over X, then $\alpha(A) \in \mathcal{FP}(Y^*)$ is a fuzzy code over Y. If B is a fuzzy code over Y, then $\alpha^{-1}(B)$ is a fuzzy code over X.*

Proof. Let $\beta : Z^* \to X^*$ be a fuzzy coding homomorphism for A. Then $\alpha(\beta(Z)) = \alpha(SuppA)$ and since $\alpha \circ \beta : Z^* \to Y^*$ is an injective homomorphism by Theorem 8.6.4, $\alpha(A)$ is a fuzzy code.

Conversely, let $A = \alpha^{-1}(B)$, $n, m \geq 1$, $x_1, \ldots, x_n, x'_1, \ldots, x'_m \in X^*$, $x_1 x_2 \ldots x_n = x'_1 \ldots x'_m = x$, then $\alpha(x_1)\alpha(x_2)\ldots\alpha(x_n) = \alpha(x'_1)\ldots\alpha(x'_m)$, as B is a fuzzy code, therefore $\wedge\{\beta(\alpha(x_1)), \ldots, \beta(\alpha(x_n)), \beta(\alpha(x'_1)), \ldots, \beta(\alpha(x'_m))\} \propto \wedge\{[n = m], [\alpha(x_1) = \alpha(x'_1)], \ldots, [\alpha(x_n) = \alpha(x'_n)]\} \geq B^+(\alpha(x))$. If $A^+(x) > 0$, then

$$x \in SuppA^+ = \left[Supp\alpha^{-1}(B)\right]^+ = \alpha^{-1}(SuppB^+)$$

and so $\alpha(x) \in Supp(B)^+$. Thus

$$\wedge([m = n], [\alpha(x_1) = \alpha(x'_1)], \ldots, [\alpha(x_n) = \alpha(x'_n)]) = 1,$$

i.e., $m = n$, $\alpha(x_i) = \alpha(x'_i)$, for $i = 1, \ldots, n$. The injectivity of α implies that $x_i = x'_i$ for $i = 1, \ldots, n$, so $\wedge\{A(x_1), .., A(x_n), A(x'_1), \ldots, A(x'_m)\} \propto$

$\wedge\{[m = n], [x_1 = x'_1], \ldots, [x_n = x'_n]\} = \wedge\{A(x_1), \ldots, A(x_n), A(x'_1), \ldots, A(x'_m)\} \propto 1 = 1 \geq A^+(x)$. If $A^+(x) = 0$, the result is obvious. Therefore, A is a fuzzy code. ■

Corollary 8.6.7 *If $A \in \mathcal{FP}(X^*)$ is a fuzzy code, then A^n is a fuzzy code for any integer $n > 0$.*

Proof. Let $\beta : Y^* \to X^*$ be a fuzzy coding homomorphism for A, then $SuppA^n = \beta(Y^n)$. But Y^n is a fuzzy code. Thus the conclusion follows from Theorem 8.6.4. ■

Definition 8.6.8 *$A \in \mathcal{FP}(X^*)$ is called **fuzzy prefix (suffix)** if for any $x, x', u \in X^*$, $x = x'u$ $(x = ux')$ implies $A(x) \wedge A(x') \leq [x = x']$. A fuzzy subset is **fuzzy biprefix** if it is both fuzzy prefix and fuzzy suffix.*

Theorem 8.6.9 *A is fuzzy prefix (suffix, biprefix) subset if and only if A_t is ordinary prefix (suffix, biprefix) subset $\forall\ t \in (0, 1]$.*

Proof. Suppose A is fuzzy prefix code. For any $x, x', u \in X^*$ such that $x = x'u$ we have $A(x) \wedge A(x') \leq [x = x']$ since A is a fuzzy prefix subset. If $x, x' \in A_t$, $t \in (0, 1]$, then $A(x) \geq t$, $A(x') \geq t$ and $[x = x'] \geq t > 0$ and so $x = x'$ and A_t is prefix set.

Suppose A_t is prefix set $\forall t \in (0, 1]$. In order to prove that A is fuzzy prefix subset, suppose the contrary. Then there exists $x, x', u \in X^*$ such that $x = x'u$ and

$$A(x) \wedge A(x') > [x = x'].$$

Thus $0 < A(x) \wedge A(x') \leq 1$ and $[x = x'] < 1$. Set

$$t = \frac{A(x) \wedge A(x')}{2} \in (0, \frac{1}{2}).$$

Then $x, x' \in A_t$ and $x \neq x'$, i.e., A_t is not a prefix set, a contradiction. Hence A is a fuzzy prefix subset. ■

Proposition 8.6.10 *Any subset of a fuzzy prefix subset is a fuzzy prefix set.*

Proposition 8.6.11 *Any fuzzy prefix (suffix, biprefix) subset $A \neq C_{\{e\}}$ is a fuzzy code.*

Proof. If A is not a fuzzy code, then there is a word w of minimal length having two factorizations

$$w = x_1 x_2 \ldots x_n = x'_1 x'_2 \ldots x'_m$$

$x_i, x_j \in X^*$ such that $A^+(w) > \wedge\{A(x_1), \ldots, A(x_n), A(x'_1), \ldots, A(x'_m)\} \propto \wedge\{[n = m], [x_1 = x'_1], \ldots, [x_n = x'_n]\}$. Thus $A(x_i) > 0$, $A(x'_j) > 0$, $i = 1, \ldots, n$, $j = 1, \ldots, m$ and

$$\wedge\{[n = m], [x_1 = x'_1], \ldots, [x_n = x'_n]\} = 0.$$

Both x_1, x_1' are nonempty, and since w has minimal length, $[x_1 = x_1'] = 0$, i.e., x_1, x_1' are distinct. Hence $x_1 = x_1'u$ or $x_1' = x_1v$ for $u, v \in X^+$. Thus $A(x_1) \wedge A(x_1') > 0 = [x_1 = x_1']$, which contradicts the fact that A is a fuzzy prefix. Therefore, A is a fuzzy code. A similar argument can be used for fuzzy suffix subsets. ■

Theorem 8.6.12 *$A \in \mathcal{FP}(X^*)$ is a fuzzy prefix code (suffix code, biprefix code) if and only if A is a fuzzy prefix subset (suffix subset, biprefix subset) and $A \neq C_{\{e\}}$.*

Proof. Suppose A is a fuzzy prefix subset and $A \neq C_{\{e\}}$. For any $x \in X^+$, if $x = x'u$ for all $x', u \in X^+$, then

$$A(x) \wedge A(x') \leq [x = x'] = 0,$$

so

$$(A \cap AX^+)(x) = A(x) \wedge (\vee\{A(x') \mid x'u = x \text{ for all } x', u \in X^+\}) = 0,$$

i.e., $A \cap AX^+ = \Phi$. Thus A is a fuzzy prefix code.

Suppose A is a fuzzy prefix code. For any $x, x', u \in X^*$ such that $x = x'u$, if $x = e$, then $x' = e$,

$$A(x) \wedge A(x') \leq [x = x'] = 1.$$

Now suppose that

$$(A \cap AX^+)(x) = A(x) \wedge (\vee\{A(x') \mid x'u = x,\ x'u \in X^+\}) = 0.$$

If $A(x) = 0$, then $A(x) \wedge A(x') = 0 \leq [x = x']$. If

$$\vee\{A(x') \mid x'u = x,\ x', u \in X^+\} = 0,$$

then $A(x') = 0$ and $A(x) \wedge A(x') = 0 \leq [x = x'] = 0$. Therefore, A is a fuzzy prefix subset and $A \neq C_{\{e\}}$. ■

Theorem 8.6.13 *A is a fuzzy prefix code (suffix code, biprefix code) if and only if A_t is prefix (suffix, biprefix) code for any $t \in (0,1]$.*

Proof. The proof follows easily from Theorems 8.6.9 and 8.6.12. ■

Definition 8.6.14 *A fuzzy code A is called a* ***maximal fuzzy code*** *over X if A is not properly contained in any other fuzzy code over X, that is, if $A \subseteq A'$, A' is fuzzy code, then $A = A'$.*

Proposition 8.6.15 *Any fuzzy code A over X is contained in some maximal fuzzy code over X.*

Proof. Let $\mathcal{F}(A)$ be the set of fuzzy codes over X containing A, ordered by set inclusion.

Consider a chain $\mathcal{C}$ of fuzzy codes containing A, then $\hat{B} = \cup_{B \in \mathcal{C}} B$ is the least upper bound of $\mathcal{C}$. Let $n, m \geq 1$, $y_1, .., y_n, y_1', \ldots, y_m' \in X^*$ be such that $y_1 y_2 \ldots y_n = y_1' y_2' \ldots y_m' = w$. If $y_1, \ldots, y_n, y_1', \ldots, y_m' \in Supp\hat{B} = \cup_{B \in \mathcal{C}} SuppB$, then each if the y_i, y_j' belongs to a support set of fuzzy code of the chain $\mathcal{C}$ and this determines $n+m$ elements of $\mathcal{C}$. One of them, say D, contains all the others. Thus $y_1, \ldots, y_n, y_1', \ldots, y_m' \in SuppD$, and since D is a fuzzy code, then $\wedge\{D(y_1), \ldots, D(y_n), D(y_1'), \ldots, D(y_m')\} \propto \wedge\{[n = m], [y_1 = y_1'], \ldots, [y_n = y_n']\} \geq D^+(w) > 0$. Thus $n = m$ and $y_i = y_i'$ for $i = 1, \ldots, n$. Since $D \subseteq \hat{B}$, $\wedge\{\hat{B}(y_1), \ldots, \hat{B}(y_n), \hat{B}(y_1'), \ldots, \hat{B}(y_m')\} \propto \wedge\{[m = n], [y_1 = y_1'], \ldots, [y_n = y_n']\} = 1 \geq \hat{B}(w)$. Therefore, $\hat{B}$ is a fuzzy code. ∎

8.7 Fuzzy Codes and Fuzzy Submonoids

Proposition 8.7.1 *Let X be an alphabet. Suppose M is a fuzzy submonoid of X^* and*

$$A = (M \backslash C_{\{e\}}) \backslash \left(M \backslash \{C_{\{e\}}\}\right)^2.$$

Then $A^* \subseteq M$ and $SuppA$ is a unique minimal set of generators of $SuppM$ (A is called ***almost minimal set of generators*** *of M).*

Proof. Set $Q = M \backslash C_{\{e\}}$. First, we verify that $SuppA$ generates $SuppM$, i.e., that $(SuppA)^* = SuppM$. Since $A \subseteq M$, for any $x \in X^*$,

$$\begin{aligned} A^2(x) &= \vee\{A(y) \wedge A(z) \mid x = yz,\ y, z \in X^* \\ &\leq \vee\{M(y) \wedge M(z) \mid x = yz,\ y, z \in X^* \\ &\leq \vee\{M(x) \mid x = yz,\ y, z \in X^*\} \\ &= M(x) \end{aligned}$$

and so $A^2 \subseteq M$. By induction on $n \geq 0$, we can show that $A^n \subseteq M$. Thus $A^* \subseteq M$ and $SuppA^* \subseteq SuppM$.

By induction on the length of the words, we show that $SuppA^* \supseteq SuppM$. Of course, $e \in SuppA^*$. Let $m \in SuppQ$. If $m \in SuppQ^2$, then $m \in SuppA$. Otherwise $m = m_1 m_2$ with $m_1, m_2 \in SuppQ$ both strictly shorter than m. Therefore, m_1, m_2 belong to $SuppA^*$. Since $SuppA^* = \cup_{n=0}^{\infty} SuppA^n$, there exist natural numbers l, k such that $A^l(m_1) > 0$, $A^k(m_1) > 0$ and so

$$\begin{aligned} A^{l+k}(m) &= \vee\{A^l(y) \wedge A^k(m) \mid m = yz,\ y, z \in X^*\} \\ &\geq A^l(m_1) \wedge A^k(m_2) \\ &> 0, \end{aligned}$$

and $A^*(m) \geq A^{l+k}(m) > 0$, i.e., $m \in SuppA^*$. Hence

$$(SuppA)^* = \cup_{n=0}^{\infty}(SuppA)^n = \cup_{n=0}^{\infty} SuppA^n = SuppA^* = SuppM.$$

Now let B be fuzzy subset of X^* such that $SuppB$ is a set of generators of $SuppM$. We may assume that $C_{\{e\}} \nsubseteq B$. Then each $x \in SuppA$ is in $SuppB^*$ and therefore can be written as $x = y_1 y_2 \ldots y_n$ ($y_i \in SuppB$, $n \geq 0$). The fact that $x \neq e$ and $x \notin SuppQ^2$ implies $n = 1$ and $x \in SuppB$. This shows that $SuppA \subseteq SuppB$. Thus $SuppA$ is a minimal set of generators and such a set is unique. ∎

Proposition 8.7.2 *If M is a fuzzy free submonoid of X^*, then a fuzzy subset whose support set is a minimal generating set of SuppM is a fuzzy code. Conversely, if $A \in \mathcal{FP}(X^*)$ is a fuzzy code, then the fuzzy submonoid A^* of X^* is fuzzy free and A is its minimal fuzzy set of generators.*

Proof. Let $\alpha : Y^* \to SuppM$ be an isomorphism. By Theorem 8.6.4, a fuzzy subset A of X^* which is such that $SuppA = \alpha(Y)$ is a fuzzy code. Now

$$SuppM = \alpha(Y^*) = (\alpha(Y))^* = (SuppA)^*.$$

Thus $SuppA$ generates $SuppM$. Furthermore, $Y = Y^+ \backslash Y^+ Y^+$ and $\alpha(Y^+) = Supp\left(M \backslash C_{\{e\}}\right)$. Consequently,

$$SuppA = Supp\{(M \backslash C_{\{e\}}) \backslash \left(M \backslash C_{\{e\}}\right)^2\},$$

showing that $SuppA$ is the minimal set of generators of $SuppM$.

Conversely, assume that $A \in \mathcal{FP}(X^*)$, is a fuzzy code and consider a fuzzy coding homomorphism $\alpha : Y^* \to X^*$ for A. Then α is injective and α is a bijection from Y into $SuppA$. Thus α is a bijection from Y^* onto $\alpha(Y^*) = (\alpha(Y))^* = (SuppA)^*$. Since A is a fuzzy code, A^* is a fuzzy code and so A^* is a fuzzy free. Now α is a bijection. Hence $Y = Y^+ \backslash Y^+ Y^+$ implies $A = A^+ \backslash A^+ A^+$. Thus by Proposition 8.7.1, $SuppA$ is the minimal set of generators of $SuppA^*$. Since A generates A^*, A is the minimal fuzzy set of generators of A^*. ∎

Corollary 8.7.3 *Let A and B be fuzzy codes over X. If $A^* = B^*$, then $A = B$.*

Definition 8.7.4 *A fuzzy submonoid N of monoid S is called **fuzzy stable** if for all $u, v, w \in S$, $N(u) \wedge N(u) \wedge N(uw) \wedge N(wu) > 0$ implies $N(w) > 0$.*

Proposition 8.7.5 *If a fuzzy submonoid N of X^* is fuzzy stable, then $A = (N \backslash C_{\{e\}}) \backslash \left(N \backslash C_{\{e\}}\right)^2$ is a fuzzy code.*

Proof. To prove that A is a fuzzy code, suppose the contrary. Then there is a word $z \in SuppA^+$ of minimal length, having two distinct factorizations in $SuppA$, $z = x_1x_2\ldots x_n = y_1y_2\ldots y_m$, where $x_1,\ldots,x_n$, $y_1,\ldots,y_m \in SuppA$. We may assume $|x_1| < |y_1|$. Then $y_1 = x_1w$ for some nonempty word w. Since N is fuzzy stable, $\wedge\{N(x_1), N(y_2\ldots y_m), N(y_1), N(x_2\ldots x_m)\} \geq \wedge\{A(x_1),\ldots, A(x_n), A(y_1),\ldots, A(y_m)\} > 0$. Since N is fuzzy submonoid,

$$N(x_2\ldots x_n) \geq \wedge\{N(x_2),\ldots,N(x_n)\} = \wedge\{A(x_2),\ldots,A(x_n)\}.$$

Thus $N(w) > 0$ and

$$\begin{array}{rcl}(N\backslash C_{\{e\}})^2(x_1w) & = & \vee\{N(u)\wedge N(y) \mid uv = x_1w,\ u,y \in X^*\} \\ & \geq & N(x_1)\wedge N(w) \\ & > & 0.\end{array}$$

Consequently $y_1 = x_1w \notin SuppA$, a contradiction. Thus A is a fuzzy code. ■

Proposition 8.7.6 *A fuzzy free submonoid N of X^* is fuzzy stable.*

Proof. Since N is fuzzy free, it has a fuzzy base, say B. Let $u,v,w \in X^*$ and suppose that $u,v,uw,wv \in SuppN$. Set $u = x_1\ldots x_k$, $wv = x_{k+1}\ldots x_r$, $uw = y_1\ldots y_l$, $v = y_{l+1}\ldots y_s$ with x_i, y_j in $SuppB$. The equality $u(wv) = (uw)v$ implies

$$x_1\ldots x_kx_{k+1}\ldots x_r = y_1\ldots y_ly_{l+1}\ldots y_s.$$

Since B is a fuzzy base of N and N is a fuzzy submonoid,

$$\begin{array}{rl} & \wedge\{B(x_1),\ldots,B(x_r),B(y_1),\ldots,B(y_s)\} \propto \\ & \quad\wedge\{[r=s],[x_1=y_1],\ldots,[x_r=y_s]\} \\ \geq & N(uwv) \\ \geq & N(u)\wedge N(wv) \\ > & 0.\end{array}$$

Hence $r = s$ and $x_i = y_i$ $(i = 1,\ldots,s)$. Moreover, $l \geq k$ because $|uw| \geq |u|$. Thus

$$uw = x_1\ldots x_kx_{k+1}\ldots x_l = ux_{k+1}\ldots x_l.$$

Hence $w = x_{k+1}\ldots x_l$. If $w \neq e$,

$$N(w) = B^*(w) \geq B^n(w) \geq \wedge\{B(x_{k+1}),\ldots,B(x_l)\} > 0.$$

If $w = e$, $N(w) = 1$. Thus N is fuzzy stable. ■

Definition 8.7.7 *Let N be a fuzzy submonoid of X^*. N is called* ***fuzzy right (fuzzy left) unitary*** *if for all $u,v \in X^*$, $N(u)\wedge N(uv) > 0$ ($N(u)\wedge N(uv) > 0$) implies $N(u) > 0$.*

The fuzzy submonoid N of X^* is called **fuzzy biunitary** if it is both fuzzy left and fuzzy right unitary.

Proposition 8.7.8 *A fuzzy submonoid M of X^* is fuzzy right (resp., left, bi-) unitary if and only if its almost minimal set of generators is a fuzzy prefix (resp., suffix, biprefix) code.*

Proof. Let $M\backslash C_{\{e\}} = Q$ and $Q\backslash Q^2 = A$ be its almost minimal set of generators. Suppose M is fuzzy right unitary.

To show that A is fuzzy prefix, let x, xu be in $SuppA$ for some $u \in X^*$. Then $M(x) \wedge M(xu) = A(x) \wedge A(xu) > 0$. Since M is fuzzy right unitary, $M(u) > 0$. If $u \neq e$, then $u \in SuppQ$, but then $xu \in SuppQ^2$, contrary to the assumption. Thus $u = e$ and $A(x) \wedge A(xu) \leq [x = xu]$, i.e., A is fuzzy prefix.

Conversely, suppose A is fuzzy prefix. Let $u, v \in X^*$ be such that $u, u, v \in SuppM = (SuppA)^*$. Then $u = x_1x_2 \ldots x_n$, $uv = y_1y_2 \ldots y_m$ for some $x_1, \ldots, x_n, y_1, \ldots, y_m \in SuppA$. Consequently $x_1x_2 \ldots x_nv = y_1y_2 \ldots y_m$. Since A is fuzzy prefix, neither x_1 nor y_1 is a proper left factor of the other. Thus $x_1 = y_1$. Hence $x_2 \ldots x_nv = y_2 \ldots y_m$ and a similar argument yields $x_2 = y_2, \ldots, x_n = y_n$. This shows that $m \geq n$ and $v = y_{n+1} \ldots y_m$ belongs to $SuppM$. Thus M is fuzzy right unitary. ■

Proposition 8.7.9 *If M is a maximal fuzzy free submonoid of X^*, then its fuzzy base A is a maximal fuzzy code.*

Proof. Clearly, A is a fuzzy code. Let B be a fuzzy code on X with $A \subseteq B$ and $A \neq B$. Then $A^* \subseteq B^*$ and $A^* \neq B^*$ since otherwise $A = B$ by Corollary 8.7.3. Now A^* is maximal. Thus $B^* = C_{X^*}$ and $B = C_X$. Thus $A \subseteq C_{X^*}$ and $A \neq C_X$. Let $b \in X\backslash SuppA$. The fuzzy subset $C = A \cup C_{\{b^2\}}$ is a fuzzy code and $M = A^* \subset C^*$ since $b^2 \in SuppM$ and $b \notin SuppA^*$. This contradicts the maximality of M. ■

Theorem 8.7.10 *The almost minimal set of generators of the intersection of an arbitrary family of fuzzy free submonoids of X^* is a fuzzy code.*

Proof. Let $(M_i)_{i \in I}$ be a family of fuzzy free submonoids of X^*, and set $M = \cap_{i \in I} M_i$. Clearly, M is a fuzzy submonoid. It suffices to show that M is fuzzy stable. If for all $u, v, w \in X^*$, and $u, v, uw, wv \in SuppM$, then these four words belong to each of the $SuppM_i$. Since each M_i is fuzzy free, by Proposition 8.7.6, each M_i is fuzzy stable. Thus $w \in SuppM_i$ for all $i \in I$. Hence $w \in SuppM$. By Proposition 8.7.8, the almost minimal set of generators of M is a fuzzy code. ■

8.8 An Algorithm of test for Fuzzy Codes

The results in this and the next section are from [98].

Definition 8.8.1 *Let $A, B \in \mathcal{FP}(X^*)$. Define $A^{-1}B, BA^{-1} \in \mathcal{FP}(X^*)$ as follows:*

$$(A^{-1}B)(x) = \vee\{A(a) \wedge B(b) \mid ax = b,\ a, b \in X^*\}$$

and

$$(BA^{-1})(x) = \vee\{A(a) \wedge B(b) \mid ax = b,\ a, b \in X^*\}$$

for all $x \in X^*$.

Lemma 8.8.2 *Let $A \in \mathcal{FP}(X^+)$ and define the fuzzy subsets U_n of X^+, $n = 1, 2, \dots,$ as follows: $U_1 = A^{-1}A \backslash C_{\{e\}}, U_{n+1} = A^{-1}U_n \cup U_n^{-1}A$ for $n \geqslant 1$. For all $n \geqslant 1$ and $k \in \{1, \dots, n\}$, we have $e \in SuppU_n$ if and only if there is a word $u \in SuppU_k$ and integers $i, j \geqslant 0$ such that $A^i \cap A^j \neq \Phi$, $i + j + k = n$.*

Proof. We prove the statement for all n by descending induction on k. Assume, first $k = n$. If $e \in SuppU_n$, taking $n = 1, i = j = 0$, where we have that the condition $A^i \cap A^j \neq \Phi$ is satisfied. Conversely, if we have that the condition $A^i \cap A^j \neq \Phi$ holds, then $i = j = 0$. This implies $u = e$ and consequently $e \in (U_n)_0$.

Now let $n > k \geqslant 1$, and suppose that the equivalence holds for $n, n-1, \dots, k+1$. If $e \in SuppU_n$, then by induction hypothesis, there exists $u \in SuppU_{k+1}$ and two integers $i, j \geqslant 0$ such that $A^i \cap A^j \neq \Phi$, $i+j+k+1 = n$.

Thus there are words $x \in SuppA^i, y \in SuppA^j$ such that $ux = y$. Now $u \in SuppU_{k+1}$. Hence $U_{k+1}(v) = \vee\{(\vee(A(a) \wedge U_k(b))) \vee (\vee(A(a) \wedge U_k(b)) \mid a, b \in X^*, av = b\} > 0$, i.e.,

$$\vee\{A(a) \wedge U_n(b) \mid a, b \in X^*, av = b\} > 0$$

or

$$\vee\{A(a) \wedge U_n(b) \mid a, b \in X^*, av = b\} > 0.$$

Consequently, there are two cases. Either there is a word $z \in A_0$ such that $zv = u \in SuppU_k$ or there exist $z \in A_0, u \in SuppU_k$ such that $z = uv$.

In the first case, we have $ux = zy$. Thus

$$\begin{aligned}
(uA^i \cap A^{j+1})(zy) &= (uA^i(zy) \wedge A^{j+1}(zy) \\
&= (\vee\{A^i(x') \mid x' \in X^+,\ ux' = ux\}) \\
&\quad \wedge(\vee\{(A(z') \wedge A^j(y')) \mid z', y' \in X^+,\ z'y' = zy\}) \\
&\geqslant A^i(x) \wedge A(z) \wedge A^j(y) \\
&> 0,
\end{aligned}$$

so $A^i \cap A^{j+1} \neq \Phi$, $u \in SuppU_k$.

In the second case, we have $zx = uvx = uy$ and similarly, $uA^j \cap A^{i+1} \neq \Phi$, $u \in SuppU_k$. In both cases, the condition is satisfied.

Conversely, assume that there are $u \in SuppU_k$ and $i, j \geqslant 0$ with $A^i \cap A^j \neq \Phi$, $i+j+k=n$. Then $ux_1x_2\ldots x_i = y_1y_2\ldots y_j$; for some x_r, y in A_0. If $j=0$, then $i=0$ and $k=n$. Thus $j \geqslant 1$. Once more, we distinguish cases according to the length of u compared to the length of y_1.

If $u = y_1v$ for some $v \in SuppA^i$, then

$$\begin{aligned}(A^{-1}U_k)(v) &= \vee\{A(a) \wedge U_k(b) \mid a, b \in X^*, av = b\} \\ &\geqslant A(y_1) \wedge U_k(u) \\ &> 0,\end{aligned}$$

$X^{-1}U_k \subset U_{k+1}$, i.e., $v \in SuppU_{k+1}$ and further, $ux_1x_2\ldots x_i = y_1y_2\ldots y_j$.

Thus $vA^i \cap A^{j-1} \neq \Phi$ and by the induction hypothesis $e \in SuppU_n$.

If $y_1 = uv$ for some A_0^+, then similarly, $v \in SuppU_k^{-1}A \subseteq SuppU_{k+1}$ and $x_1x_2\ldots x_i = vy_2\ldots y_i$ showing that $A^i \cap vA^{j-1} \neq \Phi$. Thus again $e \in SuppU_n$ by the induction hypothesis. ■

Remark 8.8.3 *It may be noteworhty that the nature of U_1 did not play any role in the proof of Lemma* 8.8.2. *In other words, Lemma* 8.8.2 *remains true if U_1 is replaced by any set B.*

Theorem 8.8.4 $A \in \mathcal{FP}(X^+)$ *is a fuzzy code if and only if none of the sets U_n defined above contain $C_{\{e\}}$.*

Proof. If A is not a fuzzy code, then there is a word $\omega \in SuppA^+$ of minimal length. $\omega = x_1x_2\ldots x_\beta = y_1\ldots t_q$, $\omega, x_i, y_i \in X^+$ such that $A^+(\omega) > \{A(x_1) \wedge \ldots \wedge A(x_p) \wedge A(y_1) \wedge \ldots \wedge A(y_q)\} \propto \{[m = n \wedge [x_1 = y_1] \wedge \ldots \wedge [x_p = y_q]\}$. Thus

$$A(x_i) > 0, A(Y_j) > 0, i = 1, \ldots, p, j = 1, \ldots, q$$

and

$$[p = q] \wedge [x_1 = y_1] \wedge \ldots \wedge [x_p = y_q] = 0.$$

Both x_1, y_1 are nonempty, and since ω has minimal length, $[x_1 = y_1] = 0$, i.e., $x_1 \neq y_1$. Assume $|y_1| < |x_1|$. Then $x_1 = y_1u$ for some $u \in X^+$. But then $u \in SuppU_k$ and

$$\begin{aligned}(A^{p-1} \cap A^{q-1})(y_2\ldots y_q) &= A^{p-1}(ux_2\ldots x_p) \wedge A^{q-1}(y_2\ldots y_q) \\ &\geqslant A(x_2) \wedge \ldots \wedge A(x_p) \wedge A(y_2) \wedge \ldots \wedge A(y_q) \\ &> 0,\end{aligned}$$

i.e., $u \in SuppU_1$ and $A^{p-1}A^{q-1} \neq \Phi, (p-1)+(q-1)+1 = p+q+1$.

According to Lemma 8.8.2, $e \in SuppU_{pq-1}$.

Conversely, if $e \in SuppU_n$, take $k = 1$ in Lemma 8.8.2. There exists $u \in SuppU_1$ and integers $i, j \geqslant 0$, such that $A^i \cap A^j \neq \Phi, i+j+1 = n$. Now $u \in SuppU_1$ implies that

$$U_1(u) = (A^{-1}A \backslash C_{\{e\}})(u) = \vee\{A(a) \wedge A(b) \mid a, b \in X^*,\ av = b\} > 0.$$

Hence there exists $x, y \in SuppA$ such that $xu = y$.

Furthermore, $x \neq y$ since $u \neq e$. It follows from $xA^i \cap xA^j \neq \Phi$ that $yA^i \cap xA^j \neq \Phi$ since $A^i \cap A^j \neq \Phi$. There exists $\omega \in X^+$ such that $(uA^i)(\omega) > 0$, $A^j(\omega) > 0$. Thus $(xuA^i)(x\omega) = (uA^i)(\omega) > 0$ and $(xA^j)(x\omega) = A^j(\omega) > 0$.

Thus there exists $m_1, m_2, \ldots, m_i, m'_1, m'_2, \ldots, m'_j \in SuppA$ such that

$$ym_1m_2 \ldots m_i = xm'_1, m'_2 \ldots m'_j.$$

Since $x \neq y$, $A^+(ym_1m_2 \ldots m_i) \geqslant A^{i+1}(ym_1 \ldots m_i) > 0$ and $A^+(ym_1m_2 \ldots m_i) > \{A(u) \wedge A(m_1) \wedge \ldots \wedge A(m_i) \wedge A(x) \wedge A(m'_1) \wedge \ldots \wedge A(m'_j)\} \propto \{[i = j] \wedge [x = y] \wedge [m_1 = m'_1] \wedge \ldots \wedge [m_i = m'_i]\}$, showing that A is not a fuzzy code. ∎

Proposition 8.8.5 *If $A \in \mathcal{FP}(X^+)$ is a recognizable fuzzy subset, then the set of all $U_n (n \geqslant 0)$ is finite.*

Proof. Let σ be the fuzzy syntalic congruence of A, defined by for any $\omega, \omega' \in X^*, \omega = w' \bmod \sigma$ if and only if for any $u, v \in X^*$, $A(uxv) = A(yv)$.

Let μ be the fuzzy congruence of X^* with the two classes $\{e\}$ and X^+.

Let $\lambda = \sigma \cap \mu$. We use the following general fact.

If $L \in \mathcal{FP}(A^*)$ is a union of equivalence classes of a fuzzy congruence θ, then for any fuzzy subset $B \in \mathcal{FP}(X^*)$, $B^{-1}L$ is a union of fuzzy congruence classes $\bmod \theta$. (Indeed, let $x \in SuppB^{-1}L$ and $x' \equiv x \bmod \theta m$. Then $yx \in SuppL$ for some $y \in B_0$, whence $yz' \in L_0$. Thus $z' \in SuppB^{-1}L$).

We prove that each U_n is a union of equivalence classes of λ be induction on $n \geqslant 1$. For $n = 1$, A is a union of classes of σ. Thus $A^{-1}A$ also is a union of classes for σ, and finally $A^{-1}A \backslash C_{\{e\}}$ is a union of classes of λ. Next, if U_n is a union of classes of λ, then by the previous fact both $U_n^{-1}A$ and $A^{-1}U_n$ are unions of classes of λ. Thus U_{n+1} is a union of classes of λ. The fact that A is recognizable implies that λ has finite index. ∎

Remark 8.8.6 *Proposition 8.8.5 shows that Theorem 8.8.4 provides an algorithm for testing whether a recognizable fuzzy subset is a code.*

Example 8.8.7 *Let $X = \{a, b\}$ and*

$$A = \{(b, 0.9), (abb, 0.3), (abbba), 0.6), (bbba, 0.1), (baabb, 0.8)\},$$

i.e., $A(b) = 0.9$, $A(aab) = 0.3$, $A(abbba) = 0.6$, $A(bbba) = 0.1$, $A(baabb) = 0.8$, and $A(x) = 0$, otherwise.

We obtain

$$U_1 = \{ba, 0.3\}, (bba, 0.1), (aabb, 0.8)\},$$

$$A^{-1}U_1 = \{(a, 0.3), (ba, 0.1)\},$$

$$U_1^{-1}A = \{(abb, 0.3)\};$$

$$U_2 = \{(a, 0.3), (ba, 0.1), (abb, 0.3)\},$$

$$A^{-1}U_2 = \{(a, 0.1), (e, 0.3)\},$$

$$U_2^{-1}A = \{(bb, 0.3), (bbba, 0.3), (abb, 0.1), (e, 0.3), (ba, 0.3)\},$$

$$U_3 = \{(a, 0.1), (e, 0.3), (bb, 0.3), (bbba, 0.3), (abb, 0.1), (ba, 0.3)\}.$$

Thus $e \in SuppU_3$ *and* A *is not fuzzy code according to Theorem 8.8.4.*

Remark 8.8.8 *If* $A \in \mathcal{FP}(X^+)$ *is a fuzzy prefix set, then* $U_1 = A^{-1}A \backslash C_{\{e\}} = \Phi$. *Thus the algorithm ends immediately for such codes. On the other hand, if* A *is fuzzy suffix, the algorithm does not stop after one step. This symmetrical behavior is due of course to the definition of the* U_n *which favors computing from left to right.*

8.9 Measure of a Fuzzy Code

Definition 8.9.1 *A function* $\pi : X^* \to \mathbb{R}^0$ *is called a* **Bernoulli distribution** *on* X^* *if* π *is a homomorphism into the multiplicative monoid* $\mathbb{R}^0$ *of nonnegative real numbers and satisfies* $\sum_{a \in X} \pi(a) = 1$.

Definition 8.9.2 *A Bernoulli distribution is* **positive** *if* $\pi(a) > 0$ *for all* $a \in X$.

Proposition 8.9.3 *Let* π *is a Bernoulli distribution. Then*

$$\pi(e) = 1, \sum_{u \in X^n} \pi(u) = 1, n \geqslant 1.$$

Proof. Let $u \in X^n$. Since π is a homomorphism, we have

$$\sum_{a \in X} \pi(ua) = \sum_{a \in X} \pi(u)\pi(a) = \pi(u) \sum_{a \in X} \pi(a) = \pi(u).$$

Thus

$$\sum_{u \in X^{n+1}} \pi(u) = \sum_{u=va,\ v \in X^n, a \in X} \pi(va) = \sum_{v \in X^n} \pi(v) \sum_{a \in X} \pi(a) = \sum_{v \in X^n} \pi(v).$$

Hence

$$\sum_{u \in X^n} \pi(u) = \sum_{u \in X^{n-1}} \pi(u) = \ldots = \sum_{u \in X} \pi(u) = 1,$$

$$\pi(e) \sum_{a \in X} \pi(a) = \sum_{a \in X} \pi(e)\pi(a) = \sum_{a \in X} \pi(ea) = \sum_{a \in X} \pi(a) = 1.$$

■

Definition 8.9.4 *A function $\pi_f : \mathcal{FP}(X^*) \to \mathbb{R}^0 \cup \{\infty\}$, is called a* ***Bernoulli fuzzy distribution*** *on X^* if $\pi_f(A) = \sum_{A(x)>0} \pi(x)$ for any $\Phi \neq A \in \mathcal{FP}(X^*)$ and $\pi_f(\Phi) = 0$, where $\pi : X^* \to \mathbb{R}^0$ is a Bernoulli distribution. π_f is called* ***positive*** *if π is positive.*

Proposition 8.9.5 *For any family $\{A_i \mid i \in I\} \subseteq \mathcal{FP}(X^*)$*

$$\pi_f(\cup_{i\in I} A_i) \leqslant \sum_{i\in I} \pi_f(A_i).$$

If for any $i, j \in I$, $i \neq j$, $A_i \cap A_j = \Phi$, then $\pi_f(\cup_{i\in I} A_i) = \sum_{i\in I} \pi_f(A_i)$.

Proof. The proof is omitted. ■

Proposition 8.9.6 *If $A, B \in \mathcal{FP}(X^*)$ and for $n \geqslant 1$ set $s_0 = 0$, $s_n = \pi\{x \mid A(x) > 0$ and $|x| \leqslant n$ for any $x \in X^*\}$, then*

$$\pi_f(A) = \vee_{n\geqslant 0} s_n.$$

Proposition 8.9.7 *If $A, B \in \mathcal{FP}(X^*)$, then*

$$\pi_f(AB) \leqslant \pi_f(A)\pi_f(B), \pi_f(A^*) \leqslant \sum_{n\geqslant 0} \pi_f(A^n) \leqslant \sum_{n\geqslant 0} \pi(A)^n.$$

Proof. Since $AB = \cup_{A(l)>0,\ l\in X^n} \cup_{B(m)>0,\ m\in X^n} \{lm, (A(l) \wedge B(m))\}$,

$$\begin{aligned} \pi_f(AB) &\leqslant \textstyle\sum_{A(l)>0,\ l\in X^*} \sum_{B(m)>0,\ m\in X^*} \pi(lm) \\ &= \textstyle\sum_{A(l)>0, l\in X^*} \pi(l) \sum_{B(m)>0,\ m\in X^*} \pi(m) \\ &= \pi_f(A)\pi_f(B). \end{aligned}$$

By Proposition 8.9.5, we have

$$\pi_f(A^*) \leqslant \sum_{n\geqslant 0} \pi_f(A^n) \leqslant \sum_{n\geqslant 0} \pi_f(A)^n.$$

■

Theorem 8.9.8 *Let $A \in \mathcal{FP}(X^+)$ and let π_f be a Bernoulli fuzzy distribution on X^*.*

(1) If A is a fuzzy code, then $\pi_f(A^n) = \pi_f(A)^n$ for $n \geqslant 1$; $\pi_f(A^*) = \sum_{n\geqslant 0} \pi(A)^n$; and in particular $\pi_f(A^*) < \infty$ if and only if $\pi_f(A) < 1$.

(2) Conversely, if π_f is positive, if $\pi_f(A)$ is finite and if $\pi_f(A^n) = \pi_f(A)^n$ for all $n \geqslant 1$, then A is a fuzzy code.

Proof. For $n \geqslant 1$, let S_n be the n-fold Cartesian product of supp$A$.

(1) Assume that A is a fuzzy code. Then the function $(x_1, \ldots, x_n) \to x_1 \ldots x_n$ is a bijection of S_n onto suppA^n. Consequently,

$$\pi_f(A)^n = \sum_{(x_1,\ldots,x_n)\in S_n} \pi(x_1)\ldots\pi(x_n) = \sum_{A(x)>0,\ x\in X^*} \pi(x) = \pi_f(A^n).$$

Since A is a fuzzy code, the A^n, $n \geqslant 0$ are pairwise disjoint. Thus

$$\pi_f(A^*) = \sum_{n\geqslant 0} \pi_f(A^n) = \sum_{n\geqslant 0} \pi_f(A)^n.$$

By the above equality, $\pi_f(A^*) < +\infty$ implies $\pi_f(A) < 1$. By Proposition 8.9.7, $\pi_f(A) < 1$ implies $\pi_f(A^*) < +\infty$.

(2) Assume that A is not a fuzzy code. Then there is a word $u \in$ suppA^+ having two distinct factorizations in supp$A$, say $u = x_1x_2\ldots x_n = x'_1x'_2\ldots x'_m$ with $n, m \geqslant 1$ and $x_i, x'_j \in$ *SuppA*. Thus $\omega = uu$ has two factorizations:

$$\omega = x_1x_2\ldots x_nx'_1x'_2\ldots x'_m = x'_1\ldots x'_mx_1\ldots x_n$$

of the same length $k = n + m$. It follows that

$$\pi_f(A)^k = \sum_{(y_1\ldots y_k)\in S_k} \pi(y_1\ldots y_k) \geqslant \pi_f(A^k) + \pi(\omega).$$

But the finiteness of $\pi_f(A)$ and the condition $\pi_f(A)^k = \pi_f(A^k)$ imply $\pi(\omega) \leqslant 0$ and so $\pi(\omega) = 0$. This contradicts the hypothesis that π_f is positive. ∎

Theorem 8.9.9 *Let A be a fuzzy code over X. For all Bernoulli fuzzy distributions π_f on X^*, we have $\pi_f(A) \leqslant 1$.*

Proof. We prove the statement first in the case where the numbers $|x|$, for $A(x) > 0$, are bounded. Thus we assume that there is some integer $k \geqslant 1$ with $A \subseteq X \cup X^2 \cup \ldots \cup X^k$. It follows that for $n \geqslant 1$, $A^n \subseteq X \cup X^2 \cup \ldots \cup X^{nk}$. Hence

$$\pi_f(A)^n = \pi_f(A_n) \leqslant \sum_{i=1}^{nk} \pi_f(x^i) = \sum_{i=1}^{nk} \sum_{u\in X^i} \pi(u) = nk.$$

Assume now that $\pi_f(A) > 1$, i.e., $\pi_f(A) = 1 + \varepsilon$ for some $\varepsilon > 0$. Then for all $n \geqslant 1$, $(1+\varepsilon)^n \leqslant kn$. Since

$$(1+\varepsilon)^n = 1 + n\varepsilon + \frac{n(n-1)}{2}\varepsilon^2 + \ldots > \frac{n(n-1)}{2}\varepsilon^2,$$

$$0 < \frac{nk}{(1+\varepsilon)^n} < \frac{2k}{(n-1)\varepsilon^2}.$$

Taking $n \geqslant \frac{2k}{\varepsilon^2} + 1$, we have $\frac{nk}{(1+\varepsilon)^n} < 1$, i.e., $nk < (1+\varepsilon)^n$, which contradicts the fact that $(1+\varepsilon)^n \leqslant kn$ for all $n \geqslant 1$. This $\pi_f(A) \leqslant 1$.

If A is an arbitrary fuzzy code and $n \in \mathbb{N}$, define the fuzzy subset A_n of X^* by $\forall x \in X^*$, $A_n(x) = A(x)$ if $|x| \leq n$ and $A(x) = 0$ otherwise. Since $A_n \subseteq A$, the set A_n is a fuzzy code satisfying the condition of the first part of the proof. Consequently, $\pi_f(A_n) \leqslant 1$. Hence $\pi_f(A) = \vee_{n\geqslant 1}\pi_f(A_n) \leqslant 1$.

∎

Corollary 8.9.10 *Let A be a fuzzy code over an alphabet with k letters. Then*

$$\sum_{A(x)>0,\ x\in X^*} k^{-|x|} \leqslant 1.$$

Theorem 8.9.11 *Let A be a fuzzy code over X. If there exists a positive Bernoulli fuzzy distribution π_f on X^* such that $\pi_f(A) = 1$, then the fuzzy code A is maximal.*

Proof. Assume that A is not maximal. Then there is some word $y \in X^+$ such that $B = A \cup C_{\{y\}}$ is a fuzzy code. By Theorem 8.9.9, we have $\pi_f(B) \leqslant 1$, On the other hand, $\pi_f(B) = \pi_f(A) + \pi(y) = 1 + \pi(y)$. Thus $\pi(y) = 0$, which is impossible since π_f is positive. ■

Example 8.9.12 *Let $X = \{a, b\}$ and let A be the fuzzy subset of X^* defined by $A(b) = 0.1$, $A(ab) = 0.7$, $A(ba) = 0.4$, and $A(x) = 0$ otherwise. Define π by $\pi(a) = \frac{1}{3}$, $\pi(b) = \frac{2}{3}$. Then*

$$\pi_f(X) = \frac{2}{3} + \frac{2}{9} + \frac{2}{9} = \frac{10}{9}.$$

Thus A is not a fuzzy code.

Example 8.9.13 *Let $X = \{a, b\}$ and $A(a) = 0.2, A(ba) = 0.8, A(bb) = 0.9$. Then A is a fuzzy code since A is fuzzy prefix. Define π by $\pi(a) = p$, $\pi(b) = 1 - p$. We have $\pi_f(A) = p + (1-p)p + (1-p)^2 = 1$. Hence A is maximal.*

We bring to attention the paper [76] that appeared as the book was being concluded. In [76], the authors introduce the notion of a fuzzy complete set. They use this notion in the study of maximal codes.

8.10 Code Theory and Fuzzy Subsemigroups

In the remainder of the chapter, we consider the approach of Gerla [33] for the construction of codes. For this purpose free, pure, very pure, left unitary, right unitary, and unitary fuzzy subsemigroups of a free semigroup are defined. For every such fuzzy subsemigroup, there is a family of subsemigroups of the same type and so a family of codes can be associated with each one.

We give some general methods for constructing the above-mentioned fuzzy subsemigroups.

Let B be a set and B^+ the free semigroup generated by B. Recall that a code is a base C of a free subsemigroup S of B^+. Since $C = S\backslash S^2$, the search for a code is, in a sense, equivalent to the search for the corresponding subsemigroup. Moreover, the codes determined by pure, very pure, left unitary, right unitary, and unitary subsemigroups [22, 75, 109] are particularly useful.

Let $(L, \wedge)$ be a semilattice and let $\mathcal{F}_L(B^+)$ denote the set of functions of B^+ into L. Then $\mathcal{F}_L(B^+)$ is a semilattice whose elements are named *fuzzy subsets* or *L-subsets* of B^+. In L and L^+, the order relation and the *infimum* and *supremum* operators are defined as usual.

An L-subset g of B^+ is an **L-subsemigroup** or **fuzzy subsemigroup** of B if, for any $u \in L$, the u-cut $g_u = \{x \in B^+ \mid g(x) \geqslant u\}$ is a subsemigroup of B^+. It follows that g is a fuzzy subsemigroup if and only if for any $x, y \in B^+$,

$$g(xy) \geqslant g(x) \wedge g(y). \tag{8.1}$$

We call an L-subsemigroup **free** if and only if its nonempty level subsets are free semigroups. In the same manner, we define **pure, very pure, left unitary, right unitary**, or **unitary**, L-subsemigroups of B^+. If $L = \{0, 1\}$ such L-subsemigroups coincide with the characteristic functions of the subsemigroups with the same name.

Clearly, every method for constructing an L-subsemigroup g of a particular type gives a method for constructing a family $\{g_u \mid u \in L\}$ of very pure subsemigroups of the same type.

Moreover, if X and Y are two sets, the rectangular band on $X \times Y$ is the semigroup defined by the setting $(x, y)(x', y') = (x, y')$ for every $(x, y), (x', y') \in X \times Y$.

The following proposition shows that a very simple characterization of the above defined subsemigroups exists.

Proposition 8.10.1 *An L-subsemigroup g of B^+ is free, pure, very pure, left unitary, right unitary, or unitary if and only if for any x, $y \in B^+$:*
(1) $g(x) \geqslant g(yx) \wedge g(xy) \wedge g(y)$,
(2) $g(x) = g(x^n)$ for all $n \in \mathbb{N}$,
(3) $g(x) \geqslant g(yx) \wedge g(xy)$,
(4) $g(x) \geqslant g(yx) \wedge g(y)$,
(5) $g(x) \geqslant g(xy) \wedge g(y)$,
(6) $g(x) \geqslant g(yx) \wedge g(y)$ and $g(x) \geqslant g(xy) \wedge g(y)$,
respectively.

Proof. By Schützenberger's condition [128], a subsemigroup of B^+ is free if and only if from $yx \in S, xy \in S$, and $y \in S$, it follows that $x \in S$ for every $x, y \in B^+$. Then g is free if and only if

$$g(yx) \geqslant u, \; g(xy) \geqslant u, \; g(y) \geqslant u \qquad \text{imply} \qquad g(x) \geqslant u \tag{8.2}$$

for every $x, y \in B^+$ and $u \in L$. Now (8.2) follows from (1). Conversely, (1) follows from (8.2) by setting $u = g(yx) \wedge g(xy) \wedge g(y)$. The remaining assertions of the proposition can be proved in a similar manner. ■

The following proposition shows that, if L is $\wedge$-complete, then for every L-subset g of B^+ we can associate the free (pure, very pure, left unitary, right unitary, or unitary) L-subsemigroup g generated by g, i.e., the intersection of the free (pure, very pure, left unitary, right unitary, or unitary) L-subsemigroups containing g.

Proposition 8.10.2 *If L is a $\wedge$-complete semilattice, then any set of free (pure, very pure, left unitary, right unitary, or unitary) L-subsemigroups of B^+ is closed with respect to intersection.*

Proof. Let $\{g_i \mid i \in I\}$ be a set of free L-subsemigroups of B^+. Then for any $i \in I$, $g_i(xy) \geqslant g_i(x) \wedge g_i(y)$. Thus

$$\wedge_{i\in I} g_i(xy) \geqslant \wedge_{i\in I} g_i(x) \wedge g_i(y) = [\wedge_{i\in I} g_i(x)] \wedge [\wedge_{i\in I} g_i(y)].$$

Hence $\cap_{i\in I} g_i$ is an L-subsemigroup. Moreover, by (1) of Proposition 8.10.1,

$$\wedge_{i\in I} g_i(x) \geqslant \wedge_{i\in I}[g_i(yx) \wedge g_i(xy) \wedge g_i(y)] = [\wedge_{i\in I} g_i(xy)] \wedge [\wedge_{i\in I} g_i(y)].$$

Thus $\cap_{i\in I} g_i$ is free. One proceeds in a similar manner if every g_i is pure, very pure, left unitary, right unitary, or unitary. ∎

8.11 Construction of Examples by Closure Systems

The following result provides the first class of examples of fuzzy subsemigroups of B^+. Recall that a *closure system* for a semigroup S is an $\cap$-complete class C of subsemigroups of S such that $S \in C$. For any subset X of S, let $\langle X \rangle = \cap\{Z \in C \mid Z \supseteq X\}$. If X is the finite set $\{a_1, \ldots, a_n\}$, then $\langle X \rangle$ is denoted also by $\langle a_1, \ldots, a_n \rangle$. Every closure system C is a semilattice with respect to the join operation $\vee$ defined by setting $X \vee Y = \langle X \cup Y \rangle$ for every $X, Y \in C$. The order relation associated to $\vee$ is the dual of ordinary inclusion.

Proposition 8.11.1 *Let S be a semigroup, C a closure system for S, and $g : B^+ \to S$ a homomorphism from B^+ to S. Then the function $g' : B^+ \to C$ defined by, $\forall x \in B^+$,*

$$g'(x) = \langle g(x) \rangle \tag{8.3}$$

is a fuzzy subsemigroup of B^+. If for all $x, y \in B^+$,
(1) $x \in \langle yx, xy, y \rangle$,
(2) $x \in \langle x^n \rangle$ for all $n \in \mathbb{N}$,
(3) $x \in \langle yx, xy \rangle$,
(4) $x \in \langle yx, y \rangle$,
(5) $x \in \langle xy, y \rangle$,
(6) $x \in \langle yx, y \rangle \cap \langle xy, y \rangle$,
then g' is free, pure, very pure, left unitary, right unitary, or unitary, respectively. Moreover, $g'_X = g^{-1}(X)$ for every $X \in C$.

Proof. For all $x, y \in B^+$, $g'(x,y) = \langle g(xy)\rangle = \langle g(x)g(y)\rangle \subseteq \langle g(x), g(y)\rangle = \langle g(x)\rangle \vee \langle g(y)\rangle = g'(x) \vee g'(y)$. Thus g' is a fuzzy subsemigroup. Moreover, g' is free if and only if $g'(x) \subseteq g'(yx) \vee g'(xy) \vee g'(y)$, i.e., $\langle g(x)\rangle \subseteq \langle g(yx)\rangle \vee \langle g(xy)\rangle \vee \langle g(y)\rangle = \langle g(yx), g(xy)g(y)\rangle$. Hence from (1), it follows that g' is free.

The remaining assertions of the proposition follow in a similar manner. ■

It follows that if S is finite, then every cut of g' is a *recognizable* subsemigroup (see [75], Proposition 1.5).

Corollary 8.11.2 *There exists an L-subsemigroup g of B^+ such that $\{g_u \mid u \in L\}$ is the family of all free (pure, very pure, left unitary, right unitary, or unitary) subsemigroups of B^+.*

Proof. Let C be the class of all free (pure, very pure, left unitary, right unitary, or unitary) subsemigroups of B^+. Then from Proposition 8.10.2, it follows that C is closed with respect to intersection. Since $B^+ \in C$, it follows that C is a closure system. Thus the corollary follows from Proposition 8.11.1 by assuming $S = B^+$ and that g is the identity function. ■

Clearly, Corollary 8.11.2 is not useful to build, in an effective way, examples of desired subsemigroups of B^+. Also, since every function from B to a semigroup S can be extended to a homomorphism from B^+ to S, a difficulty in utilizing Proposition 8.11.1 is finding semigroups S with suitable closure systems.

Corollary 8.11.3 *Let S be a semigroup, C a class of subsemigroups of S, and $\overline{C}$ the closure system generated by C. Then if $g : B^+ \to S$ is a homomorphism, the function g', defined by Eq. (8.3), is a unitary $\overline{C}$-subsemigroup of B^+.*

Proof. Every element X of $\overline{C}$ is equal to S or to an intersection of elements of C. Thus for every $x, y \in S$, $\langle yx, y\rangle = S$, or $\langle yx, y\rangle = \cap\{X \mid X \in C, yx \in X, y \in X\}$. In the first case, that $x \in \langle yx, y\rangle$ is clear. In the second, we have that for every $X \in C$, from $yx \in X$ and $y \in X$, it follows that $x = y^{-1}\, yx \in X$. Thus $x \in \langle yx, y\rangle$. In the same manner, it follows that $x \in \langle xy, y\rangle$. ■

One way to use Corollary 8.11.3 is to start from a group G and the class of all subgroups of G. Clearly, it is convenient to examine groups whose lattice of subgroups is known. For example, let $S = \mathbb{Z}/(m)$ be the additive group of residue classes modulo m and X the subgroup of S generated by the class $[r]_m, r \in \mathbb{Z}$. Then

$$g'_X = \{b_1 \ldots b_p \in B^+ \mid g(b_1) + \ldots + g(b_p) \equiv rs \bmod m \text{ for a suitable } s \in \mathbb{Z}\}$$

provides an example of a unitary subsemigroup for any function $g : B \to \mathbb{Z}$ and $r, m \in \mathbb{Z}$. Such a procedure can be generalized in the following way.

Let $(A, \cdot, +)$ be a near field, i.e., $(A, +)$ is a group, $(A, \cdot)$ is a semigroup, and that the right distribution of $\cdot$ with respect to $+$ holds. Moreover, let C be the set of subgroups of $(A, +)$ of the type $\{xa \mid x \in A\}$, where a is any fixed element of A. Then it follows from Corollary 8.11.3 that

$$\{b_1 \dots b_p \in B^+ \mid g(b_1) + \dots g(b_p) = xa \text{ for } x \in A\}$$

is an example of a unitary subsemigroup for any function $g : B \to A$ and $a \in A$.

If we want g' to also be pure, it suffices to assume that the elements of C have an algebraic closure property. For example, we can assume that S is the multiplicative group of a field F and that C is the set of the multiplicative groups of an algebraically closed subfield of F. In this case, $\langle x^n \rangle \cup \{0\}$ is an algebraically closed subfield of F and therefore $x \in \langle x^n \rangle$ for every $x \in F \backslash \{0\}$ and $n \in \mathbb{N}$.

Another way of obtaining pure fuzzy subsemigroups is to start from bands, i.e., idempotent semigroups. In this case, condition (2) of Proposition 8.11.1 is commonly true. In particular, the hypothesis that S is a rectangular band is of interest.

Corollary 8.11.4 *Let S be a rectangular band, C a closure system for S, and $g : B^+ \to S$ a homomorphism from B^+ to S. Then the function g', defined by Eq.(8.3), is a very pure fuzzy subsemigroup of B^+.*

Proof. Recall that $xyyx = x$ for every $x, y \in S$. It follows that $x \in \langle yx, xy \rangle$ and condition (3) of Proposition 8.11.1 hold. ■

Let X and Y be sets and S be the rectangular band on $X \times Y$. Then the subsemigroups of S are of type $X' \times Y'$ with X' and Y' subsets of X and Y, respectively. Moreover, let g_1 and g_2 denote the two components of a pre-fixed function $g : B \to S$. Then Corollary 8.11.4 ensures that the sets of type

$$\{b_1 \dots b_p \in B^+ \mid g_1(b_1) \in X' \text{ and } g_2(b_p) \in Y'\}$$

are examples of very pure subsemigroups of B^+.

8.12 Examples by $*$-morphisms

A $*$-semilattice is a triple $(M, *, \wedge)$ such that $(M, \wedge)$ is a semilattice with zero element 0, $(M, *)$ is a semigroup, and for every $u, v \in M$

$$u * v \geqslant u \wedge v.$$

If S is a subsemigroup of B^+, then a $*$-morphism from S into M is a function $g : B^+ \to M$ such that $g(xy) = g(x) * g(y)$ for every $x, y \in S$ and $g(x) = 0$ for every $x \in S$.

Proposition 8.12.1 *Every $*$-morphism g from a subsemigroup S of B^+ to a $*$-semilattice M is a fuzzy subsemigroup of B^+. Moreover, if for every $u, v \in M$:*

(1) $u \geqslant u * v \wedge v * u \wedge v$ and S is free,
(2) $u^n = u$ for every $n \in \mathbb{N}$ and S is pure,
(3) $u \geqslant u * v \wedge v * u$ and S is very pure,
(4) $u \geqslant u * v \wedge v$ and S is left unitary,
(5) $u \geqslant u * v \wedge v$ and S is right unitary,
(6) $u \geqslant u * v \wedge v$ and $u \geqslant u * v \wedge v$ and S is unitary,

then g is free, pure, very pure, left unitary, right unitary, or unitary, respectively.

Proof. If $x, y \in S$, then $g(xy) = g(x) * g(y) \geqslant g(x) \wedge g(y)$. If $x \notin S$ or $y \notin S$, then $g(x) \wedge g(y) = 0$ and (1) is proved.

Assume (1). Then from $g(yx) = 0, g(xy) = 0$, or $g(y) = 0$ it follows that $g(yx) \wedge g(xy) \wedge g(y) = 0$ and condition (1) of Proposition 8.10.1 holds. If $g(yx) \neq 0, g(xy) \neq 0$, and $g(y) \neq 0$, then $yx \in S, xy \in S$, and $y \in S$. It follows that $x \in S$ and

$$g(x) \geqslant g(y) * g(x) \wedge g(x) * g(y) \wedge g(y) = g(yx) \wedge g(xy) \wedge g(y).$$

Assume condition (2) holds. Then from $x \notin S$, it follows that $x^n \notin S$ and therefore $g(x) = g(x^n) = 0$. Moreover, from $x \in S$ it follows that $g(x^n) = (g(x))^n = g(x)$. Thus g is pure. The remaining parts of the proposition can be proved similarly. ■

In general, in order to apply Proposition 8.12.1, we assume $S = B^+$. Moreover, since every function $g : B \to M$ is extendible to a $*$-morphism, the only difficulty is to find suitable $*$-semilattices. If the cardinality of M is not large, then a combinatorial approach to this question is possible. For example, if M is a set of two elements, then a semilattice structure in M is necessarily isomorphic to the semilattice $\{0, 1\}$. Then we can set $M = \{0, 1\}$. Moreover, if we define the operation $*$ in such a way that condition (4) holds, then $1 * 1 = 1$. If we also impose condition (1) of Proposition 8.12.1, then we must set $0 * 1 = 0$ or $1 * 0 = 0$. Then in order to obtain free fuzzy subsemigroups, we must examine the following cases:

(1) $0 * 1 = 0$, $1 * 0 = 1$, $0 * 0 = 0$. In this case, every $*$-morphism is very pure and right unitary.

(2) $0 * 1 = 0$, $1 * 0 = 1$, $0 * 0 = 1$. The operation $*$ is not associative.

(3) $0 * 1 = 0$, $1 * 0 = 0$, $0 * 0 = 0$. Every $*$-morphism is very pure and unitary.

(4) $0 * 1 = 0$, $1 * 0 = 0$, $0 * 0 = 0$. Every $*$-morphism is unitary.

(5) $0 * 1 = 1$, $1 * 0 = 0$, $0 * 0 = 0$. Every $*$-morphism is very pure and left unitary.

(6) $0 * 1 = 0$, $1 * 0 = 1$, $0 * 0 = 1$. The operation $*$ is not associative.

Consider case (5). Then a $*$-morphism g obtains the very pure left unitary subsemigroup $g_l = \{b_1 \ldots b_p \in B^+ \mid g(b_1) \ldots g(b_p) = 1\}$. It follows that the relative code C is $B_1 \cup B_0^+ \times B_1$, where $B_0 = \{b \in B \mid g(b) = 0\}$ and $B_1 = \{b \in B \mid g(b) = 1\}$.

Clearly, the above combinatorial approach is not possible in general. However, the search for suitable $*$-semilattices can be simplified by some elementary considerations. For example, it follows from Proposition 8.12.1 that, in order to obtain pure fuzzy subsemigroups, we are forced to assume that $(M, *)$ is a band. This suggest the following corollary.

Corollary 8.12.2 *Let L_1 and L_2 be two semilattices, $(L, \wedge)$ the direct product of L_1 and L_2, and $(L, *)$ the rectangular band defined on $L_1 \times L_2$. Then $(L, *, \wedge)$ is a $*$-semilattice and every $*$-morphism from B^+ into $(L, *, \wedge)$ is a very pure fuzzy subsemigroup.*

Proof. Consider (4). For any $(u, u'), (v, v') \in L$,

$$(u, u') * (v, v') = (u, v') \geqslant (u \wedge v, u' \wedge v') = (u, u') \wedge (v, v').$$

Moreover,

$$\begin{aligned} (u, u') &\geqslant (u \wedge v, u' \wedge v') \\ &= (u, v') \wedge (v, u') \\ &= (u, u') * (v, v') \wedge (v, v') * (u, u'), \end{aligned}$$

and condition (3) of Proposition 8.12.1 holds. ■

Other examples are obtained if $(L, \vee, \wedge)$ is any lattice, $(M, \wedge) = (L, \wedge)$ and $(M, *) = (M, \vee)$ or $(M, *) = (L, \wedge)$. In this case every $*$-morphism is a pure or very pure fuzzy subsemigroup, respectively.

Corollary 8.12.3 *Let $B = B^+ \cup \{1\}$ and $S = \{xp \in B^+ \mid x \in B^*\}$, where p is a word whose right factors are not left factors. Moreover, let L be any lattice and $(L, *)$ the right zero semigroup on L. Then $(L, *, \wedge)$ is a $*$-semilattice and every $*$-morphism from S into $(L, *, \wedge)$ is a very pure left unitary fuzzy subsemigroup.*

Proof. For all $u, v \in L$, $v * u = u$. Thus $u \geqslant u * v \wedge u = u * v \wedge v * u$ and $u \geqslant u \wedge v = v * u \wedge v$. We show that S is very pure and left unitary. Assume that $xy \in S$ and $yx \in S$, i.e., $xy = ap$ and $yx = bp$ for suitable $a, b \in B^*$. If $x \notin S$, then p is not a right divisor of x. Hence from the equality $yx = bp$, it follows that there exist $y_1, y_2 \in B^+$ such that $y = y_1 y_2$ and $p = y_2 x$. Thus y_2 is a left factor of p and that $|p| > |y_2|$. Hence from $xy = xy_1y_2 = ap$, it follows that y_2 is also a right factor of p, a contradiction. This proves that S is very pure. Similarly, it follows that S is left unitary.

It now suffices to apply condition (3) and condition (4) of Proposition 8.12.1. ■

The fuzzy subsemigroup g of Corollary 8.12.3 can be defined by starting with a function α from the code $C = S \backslash S^2 = \{xp \mid x \in B^*$ and p does not occur in $x\}$ in a semilattice L. g can be precisely computed by setting

$g(x) = 0$ if p is not a right factor of x and $g(x) = \alpha(c_n)$ if $x = c_1 \dots c_n$ for suitable $c_1, \dots, c_n \in C$.

The class of $*$-semilattices ($*$-semialttices satisfying (1), (2), (3), (4), (5), or (6), respectively) is equational.

This allows for the construction of free $*$-semilattice of the desired type. Also, direct products of previously constructed $*$-semilattices can be used to obtain new suitable $*$-semilattices.

9

Generalized State Machines

9.1 T-generalized State Machines

For a finite state machine (Q, X, σ), the transition function $\sigma : Q \times X \to Q$ can be regarded as a fuzzy subset τ of $Q \times X \times Q$ defined by $\tau(p, a, q) = 1$ if $\sigma(p, a) = q$ and $\tau(p, a, q) = 0$ otherwise, and $\sum_{q \in Q} \tau(p, a, q) \leqslant 1$ for all $p \in Q$ and $a \in X$. Conversely, for a triple (Q, X, τ) with a fuzzy subset τ of $Q \times X \times Q$ such that $\tau(Q \times X \times Q) \subseteq \{0, 1\}$ and $\sum_{q \in Q} \tau(p, a, q) \leqslant 1$ for all $p \in Q$ and $a \in X$, τ can be regarded as a partial function $\sigma : Q \times X \to Q$ defined by $\sigma(p, a) = q$ if $\tau(p, a, q) = 1$.

Definition 9.1.1 *A triple $M = (Q, X, \tau)$, where Q and X are finite nonempty sets and τ is a fuzzy subset of $Q \times X \times Q$ is called a* ***generalized state machine*** *if $\sum_{q \in Q} \tau(p, a, q) \leqslant 1$ for all $p \in Q$ and $a \in X$. If $\sum_{q \in Q} \tau(p, a, q) = 1$ for all $p \in Q$ and $a \in X$, then M is said to be* ***complete****.*

We present the approach of [48] for the fuzzification of finite state machines. This approach differs from the one introduced in [86]. The latter approach is presented in detail in [85, 100]. The opinion given in [48] is that the concept of a generalized state machine is a more adequate fuzzification of a finite state machine than the one given in [86]. The reasoning in [48] is as follows: While a generalized state machine (Q, X, τ) with $\tau(Q \times X \times Q) \subseteq \{0, 1\}$ can always be regarded as a deterministic finite state machine (with a partial function for a transition function), a fuzzy finite state machine (Q, X, τ) as defined in [86] with $\tau(Q \times X \times Q) \subseteq \{0, 1\}$ cannot be regarded as a deterministic finite state machine generally. We examine this approach a little more deeply in Sections 9.6 - 9.10. However, it should be recalled that a nondeterministic finite-state automaton is equivalent to a deterministic one in the sense of language recognition.

Let $M = (Q, X, \tau)$ be a generalized state machine. Then Q is called the set of states and X is called the set of input symbols. Let X^+ denote the set of all words of elements X of (positive) finite length.

Every incomplete generalized state machine can be extended to a complete generalized state machine as can be seen from the following definition.

Definition 9.1.2 *Let $M = (Q, X, \tau)$ be an incomplete generalized state machine. Let z be a state not in Q. The completion M^c of M is the complete generalized state machine (Q', X, τ') given by $Q' = Q \cup \{z\}$ and*

$$\tau'(p', a, q') = \begin{cases} \tau(p', a, q') & \text{if } p', q' \in Q, \\ 1 - \sum_{q\in Q} \tau(p', a, q) & \text{if } p' \in Q \text{ and } q' = z, \\ 0 & \text{if } p' = z \text{ and } q' \in Q, \\ 1 & \text{if } p' = z \text{ and } q' = z \end{cases}$$

for all $a \in X$. The new state z is called the ***sink state*** *of M^c. If M is complete, then we take M itself as M^c.*

Definition 9.1.3 (*Schweizer and Sklar* [129]). *A binary operation T on $[0,1]$ is called a t-norm if*
(1) $T(a, 1) = a$,
(2) $T(a, b) \leqslant T(a, c)$ whenever $b \leqslant c$,
(3) $T(a, b) = T(b, a)$,
(4) $T(a, T(b, c)) = T(T(a, b), c)$
for all $a, b, c \in [0,1]$.

For $\vee$ representing maximum, T is clearly $\vee$-distributive, i.e., $T(a \vee b, c) = T(a, c) \vee T(b, c)$ for all $a, b, c \in [0,1]$. Define T_0 on $[0,1]$ by $T_0(a, 1) = a = T_0(1, a)$ and $T_0(a, b) = 0$ if $a \neq 1$ and $b \neq 1$ for all $a, b \in [0,1]$. Then $\wedge$ is the greatest t-norm on $[0,1]$ and T_0 is the least t-norm on $[0,1]$, i.e., for any t-norm T, $\wedge(a, b) \geqslant T(a, b) \geq T_0(a, b)$ for all $a, b \in [0,1]$.

T will always mean a t-norm on $[0,1]$. Since T is associative, we denote $T(a_1, T(a_2, T(\ldots, T(a_{n-1}, a_n)\ldots)))$ by $T(a_1, \ldots, a_n)$, where $a_1, \ldots, a_n \in [0,1]$.

Definition 9.1.4 *Let $M = (Q, X, \tau)$ be a generalized state machine. Define $\tau^+ : Q \times X^+ \times Q \to [0,1]$ by*

$$\begin{aligned} \tau^+(p, a_1 \ldots a_n, q) &= \vee\{T(\tau(p, a_1, r_1), \tau(r_1, a_2, r_2), \ldots, \\ &\quad \tau(r_{n-2}, a_{n-1}, r_{n-1}), \tau(r_{n-1}, a_n, q)) \\ &\quad \mid r_i \in Q,\ i = 1, \ldots, n-1\}, \end{aligned}$$

where $p, q \in Q$ and $a_1, \ldots, a_n \in X$. When T is applied to M as above, M is called a ***T-generalized state machine.***

Unless otherwise specified, a generalized state machine will always be written as a T-generalized state machine since a generalized state machine always induces a T-generalized state machine as in Definition 9.1.4.

Proposition 9.1.5 *Let* (Q, X, τ) *be a* T*-generalized state machine. Then*

$$\tau^+(p, xy, q) = \vee\{T(\tau^+(p, x, r), \tau^+(r, y, q)) \mid r \in Q\}$$

for all $p, q \in Q$ *and* $x, y \in X^+$.

Proof. Let $p, q \in Q$. Let $x = a_1 \ldots a_n$ and $y = b_1 \ldots b_m$ with $a_1, \ldots, a_n$, $b_1, \ldots, b_m \in X$. Then $\vee\{T(\tau^+(p, x, r), \tau^+(r, y, q)) \mid r \in Q\} = \vee\{T(\tau^+(p, a_1 \ldots a_n, r), \tau^+(r, b_1 \ldots b_m, q)) \mid r \in Q\} = \vee\{T(\vee\{T(\tau(p, a_1, q), \ldots, \tau(q_{n-1}, a_n, r)) \mid q_1, \ldots, q_{n-1} \in Q\}, \vee\{T(\tau(r, b_1, q_n), \ldots, \tau(q_{n+m-1}, b_m, q)) \mid q_n, \ldots, q_{n+m-1} \in Q\}) \mid r \in Q\} = \vee\{T(\tau(p, a_1, q_1), \ldots, \tau(q_{n-1}, a_n, r), \tau(r, b_1, q_n), \ldots, \tau(q_{n+m}, b_m, q)) \mid q_1, \ldots, q_{n+m-1}, r \in Q\} = \tau^+(p, a_1 \ldots a_n b_1 \ldots b_m, q) = \tau^+(p, xy, q)$. ∎

For a T-generalized state machine, define the relation $\equiv$ on X^+ by $\forall x, y \in X^+$, $x \equiv y$ if $\tau^+(p, x, q) = \tau^+(p, y, q)$ for all $p, q \in Q$.

Lemma 9.1.6 *Let* (Q, X, τ) *be a* T*-generalized state machine. Then* $\equiv$ *is a congruence relation on* X^+.

Proof. Clearly, $\equiv$ is an equivalence relation on X^+. Let $z \in X^+$ and $x \equiv y$. Then for all $p, q \in Q$, $\tau^+(p, xz, q) = \vee\{T(\tau^+(p, x, r), \tau^+(r, z, q)) \mid r \in Q\} = \vee\{T(\tau^+(p, y, r), \tau^+(r, z, q)) \mid r \in Q\} = \tau^+(p, yz, q)$ by Proposition 9.1.5. Hence $xz \equiv yz$. Similarly, $zx \equiv zy$. Thus $\equiv$ is a congruence relation on X^+. ∎

Given a T-generalized state machine $M = (Q, X, \tau)$, let $[x] = \{y \in X^+ \mid x \equiv y\}$, where $x \in X^+$ and let $X^+/\equiv\ = \{[x] \mid x \in X^+\}$ be denoted by $S(M)$.

Theorem 9.1.7 *Let* $M = (Q, X, \tau)$ *be a* T*-generalized state machine. Then* $S(M)$ *is a semigroup, where the binary operation on* $S(M)$ *is defined by* $[x][y] = [xy]$.

Proof. Clearly, the operation is well-defined since $\equiv$ is a congruence relation by Lemma 9.1.6. Since concatenation is associative, it follows that the operation is associative. Thus $S(M)$ is a semigroup. ∎

9.2 T-generalized Transformation Semigroups

We now generalize the concept of a transformation semigroup.

Definition 9.2.1 *A* T*-generalized state machine* (Q, S, ρ) *is called a* ***T-generalized transformation semigroup*** *if* S *is a finite semigroup and the following conditions hold:*

(1) $\rho(p, uv, q) = \vee\{T(\rho(p, u, r), \rho(r, v, q)) \mid r \in Q\}$ *for all* $p, q \in Q$ *and* $u, v \in S$.

(2) For $u, v \in S$, *if* $\rho(p, u, q) = \rho(p, v, q)$ *for all* $p, q \in Q$, *then* $u = v$.

When a T-generalized transformation semigroup $\mathcal{S} = (Q, S, \rho)$ is regarded as a T-generalized state machine (Q, S, τ_p) by taking $\tau_p = \tau_p^+ = \rho$, we denote it by $SM(\mathcal{S})$.

Definition 9.2.2 *A* T*-generalized transformation semigroup* (Q, S, ρ) *is called a* T***-generalized transformation monoid*** *if* S *is a monoid with identity* e *such that the following condition holds:*

$$\rho(p, e, q) = \begin{cases} 1 & \text{if } p = q, \\ 0 & \text{if } p \neq q. \end{cases}$$

We now consider T-generalized transformation semigroups induced by T-generalized state machines.

Proposition 9.1.5 and Theorem 9.1.7 seem to suggest that a T-generalized state machine $M = (Q, X, \tau)$ naturally induces a T-generalized transformation semigroup $(Q, S(M), \rho_\tau)$, where ρ_τ is defined by $\rho_\tau(p, [x], q) = \tau^+(p, x, q)$. However, there are two obstructions. One of them is that $S(M)$ is not finite in general and the other is that ρ_τ, i.e., τ^+, does not satisfy the condition of Definition 9.1.1. These obstructions are illustrated in the following examples.

Example 9.2.3 *Let* T *be the ordinary product t-norm on* $[0, 1]$, *i.e.*, $T(a, b) = ab\ \forall a, b \in [0, 1]$. *Let* $Q = \{s, t\}$ *and* $X = \{a\}$. *Define a fuzzy subset* τ *of* $Q \times X \times Q$ *by* $\tau(p, a, q) = \frac{1}{2}$ *for all* $p, q \in Q$. *Then* $M = (Q, X, \tau)$ *is a* T*-generalized state machine. However,* $S(M)$ *is not finite.*

Example 9.2.4 *Let* $T = \wedge$, $Q = \{p_1, p_2, \dots, p_6\}$ *and* $X = \{a\}$. *Let* $\tau(p_1, a, p_2) = 0.7$, $\tau(p_1, a, p_3) = 0.3$ *and* $\tau(p_1, a, p_n) = 0$, *where* $n = 1, 4, 5, 6$. *Let* $\tau(p_2, a, p_4) = 0.6$, $\tau(p_2, a, p_5) = 0.4$ *and* $\tau(p_2, a, p_m) = 0$, *where* $m = 1, 2, 3, 6$. *Let* $\tau(p_3, a, p_6) = 0.1$ *and* $\tau(p_3, a, p_1) = 0$, *where* $l = 1, 2, 3, 4, 5$. *Then* (Q, X, τ) *is a* T*-generalized state machine. However,* $\sum_{q \in Q} \tau^+(p_1, aa, q) = \tau^+(p_1, aa, p_4) + \tau^+(p_1, aa, p_5) + \tau^+(p_1, aa, p_6) = .6 + 0.4 + 0.1 > 1$.

For the former obstruction, since Q is finite, $S(M)$ is finite if and only if the image of τ^+ is finite. Hence for example, if T is $\wedge$, T_0 or T_m, where T_m is defined by $T_m(a, b) = \vee\{a + b - 1, 0\}$, then $S(M)$ is finite for every T-generalized state machine M. For the latter obstruction, if T_1 and T_2 are t-norms with $T_1 \geqslant T_2$ and ρ_τ satisfies the condition under T_1, then ρ_τ satisfies the condition under T_2 by the definition of τ^+. Hence the two obstructions seem to depend on t-norms and not on generalized state machines. This leads us to the following definition.

Definition 9.2.5 *Let T be a t-norm. Then T is said to be T-**generalized transformation semigroup inducible** if $S(M)$ is finite and*

$$\sum_{q\in Q} \tau^+(p,x,q) \le 1$$

for all $p \in Q$ and $x \in X^+$ for every T-generalized state machine $M = (Q,X,\tau)$.

Let T be T-generalized transformation semigroup inducible. Then a T-generalized state machine $M = (Q,X,\tau)$ naturally induces a T-generalized transformation semigroup $(Q,S(M),\rho_\tau)$, where ρ_τ is defined by

$$\rho_\tau(p,[x],q) = \tau^+(p,x,q),$$

by Proposition 9.1.5 and Theorem 9.1.7. We call $(Q,S(M),\rho_\tau)$ the **T-generalized transformation semigroup induced by** M and denote it by $TS(M)$.

Proposition 9.2.6 *There exists a T-generalized transformation semigroup inducible t-norm T.*

Proof. Let $T = T_0$ and let $M = (Q,X,\tau)$ be a T-generalized state machine. Then $S(M)$ is finite since the image of τ^+ is clearly finite. Let $p \in Q$ and $a_1, a_2 \in X$. If there exists $r' \in Q$ such that $\tau(p,a_1,r') = 1$, then $\tau(p,a_1,r) = 0$ for all $r \in Q\backslash\{r'\}$, and so

$$\begin{aligned}
\textstyle\sum_{q\in Q} \tau^+(p,a_1a_2,q) &= \textstyle\sum_{q\in Q} \vee\{T(\tau(p,a_1,r),\tau(r,a_2,q)) \mid r \in Q\} \\
&= \textstyle\sum_{q\in Q} T(\tau(p,a_1,r'),\tau(r',a_2,q)) \\
&= \textstyle\sum_{q\in Q} \tau(r',a_2,q) \\
&\leqslant 1
\end{aligned}$$

since M is a T-generalized state machine. Thus suppose that $\tau(p,a_1,r) < 1$ for all $r \in Q$. Let $Q' = \{q \in Q \mid \tau(r,a_2,q) = 1 \text{ for some } r \in Q\}$ and

$$Q_q = \{r \in Q \mid \tau(r,a_2,q) = 1\}$$

in what follows. Then, since $T = T_0$,

$$\begin{aligned}
\textstyle\sum_{q\in Q} \tau^+(p,a_1a_2,q) &= \textstyle\sum_{q\in Q} \vee\{T(\tau(p,a_1,r),\tau(r,a_2,q)) \mid r \in Q\} \\
&= \textstyle\sum_{q\in Q} \vee\{T(\tau(p,a_1,r),\tau(r,a_2,q)) \mid r \in Q_q\}, \\
&= \textstyle\sum_{q\in Q'} \vee\{\tau(p,a_1,r) \mid r \in Q_q\} \\
&= \textstyle\sum_{q\in Q'} \tau(p,a_1,r_q), \text{ where } r_q \in Q_q \\
&\qquad \text{and } \tau(p,a_1,r_q) \geqslant \tau(p,a_1,r) \text{ for all } r \in Q_q \\
&\leqslant \textstyle\sum_{q\in Q} \tau(p,a_1,q) \text{ since the } Q_q \text{ are disjoint} \\
&\leqslant 1 \text{ since } M \text{ is a } T\text{-generalized state machine.}
\end{aligned}$$

Thus $\sum_{q\in Q} \tau^+(p,a_1a_2,q) \leqslant 1$. Hence by Proposition 9.1.5, it can be shown by induction on the length of x that $\sum_{q\in Q} \tau^+(p,x,q) \leqslant 1$ for all $p \in Q$ and $x \in X^+$. Therefore, T_0 is T-generalized transformation semigroup inducible. ∎

9.3 Coverings

In Sections 9.3-9.5, we always assume that T is a T-generalized transformation semigroup inducible whenever we deal with T-generalized transformation semigroups induced by T-generalized state machines.

Let ξ be a function from a set X_1 into a set X_2. Let ξ^+ be the function of X_1^+ into X_2^+ defined by

$$\xi^+(x_1 \dots x_n) = \xi(x_1) \dots \xi(x_n)$$

$\forall x_i \in X_1$, $i = 1, 2, \dots, n$ and $\forall n \in \mathbb{N}$.

Definition 9.3.1 *Let $M_1 = (Q_1, X_1, \tau_1)$ and $M_2 = (Q_2, X_2, \tau_2)$ be T-generalized state machines. If $\xi : X_1 \to X_2$ is a function and $\eta : Q_2 \to Q_1$ is a surjective partial function such that $\tau_1^+(\eta(p), x, \eta(q)) \leqslant \tau_2^+(p, \xi^+(x), q)$ for all p, q in the domain of η and $x \in X_1^+$, then we say that (η, ξ) is a **covering** of M_1 and M_2 and that M_2 covers M_1, written $M_1 \leqslant M_2$. Moreover, if the inequality is equality whenever the left-hand side of the inequality is not zero (respectively, the inequality always is equality), then we say that (η, ξ) is a **strong covering** (respectively, a **complete covering**) of M_1 and M_2 and that M_2 **strongly covers** (respectively, **completely covers**) M_1 and we write $M_1 \leqslant_s M_2$ (respectively, $M_1 \leqslant_c M_2$).*

Example 9.3.2 *Let $M = (Q, X, \tau)$ be a T-generalized state machine. Define an equivalence relation $\sim$ on X by $\forall a, b \in X$, $a \sim b$ if and only if $\tau(p, a, q) = \tau(p, b, q)$ for all $p, q \in Q$. Form the T-generalized state machine $M_1 = (Q, X/\sim, \tau^\sim)$ by defining $\tau^\sim(p, [a], q) = \tau(p, a, q)$ $\forall p, q \in Q$, $a \in X$. Now define $\xi : X \to X/\sim$ by $\xi(a) = [a]$ $\forall a \in X$ and let η be the identity map on Q. Then (η, ξ) is a complete covering of M by M_1.*

Definition 9.3.3 *Let $\mathcal{S}_1 = (Q_1, S_1, \rho_1)$, $\mathcal{S}_2 = (Q_2, S_2, \rho_2)$ be T-generalized transformation semigroups. If $\eta : Q_2 \to Q_1$ is a surjective partial function and for each $s \in S_1$ there exists $t_s \in S_2$ such that $\rho_1(\eta(p), s, \eta(q)) \leqslant \rho_2(p, t_s, q)$ for all p, q in the domain of η, then we say that η is a **covering** of $\mathcal{S}_1$ by $\mathcal{S}_2$ and that $\mathcal{S}_2$ covers $\mathcal{S}_1$ and we write $\mathcal{S}_1 \leqslant \mathcal{S}_2$. Moreover, if the inequality is equality whenever the left-hand side of the inequality is not zero (respectively, the inequality always is equality), then we say that η is a **strong covering** (respectively, a **complete covering**) of $\mathcal{S}_1$ by $\mathcal{S}_2$ and that $\mathcal{S}_2$ **strongly covers** (respectively, **completely covers**) $\mathcal{S}_1$ and we write $\mathcal{S}_1 \leqslant_s \mathcal{S}_2$ (respectively, $\mathcal{S}_1 \leqslant_c \mathcal{S}_2$).*

Proposition 9.3.4 *(1) Let M_1, M_2 and M_3 be T-generalized state machines. If $M_1 \leqslant M_2$ (resp. $M_1 \leqslant_s M_2, M_1 \leqslant_c M_2$) and $M_2 \leqslant M_3$ (resp. $M_2 \leqslant_s M_3, M_2 \leqslant_c M_3$), then $M_1 \leqslant M_3$ (resp. $M_1 \leqslant_s M_3, M_1 \leqslant_c M_3$).*

(2) Let $\mathcal{S}_1, \mathcal{S}_2$ and $\mathcal{S}_3$ be T-generalized transformation semigroups. If $\mathcal{S}_1 \leqslant \mathcal{S}_2$ (resp. $\mathcal{S}_1 \leqslant_s \mathcal{S}_2, \mathcal{S}_1 \leqslant_c \mathcal{S}_2$) and $\mathcal{S}_2 \leqslant \mathcal{S}_3$ (resp. $\mathcal{S}_2 \leqslant_s \mathcal{S}_3, \mathcal{S}_2 \leqslant_c \mathcal{S}_3$), then $\mathcal{S}_1 \leqslant \mathcal{S}_3$ (resp. $\mathcal{S}_1 \leqslant_s \mathcal{S}_3, \mathcal{S}_1 \leqslant_c \mathcal{S}_3$).

Proof. The proof is straightforward. ∎

Theorem 9.3.5 *Let $M_1 = (Q_1, X_1, \tau_1)$ and $M_2 = (Q_2, X_2, \tau_2)$ be T-generalized state machines such that $M_1 \leqslant M_2$ with the covering (η, ξ). Then $TS(M_1) \leqslant TS(M_2)$. Moreover, if $M_1 \leqslant_c M_2$ and η is a function, then $TS(M_1) \leqslant_c TS(M_2)$.*

Proof. Let $a_1, \ldots, a_n \in X_1$. Let Q denote the domain of η. Since η is surjective, $\rho_{\tau_1}(\eta(p), [a_1 \ldots a_n], \eta(q)) = \tau_1^+(\eta(p), a_1 \ldots a_n, \eta(q)) = \vee\{T(\tau_1(\eta(p), a_1, r_1'), \tau_1(r_1', a_2, r_2'), \ldots, \tau_1(r_{n-1}', a_n, \eta(q))) \mid r_i' \in Q_1\} \leqslant \vee\{T(\tau_2(p, \xi(a_1), r_1), \tau_2(r_1, \xi(a_2), r_2), \ldots, \tau_2(r_{n-1}, \xi(a_n), q)) \mid r_i \in Q\} \leqslant \vee\{T(\tau_2(p, \xi(a_1), r_1), \tau_2(r_1, \xi(a_2), r_2), \ldots, \tau_2(r_{n-1}, \xi(a_n), q)) \mid r_i \in Q_2\} = \tau_2^+(p, \xi(a_1) \ldots \xi(a_n), q) = \tau_2^+(p, \xi^+(a_1 \ldots a_n), q) = \rho_{\tau_2}(p, [\xi^+(a_1 \ldots a_n)], q)$ for all p, q in the domain of η. Thus η is a covering of $TS(M_1)$ by $TS(M_2)$. Now let $M_1 \leqslant_c M_2$ and η be a function. Then the first inequality in the first part of the proof becomes equality since the domain of η is Q_2. ∎

9.4 Direct Products

In this section, we consider restricted direct products and full direct products of T-generalized state machines and T-generalized transformation semigroup, where T is less than or equal to the ordinary product.

Definition 9.4.1 *Let $M_1 = (Q_1, X, \tau_1)$ and $M_2 = (Q_2, X, \tau_2)$ be T-generalized state machines. The* **restricted direct product** *$M_1 \wedge_T M_2$ of M_1 and M_2 is defined to be the T-generalized state machine $(Q_1 \times Q_2, X, \tau_1 \wedge_T \tau_2)$, where*

$$(\tau_1 \wedge_T \tau_2)((p_1, p_2), a, (q_1, q_2)) = T(\tau_1(p_1, a, q_1), \tau_2(p_2, a, q_2))$$

for all $p_1, q_1 \in Q_1$, $p_2, q_2 \in Q_2$, and $a \in X$.

In Definition 9.4.1, $(Q_1 \times Q_2, X, \tau_1 \wedge_T \tau_2)$ is a T-generalized state machine since

$$\begin{aligned} & \textstyle\sum_{(q_1,q_2)\in Q_1\times Q_2} (\tau_1 \wedge_T \tau_2)((p_1, p_2), a, (q_1, q_2)) \\ = & \textstyle\sum_{(q_1,q_2)\in Q_1\times Q_2} T(\tau_1(p_1, a, q_1), \tau_2(p_2, a, q_2)) \\ \leqslant & \textstyle\sum_{(q_1,q_2)\in Q_1\times Q_2} \tau_1(p_1, a, q_1)\tau_2(p_2, a, q_2) \\ = & (\textstyle\sum_{q_1\in Q_1} \tau_1(p_1, a, q_1))(\sum_{q_2\in Q_2} \tau_2(p_2, a, q_2)) \\ \leqslant & 1 \end{aligned}$$

for all $p_1 \in Q_1$, $p_2 \in Q_2$ and $a \in X$.

Lemma 9.4.2 *Let $M_1 = (Q_1, X, \tau_1)$ and $M_2 = (Q_2, X, \tau_2)$ be T-generalized state machines. Then*

$$(\tau_1 \wedge_r \tau_2)^+((p_1, p_2), x, (q_1, q_2)) = T(\tau_1^+(p_1, x, q_1), \tau_2^+(p_2, x, q_2))$$

for all $p_1, q_1 \in Q_1, p_2, q_2 \in Q_2$ and $x \in X^+$.

Proof. Let $a_1, \ldots, a_n \in X$. Then $(\tau_1 \wedge_T \tau_2)^+((p_1,p_2),\, a_1 \ldots a_n,\, (q_1,\, q_2)) = \vee\{T((\tau_1 \wedge_T \tau_2)((p_1,p_2),\, a_1,\, (r_{11},\, r_{12})),\, (\tau_1 \wedge_T \tau_2)((r_{11},\, r_{12}),\, a_2,\, (r_{21},\, r_{22})),\, \ldots,\, (\tau_1 \wedge_T \tau_2)((r_{(n-1)1},\, r_{(n-1)2}),\, a_n,\, (q_1,\, q_2))) \mid (r_{i1}, r_{i2}) \in Q_1 \times Q_2\} = \vee\{T(T(\tau_1(p_1,\, a,\, r_{11}),\, \tau_2(p_2,\, a_1,\, r_{12})),\, T(\tau_1(r_{11},\, a_2,\, r_{21}),\, \tau_2(r_{12},\, a_2,\, r_{22})), \ldots,\, T(\tau_1(r_{(n-1)1},\, a_n,\, q_1),\, \tau_2(r_{(n-q)2},\, a_n,\, q_2))) \mid r_{i1} \in Q_1,\, r_{i2} \in Q_2\} = T(\vee\{T(\tau_1(p_1,\, a,\, r_{11}),\, \tau_1(r_{11},\, a_2,\, r_{21}), \ldots,\, \tau_1(r_{(n-1)1},\, a_n,\, q_1)) \mid r_{i1} \in Q_1\},\, \vee\{T(\tau_2(p_2,\, a,\, r_{12}),\, \tau_2(r_{12},\, a_2,\, r_{22}),\, \ldots,\, \tau_2(r_{(n-1)2},\, a_n,\, q_2)) \mid r_{i2} \in Q_2\}) = T(\tau_1^+(p_1,\, a_1 \ldots a_n,\, q_1),\, \tau_2^+(p_2,\, a_1 \ldots a_n,\, q_2))$ for all $p_1,\, q_1 \in Q_1$ and $p_2,\, q_2 \in Q_2$. ■

Lemma 9.4.3 *Let $M_1 = (Q_1, X, \tau_1)$, $M_2 = (Q_2, X, \tau_2)$ be T-generalized state machines. Let $TS(M_1) = (Q_1, X^+/R_1, \rho_1)$, $TS(M_2) = (Q_2, X^+/R_2, \rho_2)$ and $TS(M_1 \wedge_T M_2) = (Q_1 \times Q_2,\, X^+/R,\, \rho)$. Then the following conditions hold:*

(1) $R_1 \cap R_2 \subseteq R$.

(2) Suppose that for all $x, y \in X^+$, there exist $p_1, q_1 \in Q_1$ and $p_2, q_2 \in Q_2$ such that

$$\tau_1^+(p_1, x, q_1) = \tau_1^+(p_1, y, q_1) = \tau_2^+(p_2, x, q_2) = \tau_2^+(p_2, y, q_2) = 1.$$

Then $R_1 \cap R_2 = R$.

Proof. (1) Let $[x]_{R_1 \cap R_2} = [y]_{R_1 \cap R_2}$. Then $[x]_{R_1} = [y]_{R_1}$ and $[x]_{R_2} = [y]_{R_2}$. Thus

$$\begin{aligned} & (\tau_1 \wedge \tau_2)^+\left((p_1,p_2), x, (q_1,q_2)\right) \\ = \;& T(\tau_1^+(p_1,x,q_1), \tau_2^+(p_2,x,q_2)) \text{ by Lemma 9.4.2} \\ = \;& T(\tau_1^+(p_1,y,q_1), \tau_2^+(p_2,y,q_2)) \\ = \;& (\tau_1 \wedge_T \tau_2)^+((p_1,p_2), y, (q_1,q_2)) \text{ by Lemma 9.4.2} \end{aligned}$$

for all $p_1, q_1 \in Q_1$ and $p_2, q_2 \in Q_2$. Hence $[x]_R = [y]_R$.

(2) Let $[x]_R = [y]_R$. Let $p_2, q_2 \in Q_2$ be such that $\tau_2^+(p_2, x, q_2) = \tau_2^+(p_2, y, q_2) = 1$. Then

$$\begin{aligned} & \tau_1^+(p_1,x,q_1) \\ = \;& T(\tau_1^+(p_1,x,q_1), \tau_2^+(p_2,x,q_2)) \\ = \;& (\tau_1 \wedge_T \tau_2)^+\left((p_1,p_2), x, (q_1,q_2)\right) \text{ by Lemma 9.4.2} \\ = \;& (\tau_1 \wedge_T \tau_2)^+\left((p_1,p_2), y, (q_1,q_2)\right) \\ = \;& T(\tau_1^+(p_1,y,q_1), \tau_2^+(p_2,y,q_2)) \\ = \;& \tau_1^+(p_1,y,q_1) \end{aligned}$$

for all $p_1, q_1 \in Q_1$. Therefore, $[x]_{R_1} = [y]_{R_1}$. Similarly, we have $[x]_{R_2} = [y]_{R_2}$. Thus $[x]_{R_1 \cap R_2} = [y]_{R_1 \cap R_2}$. Hence $R \subseteq R_1 \cap R_2$. The conclusion now follows in view of (1). ■

With a little modification, we may assume that any two T-generalized state machines have the condition in Lemma 9.4.3(2) without change of their intrinsic properties. This can be accomplished by adding an extra state z_i to each Q_i and then constructing a T-generalized state machine $(Q_i \cup \{z_i\}, X, \tau_i')$ by defining

$$\tau_i'(p,a,q) = \begin{cases} \tau_i(p,a,q) & \text{if } p,q \in Q_i, \\ 1 & \text{if } p = z_i \text{ and } q = z_i, \\ 0 & \text{otherwise,} \end{cases}$$

for $i = 1, 2$.

Definition 9.4.4 *Let $\mathcal{S}_1 = (Q_1, S_1, \rho_1)$ and $\mathcal{S}_2 = (Q_2, S_2, \rho_2)$ be T-generalized transformation semigroups such that there exists a free semigroup F with epimorphisms $\theta_1 : F \to S_1$ and $\theta_2 : F \to S_2$ and, for all $x, y \in F$, there exists $p_1, q_1 \in Q_1$ and $p_2, q_2 \in Q_2$ such that*

$$\rho_1(p_1, \theta_1(x), q_1) = \rho_1(p_1, \theta_1(y), q_1) = \rho_2(p_2, \theta_2(x), q_2) = \rho_2(p_2, \theta_2(y), q_2) = 1.$$

The ***restricted direct product*** *$\mathcal{S}_1 \wedge_T \mathcal{S}_2$ of $\mathcal{S}_1$ and $\mathcal{S}_2$ (with respect to θ_1 and θ_2) is the T-generalized transformation semigroup $(Q_1 \times Q_2, S, \rho_1 \wedge_T \rho_2)$ with $S = F/(R_1 \cap R_2)$, where R_1 and R_2 are the equivalence relations on F defined by θ_1 and θ_2, respectively, and*

$$(\rho_1 \wedge_T \rho_2)((p_1,p_2),[x],(q_1,q_2)) = T(\rho_1(p_1,[x]_{R_1},q_1), \rho_2(p_2,[x]_{R_2},q_2))$$

$\forall p_1, q_1 \in Q_1$, $p_2, q_2 \in Q_2$, and $x \in X^+$.

The legitimacy of Definition 9.4.4 is established as follows: For all $x, y \in F, p_1, q_1 \in Q_1$ and $p_2, q_2 \in Q_2$, we have

$$\begin{aligned} & (\rho_1 \wedge_T \rho_2)((p_1,p_2),[x][y],(q_1,q_2)) \\ = \;& (\rho_1 \wedge_T \rho_2)((p_1,p_2),[xy],(q_1,q_2)) \\ = \;& T(\rho_1(p_1,[xy]_{R_1},q_1), \rho_2(p_2,[xy]_{R_2},q_2)) \\ = \;& T(\rho_1(p_1,[x]_{R_1},[y]_{R_1},q_1), \rho_2(p_2,[x]_{R_2},[y]_{R_2},q_2)) \\ = \;& T(\vee\{T(\rho_1(p_1,[x]_{R_1},r_1), \rho_1(r_1,[y]_{R_1},q_1)) \mid r_1 \in Q_1\}, \\ & \vee\{T(\rho_2(p_2,[x]_{R_2},r_2), \rho_2(r_2,[y]_{R_2},q_2)) \mid r_2 \in Q_2\}) \\ = \;& \vee\{T(\rho_1(p_1,[x]_{R_1},r_1), \rho_1(r_1,[y]_{R_1},q_1), \rho_2(p_2,[x]_{R_2},r_2), \\ & \rho_2(r_2,[y]_{R_2},q_2)) \mid r_1 \in Q_1, r_2 \in Q_2\}) \\ = \;& \vee\{T((\rho_1 \wedge_T \rho_2)((p_1,p_2),[x],(r_1,r_2)), \\ & (\rho_1 \wedge_T \rho_2)((r_1,r_2),[y],(q_1,q_2))) \mid (r_1,r_2) \in Q_1 \times Q_2\}. \end{aligned}$$

Let

$$(\rho_1 \wedge_T \rho_2)((p_1,p_2),[x],(q_1,q_2)) = (\rho_1 \wedge_T \rho_2)((p_1,p_2),[y],(q_1,q_2))$$

for all $p_1, q_1 \in Q_1$ and $p_2, q_2 \in Q_2$. Let $p_2, q_2 \in Q_2$ be such that

$$\rho_2(p_2, \theta_2(x), q_2) = \rho_2(p_2, \theta_2(y), q_2) = 1.$$

Then

$$\begin{aligned}
\rho_1(p_1, [x]_{R_1}, q_1) &= T(\rho_1(p_1, [x]_{R_1}, q_1), \rho_2(p_2, \theta_2(x), q_2)) \\
&= T(\rho_1(p_1, [x]_{R_1}, q_1), \rho_2(p_2, [x]_{R_2}, q_2)) \\
&= (\rho_1 \wedge_T \rho_2)((p_1, p_2), [x], (q_1, q_2)), \\
&\quad (\rho_1 \wedge_T \rho_2)((p_1, p_2), [y], (q_1, q_2)) \\
&= T(\rho_1(p_1, [y]_{R_1}, q_1), \rho_2(p_2, [y]_{R_1}, q_2)) \\
&= T(\rho_1(p_1, \theta_{1(y)}, q_1), \rho_2(p_2, \theta_2(y), q_2)) \\
&= \rho_1(p_1, [y]_{R_1}, q_1)
\end{aligned}$$

for all $p_1, q_1 \in Q_1$, and so $[x]_{R_1} = [y]_{R_1}$. Similarly, we have $[x]_{R_2} = [y]_{R_2}$. Hence $[x] = [y]$.

As in the paragraph following Lemma 9.4.3, we may assume that any two T-generalized transformation semigroups have the condition in Definition 9.4.4 that is quite similar to the assumption of Lemma 9.4.3(2). Note that if two T-generalized state machines have the condition in Lemma 9.4.3(2), then the T-generalized transformation semigroups induced by them clearly have the condition in Definition 9.4.4 just mentioned above, and vice versa. Unless otherwise specified, we assume that any two T-generalized state machines have the condition in Lemma 9.4.3(2) whenever we consider T-generalized transformation semigroups induced by T-generalized state machines.

The following theorem follows from Lemma 9.4.3.

Theorem 9.4.5 *Let M_1 and M_2 be T-generalized state machines. Then $TS(M_1 \wedge_T M_2) = TS(M_1) \wedge_T TS(M_2)$.*

Definition 9.4.6 *Let $M_1 = (Q_1, X_1, \tau_1)$ and $M_2 = (Q_2, X_2, \tau_2)$ be T-generalized state machines. The **full direct product** $M_1 \times_T M_2$ of M_1 and M_2 is the T-generalized state machine $(Q_1 \times Q_2,\ X_1 \times X_2,\ \tau_1 \times \tau_2)$, where*

$$(\tau_1 \times \tau_2)\left((p_1, p_2), (a, b), (q_1, q_2)\right) = T(\tau_1(p_1, a, q_1), \tau_2(p_2, b, q_2))$$

for all $p_1, q_1 \in Q_1$, $p_2, q_2 \in Q_1$, and $a \in X$.

The legitimacy of Definition 9.4.6 can be checked similarly to that of Definition 9.4.1

Lemma 9.4.7 *Let $M_1 = (Q_1, X_1, \tau_1)$, $M_2 = (Q_2, X_2, \tau_2)$ be T-generalized state machines. Then $(\tau_1 \times_T \tau_2)^+ ((p_1, p_2), (a_1 \ldots a_n, b_1 \ldots b_n), (q_1, q_2)) = T(\tau_1^+(p_1, a_1 \ldots a_n, q_1), \tau_2^+(p_2, b_1 \ldots b_n, q_2))$ for all $a_1, \ldots, a_n \in X_1$, $b_1, \ldots, b_n \in X_2$, p_1, $q_1 \in Q_1$ and $p_2, q_2 \in Q_2$.*

Proof. Let $a_1, \ldots, a_n \in X_1$ and $b_1, \ldots, b_n \in X_2$. Then

$$\begin{aligned}
& (\tau_1 \times \tau_2)^+ ((p_1, p_2), (a_1 \ldots a_n, b_1 \ldots b_n), (q_1, q_2)) \\
= \; & \vee\{T((\tau_1 \times_T \tau_2)((p_1, p_2), (a_1, b_1), (r_{11}, r_{12})), (\tau_1 \times_T \tau_2)(r_{11}, r_{12}), \\
& (a_2, b_2), (r_{21}, r_{22})), \ldots, (\tau \times \tau_2)((r_{(n-1)1}, r_{(n-1)2}), (a_n, b_n), (q_1, q_2))) \\
& \mid (r_{i1}, r_{i2}) \in Q_1 \times Q_2\} \\
= \; & \vee\{T(T(gt_1(p_1, a_1, r_{11}), \tau_2(p_2, b_1, r_{12})), T(\tau_1(r_{11}, a_2, r_{21}), \\
& \tau_2(r_{12}, b_2, r_{22})), \ldots, T(\tau_1(r_{(n-1)1}, a_n, q_1), \tau_2(r_{(n-1)2}, b_n, q_2))) \\
& \mid r_{i1} \in Q_1, r_{i2} \in Q_2\} \\
= \; & T(\vee\{T(\tau_1(p_1, a_1, r_{11}), \tau_1(r_{11}, a_2, r_{21}), \ldots, \tau_1(r_{(n-1)1}, a_n, q_1)) \\
& \mid r_{i1} \in Q_1\}, \; \vee\{T(\tau_2(p_2, b_1, r_{12}), \tau_2(r_{12}, b_2, r_{22}), \ldots, \\
& \tau_2(r_{(n-1)2}, b_n, q_2)) \mid r_{i2} \in Q_2\}) \\
= \; & T(\tau_1^+(p_1, a_1 \ldots a_n, q_1), \tau_2^+(p_2, b_1 \ldots b_n, q_2))
\end{aligned}$$

for all $p_1, q_1 \in Q_1$ and $p_2, q_2 \in Q_2$. ■

Definition 9.4.8 *Let $\mathcal{S}_1 = (Q_1, S_1, \rho_1)$, $\mathcal{S}_2 = (Q_2, S_2, \rho_2)$ be T-generalized transformation semigroups. The* ***full direct product*** *$\mathcal{S}_1 \times_T \mathcal{S}_2$ of $\mathcal{S}_1$ and $\mathcal{S}_2$ is defined to be the T-generalized transformation semigroup $(Q_1 \times Q_2, S_1 \times S_2, \rho_1 \times_T \rho_2)$, where*

$$(\rho_1 \times_T \rho_2)((p_1, p_2), (u, v), (q_1, q_2)) = T(\rho_1(p_1, u, q_1), \rho_2(p_2, v, q_2))$$

for all $p_1, q_1 \in Q_1$, $p_2, q_2 \in Q_1$, and $a \in X$.

The legitimacy of Definition 9.4.8 can be checked similarly to that of Definition 9.4.4.

Theorem 9.4.9 *Let $M_1 = (Q_1, X_1, \tau_1)$ and $M_2 = (Q_2, X_2, \tau_2)$ be T-generalized state machines. Then $TS(M_1 \times_T M_2) \leqslant_c TS(M_1) \times_T TS(M_2)$.*

Proof. Let $TS(M_1) = (Q_1, X_1^+/R_1, \rho_1), TS(M_2) = (Q_2, X_2^+/R_2, \rho_2)$ and $TS(M_1 \times_T M_2) = (Q_1 \times Q_2, (X_1 \times X_2)^+/R_3, \rho_3)$. Let $a_1, \ldots, a_n \in X_1$ and $b_1, \ldots, b_n \in X_2$. Then

$$\begin{aligned}
& \rho_3((p_1, p_2), [(a_1 \ldots a_n, b_1 \ldots b_n)]_{R_3}, (q_1, q_2)) \\
= \; & (\tau_1 \times_T \tau_2)^+((p_1, p_2), (a_1 \ldots a_n, b_1 \ldots b_n), (q_1, q_2)) \\
= \; & T(\tau_1^+(p_1, a_1 \ldots a_n, q_1), \tau_2^+(p_2, b_1 \ldots b_n, q_2)) \text{ by Lemma 9.4.7} \\
= \; & T(\rho_1(p_1, [a_1 \ldots a_n]_{R_1}, q_1), \rho_2(p_2, [b_1 \ldots b_n]_{R_2}, q_2)) \\
= \; & (\rho_1 \times_T \rho_2)((p_1, p_2), ([a_1 \ldots a_n]_{R_1}, [b_1 \ldots b_n]_{R_2}), (q_1, q_2))
\end{aligned}$$

for all $p_1, q_1 \in Q_1$ and $p_2, q_2 \in Q_2$. ■

Proposition 9.4.10 *Let $M_1 = (Q_1, X, \tau_1)$ and $M_2 = (Q_2, X, \tau_2)$ be T-generalized state machines. Then the following conditions hold:*

(1) $M_1 \wedge_T M_2 \leqslant_c M_1 \times_T M_2$.

(2) $TS(M_1 \wedge_T M_2) \leqslant_c TS(M_1 \times_T M_2)$.

Proof. (1) Let η be the identity map on $Q_1 \times Q_2$ and define $\xi : X \to X \times X$ by $\xi(a) = (a, a)\ \forall a \in X$. Then (η, ξ) is a complete covering of $M_1 \wedge_T M_2$ by $M_1 \times_T M_2$.

(2) The result here follows by applying (1) to Theorem 9.3.5. ∎

The following propositions are direct consequences of the associativity of a t-norm T.

Proposition 9.4.11 *Let M_1, M_2 and M_3 be T-generalized state machines. Then the following assertions hold:*

(1) $(M_1 \wedge_T M_2) \wedge_t M_3 = M_1 \wedge_T (M_2 \wedge_T M_3)$.

(2) $(M_1 \times_T M_2) \times_t M_3 = M_1 \times_T (M_2 \times_T M_3)$.

Proposition 9.4.12 *Let $\mathcal{S}_1, \mathcal{S}_2$ and $\mathcal{S}_3$ be T-generalized transformation semigroups. Then the following assertions hold:*

(1) $(\mathcal{S}_1 \wedge_T \mathcal{S}_2) \wedge_T \mathcal{S}_3 = \mathcal{S}_1 \wedge_T (\mathcal{S}_2 \wedge_T \mathcal{S}_3)$.

(2) $(\mathcal{S}_1 \times_T \mathcal{S}_2) \times_T \mathcal{S}_3 = \mathcal{S}_1 \times_T (\mathcal{S}_2 \times_T \mathcal{S}_3)$.

9.5 Decompositions of T-generalized Transformation Semigroups

Recall that if a T-generalized transformation semigroup $\mathcal{S} = (Q, S, \rho)$ is regarded as a T-generalized state machine (Q, S, τ_ρ) by taking $\tau_\rho = \tau_\rho^+ = \rho$, we write $SM(\mathcal{S})$.

We always assume that T is T-generalized transformation semigroup inducible (see Definition 9.2.5) whenever we consider T-generalized transformation semigroups induced by T-generalized state machines.

Let T be T-generalized transformation semigroup inducible. Then by Proposition 9.1.5 and Theorem 9.1.7, a T-generalized state machine $M = (Q, X, \tau)$ naturally induces a T-generalized transformation semigroup $(Q, S(M), \rho_\tau)$, where ρ_τ is defined by $\rho_\tau(p, [x], q) = \tau^+(p, x, q)$. Recall that we call $(Q, S(M), \rho_\tau)$ the T-generalized transformation semigroup induced by M and denote it by $TS(M)$.

In this section, we consider cascade products and wreath products of T-generalized state machines and T-generalized transformation semigroups, where T is less than or equal to the ordinary product. The results are from [15].

Definition 9.5.1 *Let $M_1 = (Q_1, X_1, \tau_1)$ and $M_2 = (Q_2, X_2, \tau_2)$ be T-generalized state machines. The* ***cascade product*** *$M_1 \omega_T M_2$ of M_1 and M_2 with respect to $\omega : Q_2 \times X_2 \to X_1$ is the T-generalized state machine $(Q_1 \times Q_2, X_2, \tau_1 \omega_T \tau_2)$, where*

$$(\tau_1 \omega_T \tau_2)((p_1, p_2), b, (q_1, q_2)) = T(\tau_1(p_1, \omega(p_2, b), q_1), \tau_2(p_2, b, q_2))$$

for all $p_1, q_1 \in Q_1$, $p_2, q_2 \in Q_1$, and $a \in X$.

In Definition 9.5.1, $(Q_1 \times Q_2, X_2, \tau_1\omega_T\tau_2)$ is clearly a T-generalized state machine. In fact,

$$\begin{aligned} & \textstyle\sum_{(q_1,q_2)\in Q_1\times Q_2}(\tau_1\omega_T\tau_2)((p_1,p_2),b,(q_1,q_2)) \\ = & \textstyle\sum_{(q_1,q_2)\in Q_1\times Q_2} T(\tau_1(p_1,\omega(p_2,b),q_1),\tau_2(p_2,b,q_2)) \\ \leqslant & \textstyle\sum_{(q_1,q_2)\in Q_1\times Q_2} \tau_1(p_1,\omega(p_2,b),q_1)\cdot\tau_2(p_2,b,q_2) \\ \leqslant & (\textstyle\sum_{q_1\in Q_1}\tau_1(p_1,\omega(p_2,b),q_1))(\textstyle\sum_{q_2\in Q_2}\tau_2(p_2,b,q_2)) \\ \leq & \ 1 \end{aligned}$$

for all $p_1 \in Q_1, p_2 \in Q_2$ and $b \in X_2$.

Let $M_1 = (Q_1, X_1, \tau_1)$ and $M_2 = (Q_2, X_2, \tau_2)$ be T-generalized state machines and $\omega : Q_2 \times X_2 \to X_1$. Define $\omega^+ : Q_2 \times X_2^+ \to X_1^+$ by

$$\omega^+(p_2, b_1b_2\dots b_n) = \omega(p_2,b_1)\omega(u_1,b_2)\dots\omega(u_{n-1},b_n),$$

where $p_2, u_1, u_2, \dots, u_{n-1} \in Q_2$ and $b_1, \dots, b_n \in X_2$.

Lemma 9.5.2 *Let $M_1 = (Q_1, X_1, \tau_1)$ and $M_2 = (Q_2, X_2, \tau_2)$ be T-generalized state machines. Then*

$$(\tau_1\omega_T\tau_2)^+((p_1,p_2),x,(q_1,q_2)) = T(\tau_1^+(p_1,\omega^+(p_2,x),q_1),\tau_2^+(p_2,x,q_2))$$

$\forall\ p_1, q_1 \in Q_1,\ p_2, q_2 \in Q_2$ and $x \in X_2^+$.

Proof. Let $b_1, \dots, b_n \in X$. Then $(\tau_1\omega_T\tau_2)^+((p_1,p_2), b_1\dots b_n, (q_1, q_2)) = \vee\{T((\tau_1\omega_T\tau_2)((p_1,\ p_2),\ b_1,\ (r_{11},\ r_{12})),\ (\tau_1\omega_T\tau_2)((r_{11},\ r_{12}),\ b_2,\ (r_{21},\ r_{22})),\ \dots,\ (\tau_1\omega_T\tau_2)((r_{(n-1)},\ r_{(n-1)2}),\ b_n,\ (q_1,\ q_2)))\ \mid (r_{i1},\ r_{i2}) \in Q_1 \times Q_2\} = \vee\{T(T(gt_1(p_1,\ \omega(p_2,b_1),\ r_{11}),\ \tau_2(p_2,\ b_1,\ r_{12})),\ T(\tau_1(r_{11},\ \omega(r_{12},\ b_2),\ r_{21}),\ \tau_2(r_{12},\ b_2,\ r_{22})),\ \dots,\ T(\tau_1(r_{(n-1)1},\ \omega(r_{(n-1)2},\ b_n),\ q_1),\ \tau_2(r_{(n-1)2},\ b_n,\ q_2)))\ \mid r_{i1} \in Q_1, r_{i2} \in Q_2\} = T(\vee\{\vee\{T(\tau_1(p_1,\ \omega(p_2,\ b_1),\ r_{11}),\ \tau_1(r_{11},\ \omega(r_{12},\ b_2),\ r_{21}),\dots,\ \tau_1(r_{(n-1)1},\ \omega(r_{(n-1)2},\ b_n),\ q_1))\mid r_{i1} \in Q_1\} \mid r_{i2} \in Q_2\},\ \vee\{T(\tau_2(p_2,\ b_1,\ r_{12}),\ \tau_2(r_{12},\ b_2,\ r_{22}),\ \dots,\ \tau_2(r_{(n-1)2},\ b_n,\ q_2))\mid r_{i2} \in Q_2\}) = T(\vee\{\tau_1^+(p_1,\ \omega(p_2,\ b_1)\omega(r_{12},\ b_2)\dots\omega(r_{(n-1)2},\ b_n),\ q_1) \mid r_{i2} \in Q_2\},\ \tau_2^+(p_2,\ b_1\dots b_n,\ q_2)) = T(\tau_1^+(p_1,\ \omega^+(p_2,\ b_1\dots b_n),\ q_1),\ \tau_2^+(p_2,\ b_1\dots b_n,\ q_2))$ for all $p_1, q_1 \in Q_1$ and $p_2, q_2 \in Q_2$. ∎

Unless otherwise specified, we assume that $\omega^+ : Q_2 \times X_2^+ \to X_1^+$ satisfies the following condition:

$$T(\rho_1(r_1,\omega^+(r_2,v),q_1),\rho_2(r_3,v,q_2)) = (\rho_1\omega\rho_2)((r_1,r_3),v,(q_1,q_2))$$

$\forall\ r_1, q_1 \in Q_1, r_2, r_3, q_2 \in Q_2, v \in X_2^+$.

Definition 9.5.3 *Let $\mathcal{S}_1 = (Q_1, S_1, \rho_1)$ and $\mathcal{S}_2 = (Q_2, S_2, \rho_2)$ be T-generalized transformation semigroups. The* **cascade product** *$\mathcal{S}_1\omega_T\mathcal{S}_2$ of $\mathcal{S}_1$ and $\mathcal{S}_2$ is defined to be the T-generalized transformation semigroup $(Q_1 \times Q_2,\ S_2,\ \rho_1\omega_T\rho_2)$ such that (1) $\forall\ p_1, q_1 \in Q_1,\ p_2, q_2 \in Q_2$ and $v \in S_2^+$,*

$$(\rho_1\omega_T\rho_2)((p_1,p_2),v,(q_1,q_2)) = T(\rho_1(p_1,\omega^+(p_2,v),q_1),\rho_2(p_2,v,q_2))$$

and (2) $\forall\ p_2 \in Q_2$ and $u, v \in S_2^+$, there exist $p_1, q_1 \in Q_1$ such that

$$\rho_1(p_1, \omega^+(p_2, u), q_1) = \rho_1(p_1, \omega^+(p_2, v), q_1) = 1.$$

The legitimacy of Definition 9.5.3 is established as follows: For all $u, v \in S_2$, $p_1, q_1 \in Q_1$ and $p_2, q_2 \in Q_2$, we have $(\rho_1 \omega_T \rho_2)((p_1, p_2), uv, (q_1, q_2)) = T(\rho_1(p_1, \omega^+(p_2, uv, q_1), \rho_2(p_2, uv, q_2)) = T(\vee\{\rho_1(p_1, \omega^+(p_2, u)\omega^+(r_2, v), q_1) \mid r_2 \in Q_2\}, \rho_2(p_2, uv, q_2)) = T(\vee\{\vee\{T(\rho_1(p_1, \omega^+(p_2, u), r_1), \rho_1(r_1, \omega^+(r_2, v), q_1)) \mid r_1 \in Q_1\} \mid r_2 \in Q_2\}, \vee\{T(\rho_2(p_2, u, r_3), \rho_2(r_3, v, q_2)) \mid r_3 \in Q_2\}) = \vee\{T(\rho 1(p_1, \omega^+(p_2, u), r_1), \rho_1(\omega^+(r_2, v), q_1)), T(\rho_2(p_2, u, r_3), \rho_2(r_3, v, q_2)) \mid r_1 \in Q_1, r_2, r_3 \in Q_2\} = \vee\{T(T(\rho_1(p_1, \omega^+(p_2, u), r_1), \rho_2(p_2, u, r_3)), T(\rho_1(r_1, \omega^+(r_2, v), q_1), \rho_2(r_3, v, q_3) \mid r_1 \in Q_1, r_2, r_3 \in Q_2\} = \vee\{T((\rho_1 \omega_T \rho_2)((p_1, p_2), u, (r_1, r_3)), (\rho_1 \omega_T \rho_2)((r_1, r_3), v, (q_1, q_2))) \mid (r_1, r_3) \in Q_1 \times Q_2\}$. Now let

$$(\rho_1 \omega_T \rho_2)((p_1, p_2), u, (q_1, q_2)) = (\rho_1 \omega_T \rho_2)((p_1, p_2), v, (q_1, q_2)).$$

Then

$$T(\rho_1(p_1, \omega^+(p_2, u), q_1), \rho_2(p_2, u, q_2)) = T(\rho_1(p_1, \omega^+(p_2, u), q_1), \rho_2(p_2, v, q_2)).$$

Let $p_1, q_1 \in Q_1$ be such that

$$\rho_1(p_1, \omega^+(p_2, u), q_1) = \rho_1(p_1, \omega^+(p_2, v)) = 1.$$

Then

$$\begin{aligned} \rho_2(p_2, u, q_2) &= T(\rho_1(\omega^+(p_2, u), q_1), \rho_2(p_2, u, q_2)) \\ &= (\rho_1 \omega_T \rho_2)((p_1, p_2), u, (q_1, q_2)) \\ &= (\rho_1 \omega_T \rho_2)((p_1, p_2), v, (q_1, q_2)) \\ &= T(\rho_1(p_1, \omega^+(p_2, v), q_1), \rho_2(p_2, v, q_2)) \\ &= \rho_2(p_2, v, q_2). \end{aligned}$$

Since S_2 is a T-generalized transformation semigroup, $u = v$. Therefore $(Q_1 \times Q_2, S_2, \rho_1 \omega_T \rho_2)$ is a T-generalized transformation semigroup.

Definition 9.5.4 *Let $M_1 = (Q_1, X_1, \tau_1)$ and $M_2 = (Q_2, X_2, \tau_2)$ be T-generalized state machines. The* ***wreath product*** *$M_1 \circ_T M_2$ of M_1 and M_2 is defined to be the T-generalized state machine $(Q_1 \times Q_2, X_1^{Q_2} \times X_2, \tau_1 \circ_T \tau_2)$, where*

$$(\tau_1 \circ_T \tau_2)((p_1, p_2), (f, b), (q_1, q_2)) = T(\tau_1(p_1, f(p_2), q_1), \tau_2(p_2, b, q_2))$$

for all $p_1, q_1 \in Q_1$, $p_2, q_2 \in Q_1$, $f \in X_1^{Q_2}$, and $b \in X_2$.

In Definition 9.5.4, $(Q_1 \times Q_2, X_1^{Q_2} \times X_2, \tau_1 \circ_T \tau_2)$ is clearly a T-generalized state machine. In fact,

$$\begin{aligned}
&\sum_{(q_1,q_2)\in Q_1\times Q_2} (\tau_1 \circ_T \tau l2)((p_1,p_2),(f,b),(q_1,q_2)) \\
= &\sum_{(q_1,q_2)\in Q_1\times Q_2} T(\tau_1(p_1,f(p_2),q_1),\tau_2(p_2,b,q_2)) \\
\leqslant &\sum_{(q_1,q_2)\in Q_1\times Q_2} \tau_1(p_1,f(p_2),q_1)\cdot\tau_2(p_2,b,q_2) \\
\leqslant &\ (\sum_{q_1\in Q_1} \tau_1(p_1,f(p_2,q_1)) (\sum_{q_2\in Q_2} \tau_2(p_2,b,q_2)) \\
\leqslant &\ 1
\end{aligned}$$

for all $p_1 \in Q_1, p_2 \in Q_2$ and $b \in X_2$.

Lemma 9.5.5 *Let $\mathcal{S}_1 = (Q_1, S_1, \rho_1)$ and $\mathcal{S}_2 = (Q_2, S_2, \rho_2)$ be T-generalized transformation semigroups. Then $(S_1^{Q_2} \times S_2, *)$ is a semigroup where*

$$(f,s)*(g,t) = (fg, st).$$

Proof. Let $f,g,h \in S_1^{Q_2}$ and $s,t,u \in Q_2$. Then $((f,s)*(g,t))*(h,u) = (fg,st)*(h,u) = ((fg)h,(st)u) = (f(gh),s(tu)) = (f,s)*(gh,tu) = (f,s)*((g,t)*(h,u))$. ∎

Lemma 9.5.6 *Let $\mathcal{S}_1 = (Q_1, S_1, \rho_1)$ and $\mathcal{S}_2 = (Q_2, S_2, \rho_2)$ be T-generalized transformation semigroups satisfying the following conditions:*

(1) For any $f : Q_2 \to S_1$ and $p \in Q_2$,

$$T(\rho_1(p_1,f(p),q_1),\rho_2(p_2,s,q_2)) = T(\rho_1(p_1,f(p_2),q_1),\rho_2(p_2,s,q_2)),$$

where $p_1, q_2 \in Q_1$, $p_2, q_2 \in Q_2$ and $s \in S_2$.

(2) For any $(f,s),(g,t) \in S_1^{Q_2} \times S_2$, there exist $p_1, q_1 \in Q_1$ and $p_2, q_2 \in Q_2$ such that

$$\rho_1(p_1,f(p_2),q_1) = \rho_1(p_1,g(p_2),q_1) = \rho_2(p_2,s,q_2,qa_2) = \rho_2(p_2,t,q_2) = 1.$$

If

$$(\rho_1 \circ_T \rho_2)((p_1,p_2),(f,s),q_1,q_2)) = (\rho_1 \circ_T \rho_2)((p_1,p_2),(g,t),(q_1,q_2)),$$

then $(f,s) = (g,t)$.

Proof. Let $(\rho_1 \circ_T \rho_2)((p_1,p_2),(f,s),(q_1,q_2)) = (\rho_1 \circ_T \rho_2)((p_1,p_2),(g,t),(q_1,q_2))$. Let $p_2, q_2 \in Q_2$ be such that $\rho_2(p_2,s,q_2) = \rho_2(p_2,t,q_2) = 1$. Then for any $p \in Q_2$,

$$\begin{aligned}
\rho_1(p_1,f(p),q_1) &= T(\rho_1(p_1,f(p),q_1),\rho_2(p_2,s,q_2)) \\
&= T(\rho_1(p_1,f(p_2),q_1),\rho_2(p_2,s,q_2)) \text{ by (1)} \\
&= (\rho_1 \circ_T \rho_2)((p_1,p_2),(g,t),(q_1,q_2)) \\
&= (\rho_1 \circ_T \rho_2)((p_1,p_2),(g,t),(q_1,q_2)) \\
&= T(\rho_1(p_1,g(p_2),q_1),\rho_2(p_2,t,q_2)) \\
&= T(\rho_1(p_1,g(p),q_1),\rho_2(p_2,t,q_2)) \text{ by (1)} \\
&= \rho_1(p_1,g(p),q_1).
\end{aligned}$$

Since $\mathcal{S}_1$ is a T-generalized transformation semigroup, $f(p) = g(p)$ and thus $f = g$. Similarly, $s = t$. Hence $(f, s) = (g, t)$. ■

Theorem 9.5.7 *Let $\mathcal{S}_1 = (Q_1, S_1, \rho_1)$ and $\mathcal{S}_2 = (Q_2, S_2, \rho_2)$ be T-generalized transformation semigroups satisfying the following condititons:*

(1) For any $f : Q_2 \to S_1$ and $p \in Q_2$,

$$T(\rho_1(p_1, f(p), q_1), \rho_2(p_2, s, q_2)) = T(\rho_1, p_1, f(p_2), q_1), \rho_2(p_2, s, q_2))$$

$\forall\, p_1, q_2 \in Q_1$, $p_2, q_2 \in Q_2$ and $s \in S_2$.

(2) For any $(f, s), (g, t) \in S_1^{Q_2} \times S_2$, there exist $p_1, q_1 \in Q_1$ and $p_2, q_2 \in Q_2$ such that

$$\rho_1(p_1, f(p_2), q_1) = \rho_1(p_1, g(p_2), q_1) = \rho_2(p_2, s, q_2, qa_2) = \rho_2(p_2, t, q_2) = 1.$$

Then $(Q_1 \times Q_2, S_1^{Q_2} \times S_2, \rho_1 \circ_T \rho_2)$ is a T-generalized transformation semigroup such that

$$(\rho_1 \circ_T \rho_2)((p_1, p_2), (f, s), (q_1, q_2)) = T(\rho_1(p_1, f(p_2), q_1), \rho_2(p_2, s, q_2))$$

$\forall\, p_1, q_1 \in Q_1$, $p_2, q_2 \in Q_2$, $(f, s) \in S_1^{Q_2} \times S_2$.

Proof. By Lemmas 9.5.5 and 9.5.6, we need to prove only condition (1) in Definition 9.2.1. Now

$$\begin{aligned}
& (\rho_1 \circ_T \rho_2)((p_1, p_2), (f, s)(g, t), (q_1, q_2)) \\
= \;& (\rho_1 \circ_T \rho_2)((p_1, p_2), (fg, st), (q_1, q_2)) \\
= \;& T(\rho_1(p_1, (fg)(p_2), q_1), \rho_2(p_2, st, q_2)) \\
= \;& T(\rho_1(p_1, f(p_2)g(p_2), q_1), \rho_2(p_2, st, q_2)) \\
= \;& T(\vee\{T(\rho_1(p_1, f(p_2), r_1), \rho_1(r_1, g(p_2), q_1)) \mid r_1 \in Q_1\}, \\
& \vee\{T(\rho_2(p_2, s, r_2), \rho_2(r_2, t, q_2)) \mid r_2 \in Q_2\}) \\
= \;& \vee\{T(T(\rho_1, (p_1, f(p_2), r_1), \rho_2(p_2, s, r_2)), \\
& T(\rho_1(r_1, g(p_2), q_1), \rho_2(r_2, t, q_2))) \mid (r_1, r_2) \in Q_1 \times Q_2\} \\
= \;& \vee\{T((\rho_1 \circ_T \rho_2)((p_1, p_2), (f, s), (r_1, r_2)), \\
& (\rho_1 \circ_T \rho_2)((r_1, r_2), (g, t), (q_1, q_2)) \mid (r_1, r_2) \in Q_1 \times Q_2\}.
\end{aligned}$$

■

Definition 9.5.8 *Let $\mathcal{S}_1 = (Q_1, S_1, \rho_1)$ and $\mathcal{S}_2 = (Q_2, S_2, \rho_2)$ be T-generalized transformation semigroups such that for all $f : Q_2 \to S_1$ and $p \in Q_2$,*

$$T(\rho_1(p_1, f(p), q_1), \rho_2(p_2, s, q_2)) = T(\rho_1(p_1, f(p_2), q_1), \rho_2(p_2, s, q_2))$$

for all $p_1, q_1 \in Q_1$, $p_2, q_2 \in Q_2$, $(f, s) \in S_1^{Q_2} \times S_2$.

Theorem 9.5.9 *Let $M_1 = (Q_1, X_1, \tau_1)$ and $M_2 = (Q_2, X_2, \tau_2)$ be T-generalized state machines. Then the following conditions hold:*

(1) $M_1 \omega_T M_2 \leqslant M_1 \circ_T M_2$.

(2) $TS(M_1 \omega_T M_2) \leqslant TS(M_1 \circ_T M_2)$.

Proof. (1) Let ξ_2 be the identity map of X_2. Let $\xi : X_2 \to X_1^{Q_2} \times X_2$ be a function such that $\xi(x_2) = (\xi_1(x_2), \xi_2(x_2))$, where $x_2 \in X_2$ and $\xi_1(x_2) : Q_2 \to X_1$ is a function defined by $\xi_1(x_2)(p_2) = \omega(p_2, \xi_2(x_2))$. Let η be the identity map on $Q_1 \times Q_2$. Then for all $(p_1,p_2), (q_1,q_2) \in Q_1 \times Q_2$ and $x_2 \in X_2$,

$$
\begin{aligned}
& (\tau_1\omega_T\tau_2)\,(\eta((p_1,p_2)), x_2, \eta((q_1,q_2))) \\
= \;& (\tau_1\omega_T\tau_2)((p_1,p_2), \xi_2(x_2), (q_1,q_2)) \\
= \;& T(\tau_1(p_1, \omega(p_2, \xi_2(x_2)), q_1), \tau_2(p_2, \xi_2(x_2), q_2)) \\
= \;& T(\tau_1(p_1, \xi_1(x_2)(p_2), q_1), \tau_2(p_2, \xi_2(x_2), q_2)) \\
= \;& (\tau_1 \circ_T \tau l2)((p_1,p_2), (\xi_1(x_2), \xi_2(x_2)), (q_1,q_2)) \\
= \;& (\tau_1 \circ_T \tau_2)((p_1,p_2), \xi_2(x_2), (q_1,q_2)).
\end{aligned}
$$

Hence $M_1\omega_T M_2 \leqslant M_1 \circ_T M_2$.

(2) The result here follows by applying (1) to Theorem 9.3.5. ■

Theorem 9.5.10 *Let $M_1 = (Q_1, X_1, \tau_1)$ and $M_2 = (Q_2, X_2, \tau_2)$ and $M = (Q, X, \tau)$ be T-generalized state machines. Suppose that $M \leqslant M_1\omega_T M_2$. Then the following conditions hold:*

(1) $M \leqslant M_1 \circ_T M_2$.

(2) $TS(M) \leqslant TS(M_1 \circ_T M_2)$.

Proof. (1) Define a function $\xi : X \to X_1^{Q_2} \times X_2$ by $\xi(x) = (\xi_1(x)(\cdot), \xi_2(x))$ for all $x \in X$ such that $\xi_1(x)(p_2) = \omega(p_2, \xi_2(x))$, where $p_2 \in Q_2$. Let $\eta : Q_1 \times Q_2 \to Q$ be a surjective partial function. Since ξ_2 is a function and $M \leqslant M_1\omega_T M_2$,

$$
\begin{aligned}
\tau(\eta(p_1,p_2), x, \eta(q_1,q_2)) &= T(\tau_1(p_1, \omega(p_2, \xi_2(x)), q_1), \tau l2(p_2, \xi_2(x), q_2)) \\
&= T(\tau_1(p_1, \xi_1(x)(p_2), q_1), \tau_2(p_2, \xi_2(x), q_2)) \\
&= (\tau_1 \circ_T \tau_2)((p_1,p_2), (\xi_1(x), \xi_2(x)), (q_1, a_2)) \\
&= (\tau_1 \circ_T \tau_2)((p_1,p_2), \xi(x), (q_1,q_2)).
\end{aligned}
$$

Thus $M \leqslant M_1 \circ_T M_2$.

(2) The result here follows by applying (1) to Theorem 9.3.5. ■

Theorem 9.5.11 *Let $\mathcal{S}_i = (Q_i, S_i, \rho_1)$ be T-generalized transformation semigroups, where $i = 1, 2, 3$. Suppose that $\mathcal{S}_1 \leqslant \mathcal{S}_3$. Then $\mathcal{S}_1 \circ_T \mathcal{S}_2 \leqslant \mathcal{S}_3 \circ_T \mathcal{S}_2$.*

Proof. Let $\eta_* : Q_3 \to Q_1$ be a surjective partial function and define $\eta : Q_3 \times Q_2 \to Q_1 \times Q_2$ by $\eta(p_3, p_2) = (\eta_*(p_3), p_2)$. Then η is also a surjective partial function. Define a function $\xi : S_1^{Q_2} \times S_2 \to S_3^{Q_2} \times S_2$ by $\xi(f, s) = (\alpha \circ_T f, s)$ $\forall f \in S_1^{Q_2}$ and $\forall s \in S$, where $\alpha : S_1 \to S_3$ is a function.

Then for all $p_3, p_2), (q_3, q_2) \in Q_3 \times Q_2$ and $(f, s) \in S_1^{Q_2} \times S_2$,

$$\begin{aligned}
&(\rho_1 \circ_T \rho_2)\,(\eta(p_3, p_2), (f, s), \eta(q_3, q_2)) \\
= \;& (\rho_1 \circ_T \rho_2)((\eta_*(p_3), p_2), (f, s), (\eta_*(q_3), q_2)) \\
= \;& T(\rho_1(\eta_*(p_3), f(p_2), \eta_*(q_3)), \rho_2(p_2, s, q_2)) \\
\leqslant \;& T(\rho_3(p_3, (\alpha \circ_T f)(p_2), q_3), \rho_2(p_2, s, q_2)) \quad \text{since } \mathcal{S}_1 \leqslant \mathcal{S}_3 \\
= \;& (\rho_3 \circ_T \rho_1)((p_3, p_2), (\alpha \circ_T f, s), (q_3, q_2)) \\
= \;& (\rho_3 \circ_T \rho_1)((p_3, p_2), gx(f, s), (q_3, q_2)).
\end{aligned}$$

■

Theorem 9.5.12 *Let* $\mathcal{S}_i = (Q_i, S_i, \rho_i)$ *be T-generalized transformation semigroups, where* $i = 1, 2, 3, 4$. *Then*

$$(\mathcal{S}_1 \circ_T \mathcal{S}_2) \times (\mathcal{S}_3 \circ_T \mathcal{S}_4) \leqslant (\mathcal{S}_1 \times \mathcal{S}_3) \circ_T (\mathcal{S}_2 \times \mathcal{S}_4).$$

Proof. Define $\eta : (Q_1 \times Q_3) \times (Q_2 \times Q_4) \to (Q_1 \times Q_2) \times (Q_3 \times Q_4)$ by $\eta((p_1, p_3), (p_2, p_4)) = ((p_1, p_2), (p_3, p_4))$ for $p_i \in Q_i$, where $i = 1, 2, 3, 4$. Then η is a surjective partial function. Define $h : Q_2 \times Q_4 \to S_1 \times S_3$ by $h(p_2, p_4) = (f(p_2), g(p_4))$ where $p_2 \in Q_2$ and $p_4 \in Q_4$, $f : Q_2 \to S_1$ and $g : Q_4 \to S_3$ are functions. Define a function $\xi : (S_1^{Q_2} \times X_2) \times (S_3^{Q_4} \times S_4) \to (S_1 \times S_3)^{Q_2 \times Q_4} \times (S_2 \times S_4)$ by $\xi((f, s_2), (g, s_4)) = (h, (s_2, s_4))$, where $((f, s_2), (g, s_4)) \in (S_1^{Q_2} \times S_2) \times (S_3^{Q_4} \times S_4)$. Then for $p_i, q_i \in Q_i$, where $i = 1, 2, 3, 4$ and $s_2 \in S_2$, $s_4 \in S_4$ and $f : Q_2 \to S_1$, $g : Q_4 \to S_3$, we have $((\rho_1 \circ_T \rho_2) \times (\rho_3 \circ_T p_4))(\eta(p_1, p_3), (p_2, p_4)), ((f, s_2), (g, sl_2), (g, s_4)), \eta(q_1, q_3), (q_2, q_4))) = (\rho_1 \circ_T \rho_2) \times (\rho_3 \circ_T \rho_4))(((p_1, p_2, (p_3, p_4)), ((f, s_2), (g, s_4)), ((q_1, q_2), (q_3, q_4))) = T((\rho_1 \circ_T \rho_2)((p_1, p_2), (f(p_2), s_2), (q_1, q_2)), (\rho_3 \circ_T \rho_4))(p_3, p_4), (g, (p_4), s_4), (q_3, q_4))) = T(T(\rho_1(p_1, f(p_2), q_1), \rho_2(p_2, s_2, q_2)), T(\rho_3(p_3, gp_4), q_3), \rho_4(p_4, s_4, q_4))) = T(T(\rho_1(p_1, f(p_2), q_1), \rho_3(p_3, g(p_4), q_3)), T(\rho_2(p_2, s_2, q_2), \rho_4(p_4, s_4, q_4))) = T((\rho_1 \times \rho_3)((p_1, p_3)((p_1, p_3), (f\,(p_2), g\,(p_4)), (q_1, q_3)), (\rho_2 \times \rho_4)((p_2, p_4), (s_2, s_4), (q_2, q_4))) = ((\rho_1 \times \rho_3) \circ_T (\rho_2 \times \rho_4))((p_1, p_3), (p_2, p_4)), (h, (s_2, s_4)), ((q_1, q_3), (q_2, q_4))) = ((\rho_1 \times \rho_3) \circ_T (\rho_2 \times \rho_4))((p_1, p_3), (p_2, p_4)), \xi((f, s_2), (g, s_4)), ((q_1, q_3), (q_2, q_4))).$ ■

9.6 On Proper Fuzzification of Finite State Machines

In Section 9.1-9.5, a different definition of fuzzy finite state machine was given.

A fuzzy finite state machine $M = (Q, X, \mu)$ of [86] is not necessarily reduced to a (deterministic) finite state machine if $\mu(Q \times X \times Q) \subseteq \{0, 1\}$. Hence a so-called generalized state machine was introduced by imposing the additional condition $\Sigma_{q \in Q} \mu(p, a, q) \leq 1$. This condition makes a generalized state machine a deterministic (crisp) machine when $\mathrm{Im}(\mu) \subseteq \{0,1\}$. But when $\mathrm{Im}(\mu) \nsubseteq \{0,1\}$, it can be easily seen that it can assume more than one state of the same degree for a given input in a given state, and hence it cannot be called deterministic.

It is clear that for a complete (crisp) nondeterministic finite state machine, $\Sigma_{q \in Q}\mu(p,a,q) > 1$ for some $(p,a) \in Q \times X$. Thus a crisp nondeterministic machine is not a generalized machine.

We also show that the homomorphic image of a generalized state machine is not a generalized state machine.

One purpose now is to consider these deficiencies.

We also propose an alternative construction that leads to minimal completion of fuzzy finite state machine.

The results in the remainder of the chapter are essentially from [53, 54].

9.7 Generalized Fuzzy Finite State Machines

Definition 9.7.1 *A fuzzy finite-state machine is a triple $M = (Q, X, \mu)$, where Q, X are non-empty finite sets and μ is a fuzzy subset of $Q \times X \times Q$.*

If $\mu(p,a,q) = t$, $t \in (0,1]$, then p is said to be a **fuzzy previous state**, a is said to be **fuzzy input** and q is said to be a **fuzzy next state** of p of positive degree t.

A (crisp) finite state machine $(Q, X, f))$, where $f : Q \times X \to Q$ is a relation, can be considered as a fuzzy finite state machine (Q, X, μ), if we define $\mu : Q \times X \times Q \to [0,1]$ as follows:

$$\mu(p,a,q) = \begin{cases} 1 & \text{if } f(p,a) = q, \\ 0 & \text{otherwise.} \end{cases} \tag{9.1}$$

On the other hand, a fuzzy finite state machine (Q, X, μ), when $\mu(Q \times X \times Q) \subseteq \{0,1\}$ can be regarded as a finite state machine (Q, X, f), if we define

$$f(p,a) = \begin{cases} q & \text{if } \mu(p,a,q) = 1, \\ 0 & \text{otherwise.} \end{cases} \tag{9.2}$$

However, in this situation f is not necessarily a partial function, but may happen to be a proper relation. In other words, (Q, X, f) may not be a deterministic finite state machine, but a nondeterministic machine, as discussed in [19, 26, 43, 46, 92].

If $M = (Q, X, \mu)$ is a generalized state machine and $\mu(Q \times X \times Q) \subseteq \{0,1\}$, then under a natural identification, M is a deterministic (crisp) finite state machine.

The additional condition, $\Sigma_{q \in Q}\mu(p,a,q) \leq 1$, $\forall p \in Q$, $a \in X$, ensures that if $\mu(Q \times X \times Q) \subseteq \{0,1\}$, then for given p and a there exist at the most one q such that $\mu(p, a, q) = 1$. Consequently given a state p and an input a, if $q \in Q$ exists such that $\mu(p, a, q) = 1$, then q is unique. This makes f, as defined in (9.2) above, a partial function and so makes the machine (Q, X, f) deterministic. Thus, when $\mu(Q \times X \times Q) \subseteq \{0, 1\}$, the fuzzy finite state machine of [86, 87, 89] reduces to non-deterministic finite state machine, whereas the generalized finite state machine reduces to deterministic finite state machine. Some feel that it is natural to expect that in any fuzzification of a finite state machine there should not be more than one positive "next state" of the same positive degree for any "input". Like a fuzzy finite state machine as defined in [86, 87, 89], a generalized state machine also does not meet this expectation. In fact, one can construct a generalized state machine, where for any given input there are n number of next states of the same degree.

Let (Q, X, μ) be a generalized state machine, where $|Q| = n$ and $\mu : Q \times X \times Q \to [0, 1]$ is defined by $\mu(p, a, q) = 1/n$, $\forall p, q \in Q$ and $\forall a \in X$.

Thus generalized state machines of Sections 9.1-9.5 are on par with fuzzy finite state machines discussed in [86, 87, 89], as far as nondeterminism is concerned. It is desirable to separate deterministic finite state machines from nondeterministic finite state machines in the crisp case, then it may be desirable to do so in fuzzy case also. The main purpose now is to achieve this.

Finite state machines can be divided into four categories: deterministic, nondeterministic, complete and noncomplete machines. Let $\mathcal{M}$ represent the class of (crisp) non-deterministic (deterministic, complete, noncomplete) fuzzy finite state machines and $F(\mathcal{M})$ the class of the corresponding fuzzy concept. If $M \in F(\mathcal{M})$ is such that the valuation set for M is $\{0, 1\}$, then $M \in \mathcal{M}$ under a natural identification.

There may be general agreement that in a nondeterministic model of a fuzzy finite state machine, some input, under a given state, produces more than one distinct state of the same degree. In a deterministic model, it can be argued that this should not be the case. The model represents a complete fuzzy finite state machine if every input gives a next state of positive degree, otherwise it represents an incomplete machine.

As is shown in the next section, these concepts can be described mathematically with the help of a proper fuzzification of the concept of a relation.

9.8 Fuzzy Relations and Fuzzy Finite State Machines

Definition 9.8.1 *Let A and B be sets.*

*(1) A **fuzzy relation** from A into B is a function $\mu : A \times B \to [0, 1]$.*

*(2) A fuzzy relation μ from A into B (i.e. $\mu : A \times B \to [0, 1]$) is said to be **complete** if for all $a \in A$, there exists $b \in B$ such that $\mu(a, b) > 0$.*

*(3) A fuzzy relation μ of A into B is called a **partial fuzzy function,** if $\forall a \in A$, $b, b' \in B$, $\mu(a,b) = \mu(a,b') > 0$ implies that $b = b'$.*

*(4) A fuzzy relation μ of A into B is called a **fuzzy function** if it is a complete partial fuzzy function.*

Thus we have four distinct situations:

(1) A fuzzy relation which is a complete partial fuzzy function.

(2) A fuzzy relation which is complete, but may not be a partial fuzzy function.

(3) A fuzzy relation which may not be complete, but is a partial fuzzy function.

(4) A fuzzy relation which may not be complete and may not be a partial fuzzy function.

Accordingly, four different types of fuzzy finite state machines can be defined.

Definition 9.8.2 *A fuzzy finite state machine (FFSM) is an ordered triple $M = (Q, X, \mu)$, where Q and X are finite sets and $\mu : Q \times X \times Q \to [0,1]$. The elements of Q are called **states** and those of X are called **inputs**.*

*(1) M is called a **deterministic fuzzy finite state machine (DFFSM)** if μ is a partial fuzzy function.*

*(2) M is called a **nondeterministic fuzzy finite state machine (NDFFSM)** if μ is a fuzzy relation.*

*(3) M is called a **complete deterministic fuzzy finite state machine (CDFFSM)** if μ is a fuzzy function (i.e., complete partial fuzzy function).*

*(4) M is called a **complete nondeterministic fuzzy finite state machine (CNDFFSM)** if μ is a complete fuzzy relation.*

Clearly, if $[0,1]$ is replaced by $\{0,1\}$, then M is a deterministic finite state machine when (1) holds, M is a nondeterministic finite state machine when (2) holds, M is a complete deterministic finite state machine when (3) holds, and M is complete nondeterministic finite state machine when (4) holds.

Recall that a generalized state machine, $M = (Q, X, \mu)$, is called **complete** if $\Sigma_{q \in Q}\mu(p,a,q) = 1$, $\forall p \in Q$ and $\forall a \in X$.

Clearly, a complete generalized state machine is a complete fuzzy finite state machine in the sense of Definition 9.8.2, but not vice versa. One may easily construct a fuzzy finite state machine (Q, X, μ), which is complete in the sense of Definition 9.8.2 satisfying $\Sigma_{q \in Q}\mu(p,a,q) > 1$ and hence is not a generalized state machine.

9.9 Completion of Fuzzy Finite State Machines

Definition 9.9.1 *Let $M = (Q, X, \mu)$ be a fuzzy finite state machine. A fuzzy finite state machine $M^S = (Q^S, X^S, \mu^S)$ is called a* **sub-fuzzy finite state machine** *of M, written $M^S \subseteq M$, if*
(1) $Q^S \subseteq Q$,
(2) $X^S \subseteq X$,
(3) $\mu \mid_{Q^S \times X^S \times Q^S} = \mu^S$.

If M is crisp in Definition 9.9.1, then so is M^S

Definition 9.9.2 *Let M be a fuzzy finite state machine. A fuzzy finite state machine M^c is said to be a* **completion** *of M if*
(1) M^c is complete and
(2) $M \subseteq M^c$.

Definition 9.9.3 *Let $M = (Q, X, \mu)$ be a fuzzy finite state machine. A fuzzy finite state machine $M' = (Q', X', \mu')$ is called the* **smallest completion** *M if*
(1) M' is a completion of M and
(2) if $M'' = (Q'', X'', \mu'')$ is a completion of M, then $M' \subseteq M''$.

Let $M = (Q, X, \mu)$ be a fuzzy finite state machine. If M is not complete, adjoin a state z, $z \notin Q$, to get $Q' = Q \cup \{z\}$. If M is complete, set $Q' = Q$. Define $\mu' : Q' \times X \times Q' \to [0,1]$ as follows: $\forall (p, a, q) \in Q' \times X \times Q'$,

$$\mu'(p, a, q) = \begin{cases} \mu(p, a, q), & \text{if } p, q \in Q, \text{ and } \mu(p, a, q) \neq 0, \\ 1 & \text{if either } \mu(p, a, r) = 0 \text{ for all } r \in Q \text{ and } q = z, \\ & \quad \text{or } p = z = q, \\ 0 & \text{otherwise.} \end{cases}$$

It is easy to see that $M^c = (Q', X, \mu')$ of Definition 9.1.2 is a complete machine in the sense of Definition 9.9.1, and therefore a completion of M. Moreover, if $\mu^c(p, a, z) = 1$, then $\mu'(p, a, z) = 1$ and if $\mu^c(p, a, z) \neq 1$, then $\mu'(p, a, z) = 0$. Therefore, $M' \subseteq M^c$. Hence M' is the smallest completion of M.

Example 9.9.4 *Let* $Q = \{q_1, q_2, q_3\}$, $X = \{a, b\}$ *and* $M = (Q, X, \mu)$ *be described as in Figure 9.1.*

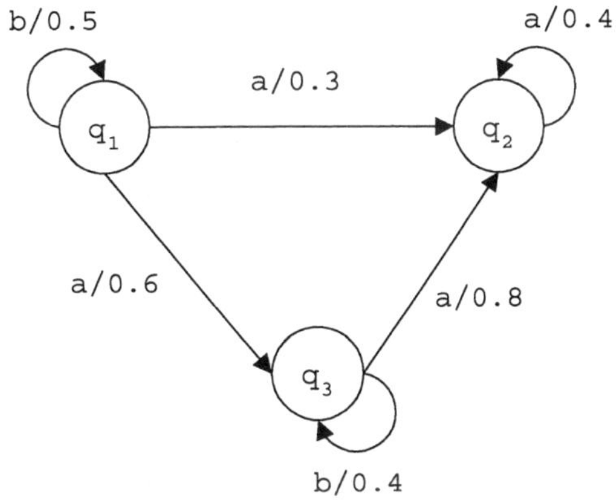

Figure 9.1

The completion of this machine by Definition 9.9.2 *is represented by Figure 9.2.*

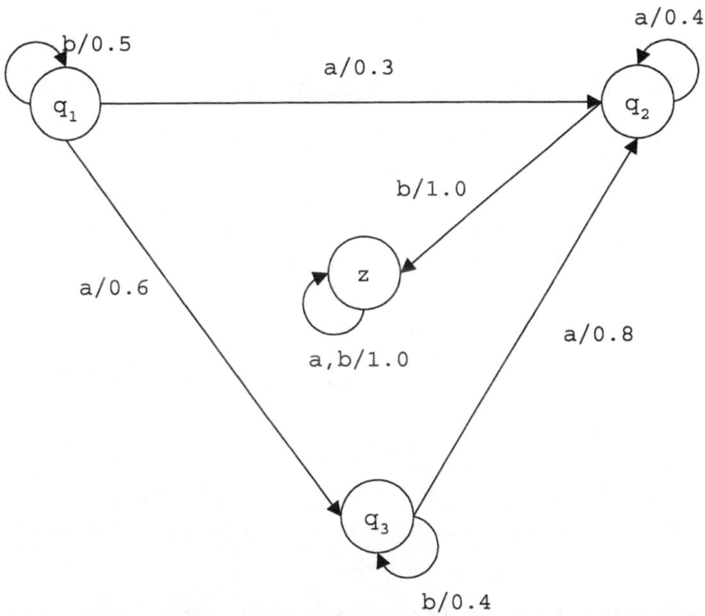

Figure 9.2

The completion by Definition 9.1.2 *is represented by Figure 9.3.*

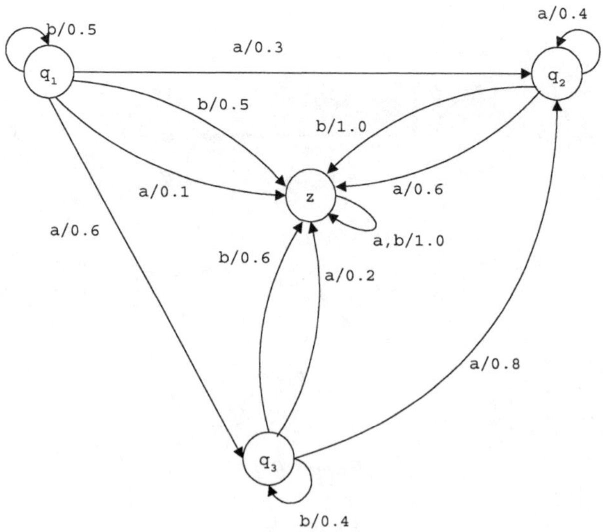

Figure 9.3

9.10 Generalized State Machines and Homomorphisms

Definition 9.10.1 *Let $M_1 = (Q_1, X_1, \mu_1)$ and $M_2 = (Q_2, X_2, \mu_2)$ be fuzzy finite state machines. An ordered pair (α, β) of functions $\alpha : Q_1 \to Q_2$ and $\beta : X_1 \to X_2$ is called a* ***fuzzy finite-state machine homomorphism*** *from M_1 into M_2, if*

$$\mu_1(p, a, q) \leq \mu_2(\alpha(p), \beta(a), \alpha(q)), \qquad \forall p, q \in Q, \ a \in X_1.$$

A homomorphic image of a fuzzy finite state machine is a fuzzy finite state machine. However, the following example shows that a homomorphic image of a generalized state machine is not a generalized state machine.

Example 9.10.2 *Let $Q_1 = \{q_1, q_2\}$, $X_1 = \{a\}$ and $Q_2 = \{q_1', q_2'\}$, $X_2 = \{a\}$ and $M_1 = (Q_1, X_1, \mu_1)$ and $M_2 = (Q_2, X_2, \mu_2)$ be as described in*

Figures 9.4 and 9.5, respectively.

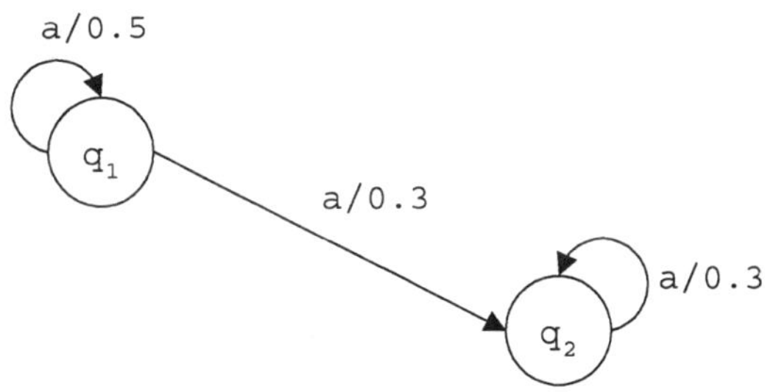

Figure 9.4

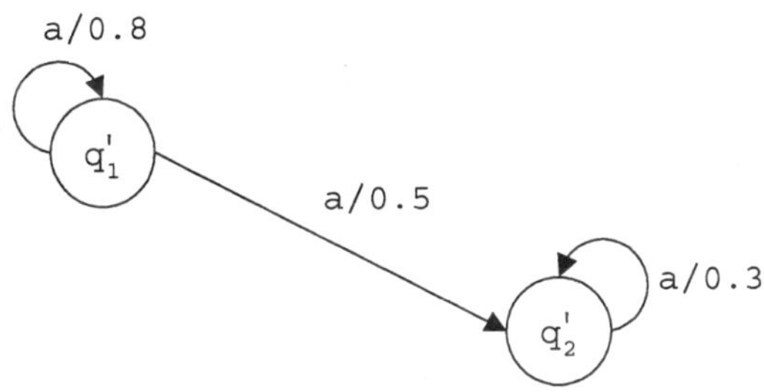

Figure 9.5

Define $\alpha : Q_1 \rightarrow Q_2$ *by* $\alpha(q_i) = q_i'$, $i = 1, 2$ *and let* β *be the identity map on* X. *Then clearly* M_2 *is a homomorphic image of* M_1, *but* M_2 *is not a generalized state machine since* $\sum_{q' \in Q_2} \mu_2(q_1', a, q') > 1$.

Examples are easily constructed showing that a homomorphic image of a complete generalized state machine in the sense of Definition 9.1.1 is not a generalized state machine.

In [54], the authors consider restricted direct products, direct products, and wreath products of fuzzy finite state machines. They show that a fuzzy reset machine with at least three states has a direct decomposition. The notion of admissible relation and partitions are studied as well as fuzzy transformation semigroups.

10
Regular Fuzzy Expressions

10.1 Regular Fuzzy Expressions

The concept of regular expressions was introduced for the purpose of representing the languages accepted by deterministic finite automata. This is necessary in the analysis and synthesis of finite automata since the languages accepted by finite automata are usually infinite.

Many approaches have been used in fuzzifying the concept of finite automata. We consider the concept of regular fuzzy expressions. This concept provides the necessary tool for the analysis and synthesis of fuzzy automata. It also provides a way for recursively generating the family of fuzzy languages accepted by fuzzy automata from certain simple fuzzy languages. The families of fuzzy languages accepted by various models of fuzzy automata is, in general, nondenumerable. Hence the concept of regular fuzzy expressions might well provide the necessary insights to the study of the structure of such families. The results of this section are mainly from [126].

The concept of an ordered semigroup is needed in order to present a concept of regular fuzzy expressions applicable to the various models of fuzzy automata.

Let $\preceq$ be a relation on a set R. Recall that $\preceq$ is called a **linear order** if the following conditions hold $\forall a, b, c \in R$:

(1) $a \preceq a$,
(2) $a \preceq b$, $b \preceq a$ implies $a = b$,
(3) $a \preceq b, b \preceq c$ implies $a \preceq c$,
(4) either $a \preceq b$ or $b \preceq a$.

If $\preceq$ is a linear order on R, then we say that R is **linearly ordered** by $\preceq$ and we call the pair $(R, \preceq)$ a **linearly ordered set.**

Let $(R, \preceq)$ be a linearly ordered set, $R_o \subseteq R$ and $a_o \in R$. Recall that a_0 is an **upper bound** of R_0 if and only if $a \preceq a_0$ for all $a \in R$. Moreover, $a_0 = \vee R_0$ if and only if (1) a_0 is an upper bound of R_0 and (2) for every upper bound b of R_0, $a_0 \preceq b$.

Definition 10.1.1 *An* ***ordered semigroup*** *is a triple $(R, \otimes, \preceq)$, where $(R, \preceq)$ is a linearly ordered set with minimal element 0, and $(R, \otimes)$ is a commutative semigroup with identity 1 such that for every $R_0 \subseteq R$ and $a_0 \in R$, $a_0 \otimes 0 = 0$ and $\vee\{a_0 \otimes a \mid a \in R_0\} = a_0 \otimes \vee R_0$.*

Clearly, for every $a, b \in R$, if $a \preceq b$, then $a \otimes c \preceq b \otimes c$ for all $c \in R$.

In what follows, unless otherwise stated, R always stands for an ordered semigroup with minimal element 0 and identity 1 for $\otimes$.

Let X be a nonempty set. Recall that X^* is the free semigroup generated by X with identity Λ.

Definition 10.1.2 *Let X be a nonempty set. A function f from X^* into R is called an R-**language** over X.*

Definition 10.1.3 *Let X be a finite nonempty set. An R-**automaton** over X is a 4-tuple $A = (S, p, h, g)$, where S is a finite nonempty set, p is a function from $S \times X \times S$ into R, and h and g are functions from S into R.*

Definition 10.1.4 *Let $A = (S, p, h, g)$ be an R-automaton over X.*

(1) Let p^* be the function from $S \times X^* \times S$ into R defined recursively as follows:

$$p^*(s, \Lambda, s') = \begin{cases} 1 & \text{if } s = s' \\ 0 & \text{if } s \neq s' \end{cases}$$

$$p^*(s, ax, s') = \vee_{s'' \in S} p(s, a, s'') \otimes p^*(s'', x, s'),$$

for all $s, s' \in S$, $a \in X$ and $x \in X^*$.

(2) Let f^A be the function from X^* into R defined by for all $x \in X^*$,

$$f^A(x) = \vee_{s \in S} \vee_{s' \in S}\, h(s) \otimes p^*(s, x, s') \otimes g(s').$$

In the above definitions, X is the set of input symbols, S is the set of states, $p(s, a, s')$ is the grade of membership that the next state is s' given that the present state is s and input a is applied, $h(s)$ is the grade of membership that s is the initial state and $g(s)$ is the grade of membership that s is an accepting state. Moreover, f^A is the R-language over X accepted by A.

Definition 10.1.4 includes many of the various existing models of fuzzy automata. These models may be obtained by appropriate choices of R, e.g., max-min automata: $R_M = (R, \otimes, \preceq)$, where R is a subset of the real number system, usually $[0, 1]$, $a \otimes b = a \wedge b$, and $a \preceq b$ means $a \leqslant b$. Other examples can be found in [37, 123, 124].

Definition 10.1.5 *An R-language f over X is called* **regular** *if $f = f^A$ for some R-automaton $A = (S, p, h, g)$ over X.*

Notation 10.1.6 *Let f, f_1 and f_2 be R-languages over X.*
(1) $f_1 \vee f_2$ is the R-language over X such that for all $x \in X^*$,

$$(f_1 \vee f_2)(x) = f_1(x) \vee f_2(x)$$

In general, if $f_1, f_2, \ldots, f_n$ are R-languages over X, then

$$\vee_{i=1}^{n} f_i = (f_1 \vee f_2 \vee \ldots \vee f_{n-1}) \vee f_n.$$

(2) $f_1 \circ f_2$ is the R-languages over X, where for all $x \in X^*$,

$$f_1 \circ f_2(x) = \vee\{f_1(y) \otimes f_2(z) \mid x = yz\}.$$

(3) Let $r \in R$. Then rf is the R-language over X, where for all $x \in X^*$,

$$(rf)(x) = r \otimes f(x).$$

(4) f^* is the R-language over X, where for all $x \in X^*$,

$$f^*(x) = \begin{cases} f(\Lambda) & \text{if } x = \Lambda, \\ \vee\{f(b_1) \otimes f(b_2) \otimes \ldots \otimes f(b_n) \mid x = b_1 b_2 \ldots b_n, & \\ \qquad b_1 \neq \Lambda \text{ for } i = 1, 2, \ldots, n \text{ and } n = 1, 2, \ldots\} & \text{if } x \neq \Lambda. \end{cases}$$

Theorem 10.1.7 *If $r \in R$ and f is a regular R-language over X, then rf is a regular R-language over X.*

Proof. Let $f = f^A$, where $A = (S, p, h, g)$ is an R-automaton over X. Let $A' = (S, p, h, g')$ be the R-automaton over X, where $g'(s) = r \otimes g(s)$ for all $s \in S$. It follows that $rf = f^{A'}$. ■

Theorem 10.1.8 *If f_1 and f_2 are regular R-languages over X, then $f_1 \vee f_2$ is a regular R-language over X.*

Proof. For $i = 1, 2$, let $f_i = f^{A_i}$, where $A_i = (S_i, p_i, h_i, g_i)$ is an R-automaton over X. Without loss of generality, assume $S_1 \cap S_2 = \emptyset$. Let $A = (S, p, h, g)$ be an R-automaton over X, where $S = S_1 \cup S_2$ and for each s, $s' \in S$ and $a \in X$,

$$p(s, a, s') = \begin{cases} p_1(s, a, s') & \text{if } s, s' \in S_1, \\ p_2(s, a, s') & \text{if } s, s' \in S_2, \\ 0 & \text{otherwise,} \end{cases}$$

$$h(s) = \begin{cases} h_1(s) & \text{if } s \in S_1, \\ h_2(s) & \text{if } s \in S_2, \end{cases}$$

$$g(s) = \begin{cases} g_1(s) & \text{if } s \in S_1, \\ g_2(s) & \text{if } s \in S_2. \end{cases}$$

It follows that $f_1 \vee f_2 = f^A$. ■

Theorem 10.1.9 *If f_1 and f_2 are regular R-languages over X, then $f_1 \circ f_2$ is a regular R-language over X.*

Proof. For $i = 1, 2$, let $f_i = f^{A_i}$, where $A_i = (S_i, p_i, h_i, g_i)$ is an R-automaton over X. Without loss of generality, assume that $S_1 \cap S_2 = \emptyset$. Let $A = (S, p, h, g)$ be the R-automaton over X, where $S = S_1 \cup S_2 \cup \{s_0\}$, $s_0 \notin S_1 \cup S_2$, and for each $s', s'' \in S$ and $a \in X$,

$$p(s', a, s'') = \begin{cases} p_1(s', a, s'') & \text{if } s', s'' \in S_1, \\ p_2(s', a, s'') & \text{if } s', s'' \in S_2, \\ \vee_{s \in S_1}[h_1(s) \otimes p_1(s, a, s'')] & \text{if } s' = s_0, s'' \in S_1, \\ f_1(\Lambda) \otimes \{\vee_{s \in S_2}[h_2(s) \otimes p_2(s, a, s'')]\} & \text{if } s' = s_0, s'' \in S_2, \\ g_1(s') \otimes \{\vee_{s \in S_2}[h_2(s) \otimes f_2(s, a, s'')]\} & \text{if } s' \in S_1, s'' \in S_2, \\ 0 & \text{otherwise}, \end{cases}$$

$$h(s') = \begin{cases} 1 & \text{if } s' = s_0, \\ 0 & \text{otherwise}, \end{cases}$$

$$g(s') = \begin{cases} f_1(\Lambda) \otimes f_2(\Lambda) & \text{if } s' = s_0, \\ f_2(\Lambda) \otimes g_1(s') & \text{if } s' \in S_1, \\ g_2(s') & \text{if } s' \in S_2. \end{cases}$$

It follows that $f_1 \circ f_2 = f^A$. ■

Theorem 10.1.10 *If f is a regular R-language over X, then f^* is a regular R-language over X.*

Proof. Let $f = f^A$ where $A = (S, p, h, g)$ is an R-automaton over X. Let $A' = (S', p', h', g')$ be an R-automaton over X, where $S' = S \cup \{s_0\}$, $s_0 \notin S$, and for every $s', s'' \in S'$ and $a \in X$.

$$p'(s', a, s'') = \begin{cases} p(s', a, s'') \vee \{g(s') \times \\ \qquad [\vee_{s \in S}\{h(s) \otimes p(s, a, s'')\}]\} & \text{if } s', s'' \in S, \\ \vee_{s \in S}[h(s) \otimes p(s, a, s'')] & \text{if } s' = s_0, s'' \in S, \\ 0 & \text{otherwise}, \end{cases}$$

$$h'(s') = \begin{cases} 1 & \text{if } s' = s_0, \\ 0 & \text{if } s' \neq s_0, \end{cases}$$

$$g'(s') = \begin{cases} g(s') & \text{if } s \in S, \\ f(e) & \text{if } s = s_0. \end{cases}$$

It follows that $f^* = f^{A'}$. ■

We now introduce some notation.

(1) If $x \in X^*$, then f_x is the R-language over X, where $f_x(x) = 1$ and $f_x(y) = 0$ for all $y \neq x$.

(2) $f_\emptyset$ is the R-language over X, where $f_\emptyset(y) = 0$ for all $y \in X^*$.

Definition 10.1.11 *Let X be a finite nonempty set. The family $\Re$ of regular R-expressions over X is defined recursively as follows:*

(1) $\emptyset \in \Re$,

(2) $\Lambda \in \Re$,

(3) $\sigma \in \Re$ for all $\Re \in X$,

(4) $r\alpha \in \Re$ for all $r \in R$ and $\alpha \in \Re$,

(5) $\alpha_1 \vee \alpha_2 \in \Re$ for all $\alpha_1, \alpha_2 \in \Re$,

(6) $\alpha_1 \circ \alpha_2 \in \Re$ for all $\alpha_1, \alpha_2 \in \Re$,

(7) $\alpha^* \in \Re$ for all $\alpha \in \Re$.

Definition 10.1.12 *For every $\alpha \in \Re$, let $\langle\alpha\rangle$ denote the R-language over X defined recursively as follows:*

(1) $\langle\alpha\rangle = f_\alpha$, for all $\alpha \in X \cup \{\emptyset, \Lambda\}$.

(2) $\langle a\alpha\rangle = a\langle\alpha\rangle$, for all $a \in R$ and $\alpha \in \Re$.

(3) $\langle\alpha_1 \vee \alpha_2\rangle = \langle\alpha_1\rangle \vee \langle\alpha_2\rangle$, for all $\alpha_1, \alpha_2 \in \Re$.

(4) $\langle\alpha_1 \circ \alpha_2\rangle = \langle\alpha_1\rangle \circ \langle\alpha_2\rangle$, for all $\alpha_1, \alpha_2 \in \Re$.

(5) $\langle\alpha^*\rangle = \langle\alpha\rangle^*$ for all $\alpha \in \Re$.

Theorem 10.1.13 *If $\alpha \in R$, then $\langle\alpha\rangle$ is a regular R-language over X.*

Proof. The proof follows from the Theorems 10.1.7-10.1.10 and the fact that for all $a \in X \cup\{\emptyset, \Lambda\}$, f_a is a regular R-language over X. ■

Theorem 10.1.14 *If f is a regular R-language over X, then $f = \langle\alpha\rangle$ for some $\alpha \in R$.*

Proof. Let $f = f^A$, where $A = (S, p, h, g)$ is an R-automaton over X with $S = \{s_1, s_2, \ldots, s_n\}$ and $X = \{a_1, a_2, \ldots, a_m\}$. For all $i, j = 1, 2, \ldots, n$, let $\alpha_{ij}^0 = (s_{ij}\Lambda) \vee [\vee_{t=1}^{p}(s_i, a_t, s_j)a_t]$ and $\alpha_{ij}^k = \alpha_{ij}^{k-1} \vee [\alpha_{ik}^{k-1} \circ (\alpha_{kk}^{k-1})^* \circ \alpha_{kj}^{k-1}]$, $k = 1, 2, \ldots, n$. It can be shown by induction on k that $\langle\alpha_{ij}^k\rangle(\Lambda) = 1$ if $i = j$ and 0 otherwise, $\langle\alpha_{ij}^k\rangle(a) = p(s_{i_1}, a, s_{i_2})$ for all $a \in X$ and that for $m \in \mathbb{N}$,

$$\begin{aligned} \langle\alpha_{ij}^k\rangle(a_{i_0}a_{i_1}\ldots a_{i_m}) &= \vee_{i_1 \leqslant k} \vee_{i_2 \leqslant k} \ldots \vee_{i_m \leqslant k} p(s_i, a_{i_0}, s_{i_1}) \otimes \\ &\quad p(s_{i_1}, a_{i_1}, s_{i_2}) \otimes \ldots \otimes p(s_{i_m}, a_{i_m}, s_j). \end{aligned}$$

Thus for every $x \in X^*$, $\langle\alpha_{ij}^n\rangle(x) = p^*(s_i, x, s_j)$. Let

$$\alpha = \vee_{i=1}^{n} \vee_{j=1}^{n} [h(s_i) \otimes g(s_j)]\alpha_{ij}^n.$$

It follows that $\langle \alpha \rangle = f$. ∎

Theorems 10.1.13 and 10.1.14 provide solutions to the synthesis and analysis problem of R-automata, respectively. They provide a generalization of Kleene's theorem for fuzzy automata.

10.2 Codes Over Languages

In [45], a restrictive structure of the customary hierarchy of formal languages attained by imposing restrictions on a set of productions and on their use by means of codewords was examined. A coded fuzzy language (CFL) is defined in order to reduce the gap between formal languages and natural languages. The fuzziness of the CFL is a consequence of the imprecision of natural languages. In the remainder of the chapter, we present some properties of CFLs as developed in [45] and their relationships to the restrictive device. A cyclic language of order n is defined to study properties of formal and fuzzy languages with regard to classes of recognizers.

Randomization of the structure of formal languages is a natural way to extend formal languages and fill the gap between them and the imprecision of natural languages. This randomization is done by means of a stochastic language based on the structure of stochastic automata and by means of fuzzy languages. The degree of artificiality of an artificial deterministic language can be measured by the amount of deviation from structures of natural languages. Time also has a very strong effect on the amount of naturalization of a language. Changes in a language over a long period of time come about naturally.

The customary hierarchy of nonfuzzy formal languages is obtained by restricting the form of rewriting rules. This hierarchy cannot quantify any fuzziness of meaning. Hence it cannot fill the gap between formal languages and natural languages. In order to reduce this gap, we impose restrictions on a set of fuzzy production rules and on their use by means of codewords. Two different approaches can be taken regarding the theory of fuzzy languages. A language L can be treated as a fuzzy binary relation from a set of terms T^* to a universe of discourse U. This definition is closer to the concept of a language on linguistics, where L, as a fuzzy relation, is characterized by a membership function

$$\mu_L : T^* \times U \to [0,1]$$

that associates with each ordered pair $(x, y) \in T^* \times U$, its grade of membership $\mu_L(x, y)$ in L.

A fuzzy formal language or a fuzzy language $\mu : T^* \to [0,1]$ can serve to indicate the degree of meaningfulness of each string in T^*, namely, for $x \in T^*$, $\mu(x)$ near 1 implies that x is meaningful and $\mu(x)$ near 0 implies that x is not meaningful.

10.3 Regulated Codes and Fuzzy Grammars

By considering a control language C for a phase structure grammar G, a method was introduced in [35, 114, 116] to impose restrictions on the use of production rules. The language generated by G under a control language C is a subset of the language generated by G. It consists of words that possess at least one derivation whose string of production rules are elements of the control C. The case when each symbol of the control word has to be applied is called Type I. We call Type II, the case when a symbol of the control word is not applicable at the corresponding step of the derivation in which case, it causes a move to the next production of the control word.

We consider only the case Type I. For a given phrase structure grammar G and control language C, $L_C(G)$, Type I is a subset (not necessarily proper) of $L_C(G)$ Type II. The results are from [45].

The problem of formalizing relations between languages and controls has been considered in [114, 116]. It was shown in [114] and [116] that if G is a type 3 grammar and C is a regular code, then $L_C(G)$ is regular and also that if G is a type 3 grammar, and C is a context-free code, then $L_C(G)$ is context free. These two results are true for both Type I and Type II.

Let C be a regular language over the alphabet $\{1, 2, \ldots, n\}$. Then the language $L_C(G)$ is defined to be a subset of $L(G)$ consisting of words that possess at least one derivation whose codeword is in C. $L_C(G)$ is called the language generated by G with code language C. If G is a context-free language, $L_C(G)$ is, in general, neither a regular nor a context-free language. This follows immediately from the following examples.

Example 10.3.1 *Let*

$$\begin{aligned}
&G = \langle N, T, P, S\rangle\,, \text{ where}\\
&N = \{S\}\\
&T = \{0, 1\}\\
&P = \{P_1, P_2\}\\
&P_1 : S \to 0S1\\
&P_2 : S \to 01\\
&C = (1)^*\,2.
\end{aligned}$$

$C = (1)^*\,2$ means that P_1 is applied $(n-1)$ times for $n \in \mathbb{N}$, followed by an application of P_2. We obtain

$$L_C(G) = \{0^n 1^n \mid n \in \mathbb{N}\}$$

which is well-known not to be a regular language.

Example 10.3.2 *Let*

$$\begin{aligned}
&G = \langle N, T, P, S \rangle\,, \text{ where}\\
&N = \{S, A, B\}\\
&T = \{0, 1, 2\}\\
&P = \{P_i \mid i = 1, 2, \ldots, 7\}\\
&P_1 : S \to SAB\\
&P_2 : S \to 0S\\
&P_3 : A \to 1A\\
&P_4 : B \to 2B\\
&P_5 : S \to 0\\
&P_6 : A \to 1\\
&P_7 : B \to 2\\
&C = 1\,(234)^*\,567.
\end{aligned}$$

Then we obtain $L_C(G) = \{0^k 1^k 2^k \mid k \geq 1\}$ *which is a context-sensitive language, but is not a context-free language.*

It is shown in [115] that $L_C(G)$ of Example 10.3.2 can be also generated by the grammar

$$G = \langle \{S, A\}\,, \{0, 1, 2\}\,, P, S \rangle\,,$$

where P consists of the productions

$$\begin{aligned}
&P_1 : S \to SA\\
&P_2 : S \to 0S1\\
&P_3 : A \to 2A\\
&P_4 : S \to 01\\
&P_5 : A \to 2
\end{aligned}$$

with the coded language consisting of words of the form

$$C = 1\,(23)^*\,45.$$

This notion of a coded language differs from the notion of a control set introduced in [35] in that [35] restricts the language to leftmost derivations only and thus restricts the result to context-free languages only.

We introduce the notion of fuzzy productions as follows: Let $P = \{P_1, P_2, \ldots, P_r\}$ be a set of productions. Define a function μ of P into the set of r-dimensional row vectors $[0,1]^r$ with elements in the interval $[0,1]$. Let μ_I be such a r-dimensional row vector such that μ_I is the grade membership signifying a fuzzy measure of applying a rewriting rule from the set of productions P, starting from the starting symbol σ. The nth term of the vector $\mu(P_j)$, where $1 \leq n, j \leq r$, is the grade of membership signifying a fuzzy measure of applying a rewriting rule from P_n after applying a rewriting rule from P_j.

Remark 10.3.3 *In a probabilistic grammar, μ and μ_I are both stochastic vectors, and μ_I is referred to as the initial designator that represents the initial distribution of the rewriting rules.*

Definition 10.3.4 *Let G be a context-free language and let C be a code structure on G. Then the system $G_f = (G, C, \mu, \mu_I)$ is called a* ***context-free coded fuzzy grammar*** *(CFCFG or $(CF)^2G$).*

Let $t \in [0,1]$. Then we can define context-free coded fuzzy languages $((CF)^2L)$, $L_f(G_f, t)$, generated by $(CF)^2G$, $G_f = (G, C, \mu, \mu_I)$, with cut-point t. This language is a subset of $L(G)$ consisting of all words x which possess at least one derivation $\sigma \xrightarrow{*} x$ such that the grade of membership of the word in $L(G)$ is greater than or equal to t.

Considering Example 10.3.2, it is clear that given a context-free grammar G and an arbitrary regular code C, the coded language $L_C(G)$ is not necessarily finite.

Remark 10.3.5 *The previous result is clearly the case for regular codes of the form $\alpha\beta^*\gamma$, or even $\beta^*\gamma$, where α, β, and γ are regular codes and $\alpha\beta^*\gamma$ denotes the code consisting of all words $\alpha\beta^j\gamma$, $j = 0, 1, 2, \ldots$. However, considering regular codes of a special form gives us a different result. If the regular code calls for applying a production that cannot be applied, or a production results in a terminal string in the middle of the regular codeword (i.e., the complete regular codeword is not utilized), then the sentences do not belong to the coded language. Hence the coded language consists of sentences that belong to T^* and that are produced under complete exhaustion of the codeword.*

Remark 10.3.6 *Complete exhaustion does not restrict the number of times that the $*$ operation is used, but refers only to the use of symbols in the codewords.*

We illustrate these remarks in the following example.

Example 10.3.7 *Let*

$$
\begin{aligned}
&G = \langle N, T, P, S \rangle \\
&N = \{A, B, S\} \\
&T = \{0, 1\} \\
&P = \{P_1, P_2, P_3, P_4, P_5\} \\
&P_1 : S \to 0B \\
&P_2 : B \to BA \\
&P_3 : B \to 1 \\
&P_4 : A \to 0 \\
&P_5 : B \to 1B \\
&C = 1(2345)^*.
\end{aligned}
$$

$L_C(G) = \emptyset$ since production P_5 cannot be used. However, if P_5 is eliminated from the set P and the code $C = 1\,(234)^*$ is used, then $L_C(G) = 010$.

Theorem 10.3.8 *Let G be a phrase structure grammar with a set P of single production rules (i.e., $p_j \in P$ if $\{p_j\}$ is a singleton) and let C be a finite sum of ultimately periodic sequences. Then the language $L_C(G)$ is finite.*

Proof. Clearly,

$$L_{\sum_i c_i}(G) = \sum_i L_{c_i}(G).$$

Hence it is suffices to consider a regular code which has an ultimately periodic structure of the form $\alpha\beta^*$, where α and β are regular codewords and $\alpha\beta^*$ denotes the code consisting of all words $\alpha\beta^j$, $j = 0, 1, 2, \ldots$. Suppose β is the empty string. Then the conclusion is immediate. Assume that β is not the empty string. It follows from the definition of an ultimately periodic sequence that only a finite number of derivations from the starting symbol possess the code string α and that the number of subterminating words in this set of finite derivations is also finite. Thus we only need to consider the effect of the code string β^* on one subterminating word γ from this set. Starting from γ, we claim that code strings in β^* will not lead to an infinite set of terminating strings. Let $l\,(r)$ denote the total number of nonterminating symbols to appear on the left (right) side of the set of productions denoted by the symbols of β, where we count every occurrence.

First, suppose $l > r$. Let q be the number of nonterminating symbols in γ. Then the code strings β^i, where $q < i$, are not applicable. For $q \geq i$, either the string is an element of the language or it is not. However, this decision does not affect the finiteness of $L_C(G)$.

Second, suppose $l \leq r$. Then the code string β^i, $i \geq 1$, will not lead to any terminating string. Thus any code of the form of an ultimately periodic sequence will yield a finite language $L_C(G)$. ■

If C is an ultimately periodic sequence and P is not a set of single production rules, then $L_C(G)$ is not necessarily finite . For example, let $G = (N, T, P, \sigma)$, where $N = \{\sigma\}$, $T = \{\alpha\}$, $P = p_1 = \{\sigma \to a\sigma, \sigma \to a\}$, and let $C = (1)^*$. Then the language $L_C(G)$ is not finite since $L_C(G) = (a)^*$. However, if, for a given set $\{P_j\}$ (not necessarily singleton), each variable on the left must be unique, then Theorem 10.3.8 and its proof are valid for arbitrary P_j such that $P = \{P_j \mid j = 1, 2, \ldots, n\}$.

Definition 10.3.9 *Let G be a context-free grammar $G = (N, T, P, S)$, where P consists of sets of production rules $P_1, P_2, \ldots, P_n$ (i.e., P_i consists of a set of production rules, and not necessarily a singleton, where $1 \leq i \leq n$.) Let C be an infinite sequence $\{1, 2, \ldots, n\}^*$. Then the* ***cyclic language*** *of order n, $L_{cn}(G)$, is generated by applying at step i a rewriting rule from P_{c_i}, where c_i is the ith element in C.*

Relationships between cyclic languages of the same order and their codes can be easily established. Consider for example the next theorem.

Theorem 10.3.10 *Let G be a context-free grammar and $L_{cn}(G)$ be a cyclic language of order n. If C_1 and C_2 are regular codes over the alphabet $\{1, 2, \dots, n\}$ such that $C_1 \subseteq C_2$, then*

$$L_{C_1 n}(G) \subseteq L_{C_2 n}(G).$$

Proof. Every sentence generated in $L_{C_2 n}(G)$ is also generated in $L_{C_1 n}(G)$ since the grammar is identical, and every codeword in C_2 is also a codeword in C_1. Equality is established by using identical codewords over the same grammar. ■

Definition 10.3.11 *A language L is defined to be a* ***sequential fuzzy language (SFL)*** *if there is a finite fuzzy automata (FFA) A_f and a cut-point t such that L is the set of coded words that yield at least one path from the initial state to a final state of A_f whose fuzzy measure is greater that t.*

Theorem 10.3.12 *Let $G_f = (G, C, \mu, \mu_I)$ be a coded fuzzy grammar and let G be a type 3 grammar. Then the SFL, $L_f(G_f, t)$, is a regular language.*

Proof. Let L be a language accepted by an FFA, $A_f = \langle Q, X, M, q_0, F \rangle$, and let G be a regular grammar, $G = \langle N, T, P, S \rangle$, where $N = Q$, $T = X$, $S = q_0$,

$$\begin{aligned} &P = \{P_1, P_2\}, \\ &P_1 : A_i \to \alpha A_j,\ A_i, A_j \in K,\ \alpha \in X \\ &P_2 : A_i \to \Lambda,\ A_i \in F,\ C = (1)^* \, 2. \end{aligned}$$

Extend G to $G_f = (G, C, \mu, \mu_I)$ by constructing a vector μ_I with 1 in the position indicating a rewriting rule with S on the left side, and 0 otherwise. The vector μ, indicating rewriting rules of the format of P_2, is the zero vector. The vector associated with P_1 has the fuzzy grade membership of A_f entering state A_j with input α from A_i, if the term of this vector corresponds to a rewriting rule with A_j on the left side, and 0 otherwise. This procedure yields all the possible fuzzy paths in A_f. Thus $L = L_f(G_f, t)$. ■

Without loss of generality we can generate $G_f = (G, C, \mu, \mu_I)$ with only two kinds of productions, namely, $A \to \alpha B$ and $A \to \Lambda$, where $A, B \in N$ and $\alpha \in T$. By extending FFA to a generalized FFA (GFFA) and by using the results of GPA as stated in [142], it follows that G_f is a regular grammar, excluding production rules of the form $A \to B$, $A, B \in N$, from the set of productions in G_f, the language $L_f(G_f, t)$ is an SFL $\forall t$.

The characterization of the family of coded fuzzy languages with regard to the family or subfamilies of context-sensitive languages should be investigated further. It is stated in [45] that the most interesting problem is the investigation of the conditions for fuzzy grammars (type 0,1,2, and 3) under which a set of derivations will define a syntactical structure of the sentence. This should be done in a similar way that, for the context-free grammars, a leftmost derivation of a sentence defines its syntactical tree.

References

1. S. Abou-Zaid, On fuzzy subnear-rings and ideals, *Fuzzy Sets and Systems*, 44(1991), 139-146.

2. J. Ahsan, M.F. Khan, and M. Shabir, Characterization of monoids by the properties of their fuzzy subsystems, *Fuzzy Sets and Systems,* 56(1993),199-208.

3. J. Ahsan, K.Y. Li, and M. Shabir, Semigroups characterized by their fuzzy bi-ideals, *J. Fuzzy Math.*, 10(2002) 221-449.

4. J. Ahsan, K. Saifullah and M.F. Khan, Semigroups characterized by their fuzzy ideals, *Fuzzy Systems and Mathematics*, 9 (1995), 29-32.

5. J. Ahsan, R.M. Latif, and M. Shabir, Fuzzy quasi-ideals in semigroups, *J. Fuzzy Math.*, 9(2001) 259-270.

6. T.M.G. Ahsanullah and F.A. Al-Thukair, Conditions on a semigroup to be a fuzzy neighborhood group, *Fuzzy Sets and Systems*, 55(1993) 333-340.

7. F.A. Al-Thukair, Fuzzy congruence pairs of inverse semigroups, *Fuzzy Sets and Systems*, 56(1993) 117-122.

8. J. Berstel and D. Perrin, *Theory of Codes*, Academic press, Inc. London 1985.

9. L. Biacino and G. Gerla, Closure systems and L-subalgebras, *Inform. Sci.*, 33(1984) 81-195.

10. G. Birkhoff, *Lattice Theory*, 3rd edition, Amer. Math. Coll. Publ. Vol. 25 (AMS, New York, 1973).

11. S. Bogdanovic, *Semigroups with a System of Subsemigroups*, New York, 1976.

12. J. Calais, Demi-groupes quasi-inverseifs, *C.S. Acad. Sci. Paris,* 252 (1961) 2357-2359.

13. Y. B. Cha and H. J. Shyr, Some Algebraic Properties of Prefix Codes, *Nanta Math.* Vol. VI No 2 (1973) 60-64.

14. M. K. Chakraborty and M. Das, Reduction of fuzzy strict order relations, *Fuzzy Sets and Systems,* 15(1985) 33-44.

15. S.J. Cho, J.G. Kim, and W.S. Lee, Decomposition of T-generalized transformatoipm semigroups, *Fuzzy Sets and Systems*, 122 (2001) 527-537.

16. S.J. Cho and J.G. Kim, Sums and joins of fuzzy finite state machines, preprint.

17. A.H. Clifford and G.B. Preston, *The Algebraic Theory of Semigroups Vol. I.* Amer. Math. Soc., Mathematical Surveys Monographs No. 7 (1961).

18. J.A. Coguen, *L*-fuzzy sets, *J. Math. Anal. Appl.,* 18(1967) 145-171.

19. D.I.A. Cohen, Introduction to Computer Theory, John Wiley Int. Inc., NY, 1991.

20. P. Das, Fuzzy regular and inverse subsemigroups, *Fuzzy Sets and Systems*, 91(1997) 99-105.

21. P. Das, Fuzzy multiplicative semigroups, *Fuzzy Sets and Systems*, 105(1999) 171-176.

22. A. De Luca, On some properties of syntactic semigroup of a very pure subsemigroup, *RAIRO Inform. Theory,* 14(1980) 39-56.

23. K.A. Dib and N. Galhum, Fuzzy ideals and fuzzy bi-ideals in fuzzy semigroups, *Fuzzy Sets and Systems*, 92(1997) 103-111.

24. V.N. Dixit, R. Kumar, N. Ajmal, Fuzzy ideals and fuzzy prime ideals of a ring, *Fuzzy Sets and Systems*, 44(1991) 127-138.

25. D. Dubois and H. Prade, *Fuzzy Sets and Systems: Theory and Applications*, Academic Press, New York, 1980.

26. S. Eilenburg, *Automata, Languages and Machines*, Vol. A, Academic Press, London, 1974.

27. K. S. Fu and T. J. Li, On stochastic automata and languages, in Proc. 3rd Annu. Conf. Information Sciences and Systems, 1969, pp. 338-377.

28. J. Garcia, The congruence extension property for algebraic semigroups, *Semgroup Forum,* 43 (1991) 1-18.

29. S. K. Geetha, On *L*-fuzzy topological semigroups, *J. Math. Anal. Appl.*, 174(1993) 147-152.

30. S.K. Geetha, On *L*-fuzzy semitopological semigroups, *J. Math. Anal. Appl.*, 175(1993) 1-9.

31. G. Gerla, Some elementary concepts of *L*-semigroups theory, *Ric. di Mat.*, 33(1984) 53-62.

32. G. Gerla, Pavelka's fuzzy logic and free *L*-subsemigroups, *Z. Math. Logik Grundlag. Math.*, 31(1985) 123-129.

33. G. Gerla, Code theory and fuzzy subsemigroups, *J. Math. Anal. Appl.*, 128(1987) 362-369.

34. G. Gerla, *L*-subsemigroups of a free semigroup, *Rend. Mat.*, in press.

35. S. Ginsburg and E.H. Spanier, Control sets on grammars, *Math. Syst. Theory*, 2(1968) 159-177.

36. K.C. Gupta and M.K. Kantroo, The nil radical of a fuzzy ideal, *Fuzzy Sets and Systems*, 59(1993), 87-93.

37. M.A. Harrison, *Introduction to Switching and Automata Theory.* McGraw-Hill, New York, 1965.

38. K. H. Hofmann and P. S. Mostert, Elements of Compact Semigroups (Charles E. Merrill, Columbus, OH, 1966).

39. W.M.L. Holcombe, *Algebraic Automata Theory*, Cambridge University press, 1982.

40. J. E. Hopcroft and J. D. Ullman, *Formal Languages and their Relation to Automata,* Reading, Mass.: Addison-Wesley, 1969.

41. J.M. Howie, *An Introduction to Semigroup Theory,* Academic Press, London, 1976.

42. K. Iseki, A characterization of regular semigroups, *Proc. Japan Acad.*, 32(1965) 676-677.

43. R. Johnsonbaugh, *Discrete Mathematics*, Maxwell MacMillon, NY, 1989.

44. Y.B. Jun, On fuzzy prime ideals of r-rings, *Soochow Journal of Mathematics*, 21(1995), 41-48.

45. A. Kandel, Codes over languages, *IEEE Trans. on Systems, Man, and Cybernetics*, SMC-4(1974) 135-138.

46. A. Kandel, and S.C. Lee, *Fuzzy Switching and Automata: Theory and its Application*, Crane Russak, 1980.

47. N. Kehayopulu, X.Y. Xie, and M. Tsingelis, A characterization of prime and semiprime ideals of semigroups in terms of fuzzy subsets, *Soochow J. Math.*, to appear.

48. Y. Kim, J. Kim, and S. Cho, Products of T-generalized state machines and T-transformation semigroups, *Fuzzy Sets and Systems*, 93(1998), 87-97.

49. R. Kumar, Certain fuzzy ideals of rings redefined, *Fuzzy Sets and Systems*, 46(1992), 251-260.

50. R. Kumar, *Fuzzy Algebra (I),* University of Delhi Publication Division, 1993.

51. H.V. Kumbhojkar amd M.S. Bapat, On prime and primary fuzzy ideals and their radicals, *Fuzzy Sets and Systems* 53(1993) 203-216.

52. H.V. Kumbhojkar and S.R. Chaudhari, Coverings of products of fuzzy finite state machines, *Fuzzy Sets and Systems*, 125(2002) 215-222.

53. H.V. Kumbhojkar and S.R. Chaudhari, On proper fuzzification of finite state machine, preprint.

54. H.V. Kumbhojkar and S.R. Chaudhari, Decomposition of fuzzy finite state machines, preprint.

55. H.V. Kumbhojkar and S.R. Chaudhari, Homomorphisms of fuzzy recognizers, preprint.

56. H.V. Kumbhojkar and S.R. Chaudhari, Fuzzy recognizers and recognizable sets, *Fuzzy Sets and Systems*, 131 (2002), 381-391.

57. N. Kuroki, Fuzzy bi-ideals in semigroups, *Comment. Math. Univ. St. Paul*, 28(1979) 17-21.

58. N. Kuroki, On fuzzy ideals and fuzzy bi-ideals in semigroups, *Fuzzy Sets and Systems*, 5(1981) 203-215.

59. N. Kuroki, T^*-pure archimedean semigroups, *Comment. Math. Univ. St. Paul*, 31(1982) 115-128.

60. N. Kuroki, Fuzzy semiprime ideals in semigroups, *Fuzzy Sets and Systems,* 8(1982) 71-80.

61. N. Kuroki, On B^*-pure semigroups, *Acta Math. Hung.*, 43(1984) 295-298.

62. N. Kuroki, On fuzzy semigroups, *Inform. Sci.*, 53(1991) 203-236.

63. N. Kuroki, Regular and intra-regular semigroups, *Tokyo Gakugei J. Math. Edu.*, 3(1991) 51-54.

64. N. Kuroki, Fuzzy congruences and fuzzy normal subgroups, *Inform. Sci.*, 60(1992) 247-259.

65. N. Kuroki, Fuzzy generalized bi-ideals in semigroups, *Inform. Sci.*, 66(1992) 235-243.

66. N. Kuroki, On fuzzy quasi-ideals in semigroups, *Advances in Fuzzy Theory and Technology Vol. 1* (Ed. Paul P. Wang) Duke University, North Carolina, (1993) 63-76.

67. N. Kuroki, Fuzzy semiprime quasi-ideals in semigroups, *Inform. Sci.,* 75(1993) 201-211.

68. N. Kuroki, Fuzzy interior ideals in semigroups, *The Journal of Fuzzy Mathematics*, 3(1995) 435-447.

69. N. Kuroki, Fuzzy congruences on T^*-pure semigroups, *Inform. Sci.*, 84(1995) 239-246.

70. N. Kuroki, Regular fuzzy duo rings, *Inform. Sci.*, 94(1996) 119-139.

71. N. Kuroki, Fuzzy congruences on inverse semigroups, *Fuzzy Sets and Systems*, 87(1997) 335-340.

72. N. Kuroki and H.V. Kumbhojkar, Fuzzy Q^*-normal semigroups, *The Journal of Fuzzy Mathematics*, 3(1995) 429-434.

73. S. Lajos, A note on semilattices of groups, *Acta Sci. Math. Szeged,* 31(1970) 179-180.

74. S. Lajos, On generalized bi-ideals in semigroups, *Algebraic Theory of Semigroups* (Ed. G. Pollak) Coll. Math. Soc. Janos Bolyai, 20(1979) 335-340.

75. G. Lallement, *Semigroup and Combinatorial Applications*, Wiley, New York, 1979.

76. S. Lan and P.J. Yin, Fuzzy complete sets, *Fuzzy Sets and Systems*, 131(2002), 393-398.

77. S.Y. Lee and K.P. Shum, Extension of certain fuzzy ideals of semigroups on strong semilattices, *J. Fuzzy Math.,* 9(2001) 151-157.

78. E. T. Lee and L. A. Zadeh, Note on fuzzy languages, *Inform. Sci.*, 1(1969) 421-434.

79. F.W. Levi, On semigroups, *Bull. Calcutta Math. Soc.*, 36(1944) 141-146.

80. W.J. Liu, Fuzzy invariant subgroups and fuzzy ideals, *Fuzzy Sets and Systems*, 8(1982) 133-139.

81. J.D. MacKnight and A.J. Storey, Equidivisible semigroups, *J. Algebra*, 12(1969) 24-48.

82. B. B. Makamba and V. Murali, Normality and congruence in fuzzy subgroups, *Inform. Sci.,* 59(1992) 121-129.

83. D.S. Malik and J.N. Mordeson, Fuzzy prime ideals of a ring, *Fuzzy Sets and Systems*, 37(1990), 93-98.

84. D.S. Malik and J.N. Mordeson, Fuzzy maximal, radical and primary ideals of a ring, *Inform. Sci.,* 53(1991) 237-250.

85. D.S. Malik and J.N. Mordeson, *Fuzzy Discrete Structures,* Physica-Verlag, Studies in Fuzziness and Soft Computing, Vol. 58, 2000.

86. D.S. Malik, J.N. Mordeson, and M.K. Sen. Semigroups of fuzzy finite state machines, in: *Advances in Fuzzy Technology*, (P.P. Wags, eds.), Bookswrite, Durham, North Carolina 1994.

87. D.S. Malik, J.N. Mordeson, and M.K. Sen, On subsystem of fuzzy finite state machines, *Fuzzy Sets and Systems,* 68(1994), 83-92.

88. D.S. Malik, J.N. Mordeson, and M.K. Sen, Submachines of fuzzy finite state machines, *J. Fuzzy Math.,* 4(1994) 781-792.

89. D.S. Malik, J.N. Mordeson, and M.K. Sen, Products of fuzzy finite state machine, *Fuzzy Sets and Systems,* 92(1997), 95-102.

90. D.S. Malik, J.N. Mordeson, and M.K. Sen, Admissible partitions of fuzzy finite state machines, *International Journal of Uncertainty Fuzziness and Knowledge-based System,* 5(6)(1997), 723-732.

91. V.L. Mannepalli, Multiplication Semigroups, *Semigroup Forum*, 11 (1975, 1976) 310-327.

92. J. C. Martin, *Introduction to Languages and the Theory of computation*, McGraw-Hill inc., NY, 1989.

93. R.C. McLean and H. Kummer, Fuzzy ideals in semigroups, *Fuzzy Sets and Systems,* 48(1992) 137-140.

94. P.P. Ming, L.Y. Ming, Fuzzy topology-I, neighbourhood structure of a fuzzy point and Moore-Smith convergence, *J. Math. Anal. Appl.,* 76(1980) 571-599.

95. Z.W. Mo and X. P. Wang, On pointwise depiction of fuzzy regularity of semigroups, *Inform. Sci.*, 74(1993) 265-274.

96. Z.W. Mo and X. P. Wang, Fuzzy ideals generated by fuzzy sets in semigroups, *Inform. Sci.*, 86(1995) 203-210.

97. Z. W. Mo and J. Y. Pen, Some algebraic properties of fuzzy prefix codes on a free monoid, preprint.

98. Z. W. Mo and J. Y. Pen, An algorithm of test for fuzzy codes, *Appl. Math. J. China Univ. Ser. B.,* 16(1) (2001) 1-7.

99. Z. W. Mo and J. Y. Pen, Some equivalent depictions of fuzzy codes, *Bull. for Studies and Exchanges on Fuzziness and Its Applications,* 81(2000) 93-101.

100. J.N. Mordeson and D.S. Malik, *Fuzzy Automata and Languages: Theory and Applications*, Chapman & Hall/CRC, Computational Mathematics Series, Series Editor Mike J. Atallah, 2002.

101. T.K. Mukherjee and M.K. Sen, On fuzzy ideals of a ring I, *Fuzzy Sets and Systems,* 21(1987) 99-104.

102. T.K. Mukeherjee and M.K. Sen, Fuzzy prime ideals in rings, *Fuzzy Sets and Systems*, 32(1989), 337-341.

103. V. Murali, Fuzzy equivalence relations, *Fuzzy Sets and Systems*, 30 (1989) 155-163.

104. C.V. Negoita and D.A. Ralescu, *Application of Fuzzy Sets to System Analysis*, Halsted press, New York, 1975.

105. W. C. Nemitz, Fuzzy relations and fuzzy functions, *Fuzzy Sets and Systems*, 19(1986) 177-191.

106. A. Paz, Some aspects of probabilistic automata, *Inform. Control*, 9(1966) 26-60.

107. M. Petrich, *Introduction to Semigroups*, Charles E. Merrill, (1973).

108. M. O. Rabin, Probabilistic automata, *Inform. Control*, 6(1963) 230-245.

109. A Restivo, On a question of McNaughton and Papert, *Inform. and Control,* 25(1974) 93-101.

110. A. Rosenfeld, Fuzzy groups, *J. Math. Anal. Appl.*, 35(1971) 512-517.

111. T. Saito, On semigroups which are semilattices of left simple semigroups, *Math. Japon.*, 18(1973) 95-97.

112. A. Salomao, On finite automata with a time-variant structure, *Inform. Control,* 11(1967) 396-401.

113. A. Salomaa, On languages accepted by probabilistic and time-variant automata, in Proc. 2nd Annu. Princeton Conf. Information Sciences and Systems, Mar. 1968.

114. A. Salomaa, On grammars with restricted use of productions, *Ann. Acad. Sci,. Fenn.*, Ser. AI 454, 1969.

115. A. Salomao, Probabilistic and weighted grammars, *Inform. Contr.*, 15(1969) 529-544.

116. A. Salomao, On some families of formal languages obtained by regulated derivations, *Ann. Acad. Sci. Fenn.*, Ser. AI 479, 1970.

117. M.A. Samhan, Fuzzy congruences on semigroups, *Inform. Sci.*, 74(1993) 165-175.

118. M. A. Samhan and T. M. G. Ahsanullah, Fuzzy congruences on groups and rings, *Internat. J. Math. Sci.,* 17(1994) 469-474.

119. M. A. Samhan and T. M. G. Ahsanullah, Fuzzy congruences on groups and rings, submitted to *Fuzzy Sets and Systems.*

120. M. A. Samhan and F. Al-Thukair, On fuzzy subalgebras and fuzzy congruence relations of a universal algebra, submitted for publication.

121. E. Sanchez, Resolution of composite fuzzy relation equations, *Inform. Control,* 30(1976) 38-48.

122. E.S. Santos, Fuzzy Sequential Functions, *J. Cybernetics,* 3(1973), 15-31.

123. E.S. Santos, Realization of Fuzzy Languages by Probabilistic, Max-Product and Maximin Automata, *Inform. Sci.*, 8(1975), 39-53.

124. E.S. Santos, Fuzzy automata and languages, *Inform. Sci.*, 10(1976) 193-197.

125. E.S. Santos, Fuzzy automata and probabilistic programs, *Inform. Sci.,* 10(1976) 331-335.

126. E.S. Santos, Regular fuzzy expressions, In: *Fuzzy Automata and Decision Processes*, Editors: M.M. Gupta, G.N. Saridis, and B.R. Gaines, Elsevier North-Holland Inc., 1977, 169-175.

127. A.A. Sardinas and C.W. Patterson, A necessary and sufficient condition for the unique decomposition of coded messages, *I.R.E. Int. Conv. Rec.*, 8(1953) 104-108.

128. M.P. Schützenberger, Sur certains sous-demigroups qui interbiennent dans un probleme de mathematiques appliques, *Publ. Sci. Univ. Alger. Ser. A,* 6(1959), 85-90.

129. B. Schweizer and A. Sklar, Statistical metric spaces, *Pacific J. Math.*, 10(1960) 313-334.

130. J. Z. Shen, On fuzzy regular subsemigroups of a semigroup, *Inform. Sci., 51(1990) 111-120.*

131. J. Z. Shen, Fuzzy language on free monoid, *Inform. Sci.,* 88(1996) 149-168.

132. J. Z. Shen, On base of fuzzy monoid, *J. Jiangxi Normal Univ.*, Vol. 3, 1998.

133. J. Z. Shen, Fuzzy codes on free monoid, *J. Jiangxi Normal University,* (1999) 189-194.

134. H. J. Shyr, Codes and factor theorems for subsets of free monoid, *Util. Math.*, 13(1973) 153-159.

135. H. J. Shyr, *Free Monoids and Languages*, Taiwan, 1991.

136. O. Steinfeld, Uber die Quasiideale con Halbgruppen, *Publ. Math. Debrece,* 4(1956) 262-275.

137. O. Steinfeld, Quasi-ideals in Rings and Semigroups, *Akad. Kiad*, Budapest, (1978).

138. K. L. N. Swamy and U.M. Swamy, Fuzzy prime ideals of rings, *J. Math. Anal. Appl.*, 134 (1988) 90-103.

139. Y. Tan, Fuzzy congruences on a regular semigroup, *Fuzzy Sets and Systems,* 117(2001) 447-453.

140. B. Tilson, The intersection of free submonoids of a free monoid is free, *Semigroup Forum,* 4(1972) 345-350.

141. P. Turakainen, On stochastic languages, *Inform. Contr.*, 12(1968) 304-313.

142. P. Turakainen, Generalized automata and stochastic languages, *Proc. Amer. Math. Soc.*, 21(1969) 303-309.

143. P. Wang and W. J. Liu, Fuzzy regular subsemigroups in semigroups, *Inform. Sci.*, 68(1993) 225-231.

144. X. P. Wang, Z.W. Mo, and W.J. Liu, Fuzzy ideals generated by fuzzy points in semigroups, *J. Sichuan Normal Univ.*, 15(1992), 17-24.

145. W.G. Wee, On generalizations of adaptive algorithm and application of the fuzzy sets concept to pattern classification, Ph.D. Thesis, Purdue University, 1967.

146. W. G. Wee and K. S. Fu, A formulation of fuzzy automata and its application as a model of learning systems, *IEEE Trans. Syst. Sci. Cybern.*, SSC-5(1969) 215-223.

147. D. Wood, A factor theorem for subsets of a free monoid, *Inform. and Control*, 21(1972) 21-26.

148. W. M. Wu, Normal fuzzy subgroups, *Fuzzy Math.*, 1(1971) 21-30.

149. X.Y. Xie, Fuzzy Rees congruences on semigroups, *Fuzzy Sets and Systems,* 102(1999) 353-359.

150. X.Y. Xie, Fuzzy ideals in semigroups, *J. Fuzzy Math.*, 7(1999) 357-365.

151. X.Y. Xie, On prime fuzzy ideals of a semigroups, *J. Fuzzy Math.*, 8(2000) 231-241.

152. X.Y. Xie, On prime quasi-prime, weakly quasi-prime fuzzy left ideals of semigroups, *Fuzzy Sets and Systems*, 123(2001) 239-249.

153. X.Y. Xie, Fuzzy ideal extensions of semigroups, *Soochow J. Math.*, 27(2001) 125-138.

154. X.Y. Xie, Fuzzy ideal extensions in semigroups, *Kyungpook Math. J., 42(2002) 39-49.*

155. Y. Yandong, J.N. Mordeson and S.C. Cheng, *Elements of L-Algebra,* Lecture Notes in Fuzzy Mathematics and Computer Science, Center for Research in Fuzzy Mathematics and Computer Science, Creighton University, 1994.

156. Zhang Yue, Prime L-fuzzy ideals and primary L-fuzzy ideals, *Fuzzy Sets and Systems,* 27(1988), 145-350.

157. L.A. Zadeh, Fuzzy Sets, *Inform. and Control,* 8(1965) 338-353.

158. L. A. Zadeh, Quantitative fuzzy semantics, Electron. Res. Lab., Univ. California, Berkeley, Memo ERL-M 281, Aug. 1970.

159. L.A. Zadeh, Similarity relations and fuzzy orderings, *Inform. Sci.*, 3(1971) 177-200.

160. C. Zhang, Fuzzy complete inner-unitary subsemigroups and fuzzy group congruences on a regular semigroup, *Fuzzy Sets and Systems,* 112(2000) 327-332.

161. C. Zhang, Fuzzy congurences on completely 0-simple semigroups, *J. Fuzzy Math.*, 10(2002) 27-48.

162. H.-J. Zimmermann, *Fuzzy Set Theory and its Applications*, 3rd edition, International Series in Management Science/Operation Research (Kluwer-Nijhoff Publishing, Dordrecht, 1988).

163. X.F. Zou and J. Shen, Fuzzifying codes and fuzzifying prefix codes, *J. Fuzzy Math.*, 8(2000) 307-313.

Index

Druck: Strauss Offsetdruck, Mörlenbach
Verarbeitung: Schäffer, Grünstadt